T0192810

Process Chemistry
of Petroleum
Macromolecules

CHEMICAL INDUSTRIES

A Series of Reference Books and Textbooks

Founding Editor

HEINZ HEINEMANN
Berkeley, California

Series Editor

JAMES G. SPEIGHT
Laramie, Wyoming

Process Chemistry of Petroleum Macromolecules

Irwin A. Wiehe

Soluble Solutions
Gladstone, New Jersey, U.S.A.

CRC Press
Taylor & Francis Group
Boca Raton   London   New York

CRC Press is an imprint of the
Taylor & Francis Group, an **Informa** business

CRC Press
Taylor & Francis Group
6000 Broken Sound Parkway NW, Suite 300
Boca Raton, FL 33487-2742

First issued in paperback 2020

ISBN-13: 978-0-367-57750-6 (pbk)
ISBN-13: 978-1-57444-787-3 (hbk)

Library of Congress Cataloging-in-Publication Data

Wiehe, Irwin A.
    Process chemistry of petroleum macromolecules / Irwin A. Wiehe. -- 1st ed.
        p. cm. -- (Chemical industries ; 121)
    Includes bibliographical references and index.
    ISBN 978-1-57444-787-3 (alk. paper)
    1. Petroleum--Refining. 2. Catalytic reforming. 3. Petroleum--Research. I. Title. II. Series.

TP690.W427 2008
665.5'38--dc22   2008002311

Visit the Taylor & Francis Web site at
http://www.taylorandfrancis.com

and the CRC Press Web site at
http://www.crcpress.com

Contents

Preface

Today, as the supply of conventional light petroleum is decreasing, the world is depending more and more on heavy and extra heavy crude oils to meet the increasing demand for transportation fuels. These heavy and extra heavy crude oils have greater amounts of larger, more aromatic molecules, the petroleum macromolecules, and lesser amounts of smaller, more paraffinic molecules that can be distilled directly into transportation fuels. Instead, the petroleum macromolecules even need to be chemically converted into a form that can be catalytically transformed into transportation fuels. However, petroleum macromolecules have not received nearly the attention of the smaller molecules in understanding and in the development of processes. This is natural because unlike petroleum macromolecules, the structure and identity of smaller molecules of petroleum can be directly determined, and in the past, conversion of petroleum macromolecules was a very minor portion of the volume of petroleum processed in a refinery.

With the present accelerating need for converting petroleum macromolecules, many are asking if and how can the yield of transportation fuels be increased from petroleum macromolecules? The present book projects the view that considerable improvement in the conversion of heavy oils is probable for those willing and able to accept the risk of new and innovative processes. In addition, this book provides the reader with the scientific understanding, the logic, and the insight to devise their own innovations. Thus, after a chapter that introduces the reader to the world of petroleum and petroleum macromolecules, separate chapters are devoted to the fundamental subjects of characterization, thermal conversion kinetics, phase behavior, and separation of petroleum macromolecules. Do not expect these to be exactly like the same subjects that you may have studied for other molecules. Petroleum macromolecules are too many and too diverse to analyze for the exact molecules, and many contain one or more polynuclear aromatics that are outside the experience of most scientists and engineers. Nevertheless, mathematical models that are not exact but still capture the important elements in a relatively simple form give surprisingly good descriptions of the behavior of petroleum macromolecules. Much of the chapters on the fundamental subjects as well as on the applications have not been previously published, except in conference proceedings.

Applications, innovations, and some case studies are inserted in the chapters on fundamental subjects for the reader to grasp the significance and direction of the research. However, these are amplified and summarized in the application chapters on fouling mitigation, coking, visbreaking, and hydroconversion with additional case studies. Much more attention is devoted to fouling mitigation than any other book on heavy oils. In part this is because fouling and coking usually either limit the conversion or the run length of heavy oil conversion processes. Another reason is that the same tools used for fouling mitigation of heavy oils can be extended to the fouling of crude oils, such as the Oil Compatibility Model. As a result, fouling mitigation, usually without requiring capital investment, can offer refineries large and immediate savings in energy, in carbon dioxide emissions, and in operating costs. Meanwhile, innovations based upon logical deduction from concepts in the fundamental chapters are given for coking, visbreaking, and hydroconversion. Some of these have been applied, but many have yet to be tried beyond the small laboratory scale. The message is loud and clear that there is significant room for improving the conversion of petroleum macromolecules and for readers devising their own innovations.

A secondary purpose of the book is to provide a glimpse of how industrial research is done by creating knowledge on a path toward innovation, as opposed to more familiar basic research by creating knowledge for knowledge sake. One approach of the author is described in differentiating the real barriers to innovation from the false barriers, which in overcoming, provide step-out improvements in technology. Another approach is described in which the author's seemingly successful mathematical models are tested by additional and different experimental data in order to determine where they fail. In this way, new understanding is created in the process of developing more robust models. All scientific concepts and models are approximations that will be eventually rejected in favor of better ones. However, those applied to petroleum macromolecules have even greater uncertainty because the system is ultracomplex. This is both the challenge and the fun because nothing is definite. Therefore, only by showing consistency with many types of experimental data can one become confident that a concept or model of petroleum macromolecules is a good enough approximation to be a basis for innovation. Many in this field have made the mistake of basing a conclusion on a single experiment or measurement.

An attempt was made by the author to humanize this book by inserting a personal view of how and why many steps of research were taken. As a result, this book greatly emphasizes the author's own research and applications over his 30+ years of experience with petroleum macromolecules. Although an attempt is made to point out where others have opposing views and where others have made contributions, this book is not a comprehensive text that attempts to survey all research and technology on petroleum macromolecules. Therefore, the author apologizes in advance for omitting the work of others.

The material for this book was initially organized as part of the notes and slides that the author provided in teaching short courses. The first course (three days) in 1999 was at the invitation of Parviz Rahimi of the National Centre of Upgrading Technology near Edmonton, Canada. The author thanks Rahimi for initiating these opportunities and for allowing the author to participate in his research as a consultant. The author also thanks C.B. Panchal of Argonne National Laboratories with whom the author collaborated in teaching short courses on refinery fouling mitigation. By the time of the writing of this book, the author had taught 23 short courses, sponsored by 17 different organizations in 7 different countries.

The author is very appreciative to Exxon (now ExxonMobil) Research and Engineering Company for providing him the opportunity, support, and wonderful environment for doing basic and applied research on petroleum macromolecules for 22 years. An attempt was made to give credit to the author's colleagues at Exxon within this book, but he offers apologies to any that he might have overlooked. Special thanks are made here to those at Exxon who provided the foundation of this book by measuring superb experimental data on difficult oils and by being partners in the design of experiments: Mike Lilga, Jerry Machusak, Walt Gerald, Kathy Greaney, Larry Kaplan, Cyntell Robertson, Ray Kennedy, Ted Jermansen, and David Jennings. The author would also like to thank his mother, Winona Wiehe, who continues to be a shining light in his life and who suggested that he humanize his technical book. Finally, the author is especially grateful to his wife, Irene, who provided the encouragement and the drive that helped enable the completion of this book after eight years of writing.

Author

Irwin (Irv) Wiehe is president of Soluble Solutions, a consulting company specializing in petroleum processing and fouling mitigation. Irv has a Ph.D. in chemical engineering from Washington University (St. Louis) and worked at Exxon Corporate Research for 22 years after holding positions at the University of Rochester and Xerox. Although originally educated in polymer science and in solution thermodynamics, his current special interests are in improving the processing of petroleum resids and heavy oils and in the solving of refinery fouling problems. He is internationally known for unique general concepts, such as the Oil Compatibility Model, the Phase-Separation Kinetic Model for Coke Formation, the Pendant-Core Building Block Model of Resids, and the Solvent-Resid Phase Diagram. In addition, Wiehe is the founder of two annual conferences that have been meeting for over 10 years: the International Conference on Refinery Processing for the American Institute of Chemical Engineers (AIChE) and the International Conference on Petroleum Phase Behavior and Fouling. Wiehe, who has held most of the elected offices in the Fuels and Petrochemicals Division of AIChE, including chairman, and has presented papers and chaired sessions for 15 consecutive years at national meetings, received the division's Distinguished Service Award and was elected a Fellow of AIChE, both in 2007.

1 Introduction to the Processing of Petroleum Macromolecules

This chapter is designed to briefly introduce the reader to petroleum, petroleum economics, petroleum processing technology, and the chemical engineering approach to research. This should help provide the reader with the background and context to understand the rest of the book, which emphasizes the application of the research approach of the author to the larger molecules in petroleum. These petroleum macromolecules are concentrated in heavy crude oils or in the least volatile fractions of light crude oils, called *resids*.

1.1 IMPORTANCE AND CHALLENGES OF PETROLEUM

Anthropologists have classified periods of ancient history by the important materials of the time, such as the Bronze Age or the Iron Age. However, fuel better classifies modern man because there is no greater dependency of modern technology than on the energy that drives it. If this is the case, then the twentieth century was clearly the Petroleum Age. It has been the fuel of choice for driving our vehicles: automobiles, trucks, airplanes, ships, and trains. In addition, the by-products of producing these fuels provided the petrochemical building blocks of the twentieth century materials: plastics, synthetic fibers, and synthetic elastomers. Although other forms of energy, such as coal, natural gas, and nuclear energy, have made their bid to overtake petroleum, the liquid state of petroleum gives it a great edge. As a result, it can be easily stored and transported in a concentrated form of chemical energy that is relatively safe. It is no wonder that, in the 1970s, when we thought the supply of petroleum was running out, the consequence was a world economic and political problem, the "energy crisis." However, the increased price encouraged more conservation and a greater search for more petroleum. By the end of the 1990s, petroleum was priced at historical lows in constant dollars. However, in the 2000s, the price of crude oil again greatly increased with the economic development of China and India and the popularity of sport utility vehicles (SUVs) in the United States reversing conservation. As a result, the recovery and upgrading of extra heavy oil has been greatly stepped up in Venezuela and in Alberta, Canada, and new discoveries of oil have been made in West Africa and offshore Brazil. Instead of a shortage, the greatest threat to continuation of the Petroleum Age well into the twenty-first century is environmental. There are concerns about the emission of sulfur oxides, nitrogen oxides, and carbon monoxide when the fuel is burned, about the dangers of accidental spills when it is transported and stored, and about air and water emissions when it is produced and refined. Even more importantly, both science and society are becoming aware that the burning of petroleum and other fossil fuels is producing enough carbon dioxide to cause global warming through a "greenhouse" effect. As a result, most of us wish neither to cause a significant climate change nor to pay the energy cost of a complete shift away from fossil fuels.

With all this history of the technological exploitation of petroleum, there is a misconception that we completely understand petroleum and have reached the technological limits of its conversion to fuels and petrochemicals. Nothing can be further from the truth. This is especially the case for the larger molecules in petroleum, the macromolecules or heavy oil.[1] The objective of the author in writing this book is to convey the richness of petroleum as a field of science and innovation. Although petroleum contains over a million different molecules, most of this diversity is in the heavy fraction, the macromolecules. Although physicists have recently initiated a new branch called the physics of complex materials,[2] by comparison petroleum is an "ultracomplex" material. Even, some of the petroleum (asphaltenes) self organizes so that it is difficult to measure molecular weight. Petroleum is neither a solution nor a colloid, but a hybrid of both. The greatest attractive interaction is between polynuclear aromatic structures that are rarely found in other materials. During thermal processing, these polynuclear aromatic structures can form a discotic liquid crystalline phase, only an example of its challenging but interesting phase behavior. Although the high viscosity of oil provides its great lubricating properties, we have little understanding why the viscosity is so high. However, with all this uncertainty, the most surprising feature of petroleum is that it can be described by simple models. Although theory tells us that reactions of complex mixtures of molecules cannot be first order, the thermal reaction of petroleum macromolecules is first order with constant activation energy over the entire temperature range that we have been able to measure. Solubility parameters are known to be at best a rough predictor of the phase behavior of pure, small molecules, but they may describe the phase behavior of petroleum better than any other system. When we ignore all that molecular complexity and represent petroleum macromolecules as either a few pseudocomponents or being composed of only two building blocks, much of the thermal chemistry can be described quantitatively as well as we can measure it. These are not merely the result of empirical correlation because they describe changes in properties, reactor type, and initial concentrations. The excitement is in exploiting this surprising simplicity with new innovations as well as to unravel the underlying reasons using modern tools for studying macromolecules. Such an endeavor may even enable us to better understand ultracomplex materials other than just petroleum.

1.2 WORLD PETROLEUM ECONOMICS

It is difficult for a technical book to capture the economics of petroleum, which historically has changed widely so frequently. However, worldwide petroleum economics has such a profound influence on the development and application of petroleum technology that it cannot be ignored. Thus, the reader should decide which of the following pertains to the time he or she is reading this section and modify the conclusions accordingly.

In 2007, when the writing of this book was completed, petroleum economics was greatly influenced by Organization of Petroleum Exporting Countries (OPEC); the occupation of Iraq by the United States and Great Britain; the booming, but emerging, economies of China and India; the strong appetite of the United States for petroleum products; the decreasing rate of discovery of sources of light crude oils; the approach to full refining capacity; and the threat that the burning of fossil fuels has begun to cause global warming. Petroleum economics is greatly influenced by the political situation in many countries because there is a delicate balance between supply and demand. A small shortage or surplus in petroleum causes wide swings in its price. However, at the time of the writing of this book, the price of light, sweet crude oil has reached above $90 per barrel, showing a greater danger of shortage than surplus.

1.2.1 OPEC

This organization is composed of 11 countries: Algeria, Indonesia, Iran, Iraq, Kuwait, Libya, Nigeria, Qatar, Saudi Arabia, the United Arab Emirates, and Venezuela. OPEC's member countries hold about two-thirds of the world's proven oil reserves (about 900 billion barrels) and supply about 40% of the world's oil production (about 50% of total exports). The objective of OPEC is to stabilize the international price range for crude oil by controlling the amount of crude oil sold in worldwide markets. Because the price of light, sweet crude oil is currently much higher than their previous target of $25 to $30 per barrel, it is clear that the demand is close to the maximum production capacity of both OPEC and petroleum-exporting countries not part of OPEC. Of course, this high price and high production rate maximizes the income of the petroleum-exporting countries as long as the petroleum-importing countries can maintain healthy economies. Thus, despite the high cost of crude oil, OPEC has less influence on the crude price than it did previously. Of course, if OPEC were to decide to cut production, the price would greatly increase. However, with all the OPEC countries producing at maximum capacity with the exception of Iraq and Saudi Arabia, there is not much OPEC can do to lower the price. Because the consumer pays the price of petroleum products, not crude oil, a shortage of refinery capacity can also raise the price consumers pay as well as local taxes. To combat the effect of a refining shortage, OPEC countries, particularly Saudi Arabia, have made plans to add refining capacity of 5.9 million barrels per day by 2012[3] and export refinery products.

The influence of OPEC also has been reduced by the increase in the amount of petroleum exported by countries not part of OPEC, particularly Russia and Norway (the latter only the seventh largest producer in the world), which are the second and third largest exporters of petroleum, respectively.[4]

Iraq with 115 billion barrels has the third largest reserves of conventional oil to that of Saudi Arabia, 267 billion barrels, and Iran, 132 billion barrels.[5] However, Iraq only ranks 14th in amount of oil produced.[4] This is because of the United Nations sanctions placed on Iraq after the 1991 Persian Gulf War, initiated by Iraq's invasion of Kuwait. More recently, Iraq's production is even less because of the obstruction by those in Iraq opposed to the occupation of the United States and Great Britain and to the present Iraq government. The facilities and equipment for the production and transport of petroleum are also much below standard in Iraq, but do not warrant upgrading until the threat of obstruction is mitigated. Obviously, if Iraq were to achieve full production, the petroleum price would greatly drop unless the rest of OPEC would cut their production accordingly. At this writing, the unrest in Nigeria, an OPEC member, also threatens to reduce production in the eighth largest exporter of petroleum.

1.2.2 CHINA AND INDIA

The recent large increase in demand for petroleum by China and supposedly by India to fuel their booming economies has been a strong contributor to the high price of petroleum. Because these are the two most populous countries in the world, the effect is enormous, has large momentum, and is expected only to increase in the future. China has become the manufacturing center of the world based on low wages of workers and capitalistic entrepreneurs who rose up in spite of government restrictions.[5] On the other hand, the economy in India has blossomed because of the convergence of computer and communication technology with a surplus of English-speaking people, well educated in technical disciplines.[6] This has made India the service center of the world. As will be discussed later, the consumption of petroleum products in India has not met projections (lower in 2005 than 2004). Although India still imports much more petroleum than previously, and it is expected to increase much more in the future, they actually export 17% of their refined fuels, with more exports expected in the future.

China is now the second largest importer of petroleum from the United States.[4] Like the United States, China has significant production of petroleum (sixth largest at 3.8 million barrels per day in 2004) of its own.[4] When the author visited two refineries in China in 2004, he was surprised to find that this increase in demand for petroleum was more because of the increase in the use of motor bicycles and motorcycles than that of automobiles. The amount and quality of roads in China are quite poor, and even professionals in China could not afford automobiles. As these obstacles are overcome, the demand for petroleum in China will go through the roof. In addition, the oil produced in China is very paraffinic. Unlike most crude oils, the nonvolatile part of Chinese crude oil can be more paraffinic than the volatile part. As a result, the Chinese process petroleum in refineries in ways never considered outside China. However, this restricts China to import only light, paraffinic crude oils. Therefore, the price of light, sweet (low-sulfur) crude oils has greatly increased relative to heavy, sour (high-sulfur) crude oils. In addition, in 2005, both China and India greatly decreased the sulfur levels required for their transportation fuels. Because both have limited hydrotreating capacity, their solution is to import lighter, sweeter crude oils, further driving up their price.[3]

1.2.3 UNITED STATES AND CANADA

Because the result of the large demand for light, sweet crude oils has been the price of heavy, sour crude oils being as low as half the cost of the former, the demand for heavy crude oils has greatly increased. This is particularly true in the North America where refineries generally have more resid conversion capacity (usually delayed cokers) and more hydrotreating capacity to reduce sulfur in transportation fuels than the rest of the world. Most of the heavy oil and extra heavy oil is imported into the United States from Canada, Mexico, and Venezuela. Unlike light oils, heavy and extra heavy oils are easy to find, but are difficult to produce from their source, difficult to transport because of high viscosity, and difficult to convert into transportation fuels. Nevertheless, with their much lower prices compared to light crude oils, the incentives for producing, transporting, and refining heavy and extra heavy oils are more than sufficient. This is particularly true for Alberta, Canada, which is connected by pipelines to many refineries in the United States and Canada. The construction of new production and upgrading facilities for Athabasca tar sands is only limited by the availability of construction workers, such as welders. Cities in the area (northern Alberta, Canada), such as Fort McMurray (actually a hamlet), cannot build homes, streets, and schools fast enough to absorb the huge increase in workers. Extra heavy oils, such as Cold Lake bitumen, need to be blended with gas condensate, a light liquid by-product of natural gas production, to lower the viscosity and density enough to meet pipeline specifications. With the increase in the production of extra heavy oil in Alberta, there is a shortage of gas condensate, and it has by far the highest price in the world. Elsewhere, gas condensate is difficult to sell. In addition, these operations are stressing the Athabasca River supply of water in the area of the tar sands. Hot water (two to three barrels per barrel of bitumen) is used to separate the fine minerals from the bitumen and require giant "tailings ponds" to separate oil-coated fine clays from the waste water. These are just some of the symptoms of the large increase in the production of heavy and extra heavy oils that contain, high proportions of petroleum macromolecules. Large facilities have also been built or expanded for converting extra heavy oils into light, sweet synthetic crude oils near the production sites in Alberta, and in Venezuela. Most of these synthetic crude oils have no vacuum-resid (low-volatility) fraction and are attractive to refineries without sufficient resid conversion capacity. However, blends of synthetic crudes and unprocessed extra heavy oils are also marketed as a way to circumvent the shortage of gas condensate.

The United States provides by far the greatest demand for petroleum. The 2004 statistics[4] indicate that the United States is the third largest producer of petroleum at 8.4 million barrels per day, but consumes by far the most petroleum at 20.6 million barrels per day. Thus, 11.8 million barrels per day need to be imported. This is greater than the total oil production of Saudi Arabia at 10.7 million barrels per day. Therefore, when Americans are looking for whom to blame for the high cost of gasoline, they only need to look in the mirror. With the typical suburban American family owning a car for every driver and at least one SUV, we are consuming petroleum at an alarming rate. With 5% of the population, the United States consumes 45% of the world's production of gasoline[7] and 25% of the petroleum. It is the author's opinion that gasoline and diesel fuel are priced much too low in the United States. As in most other countries, gasoline and diesel fuel should be taxed to make them cost two or three times the present cost. This would greatly encourage conservation by forcing Americans to use mass transit, to purchase fewer cars, to car pool, and to operate cars with high mileage per gallon.

Who provides the petroleum the United States imports? In 2005, Canada led with 18%, Mexico was second with 15%, Saudi Arabia was third with 12%, and Venezuela was fourth with 10%.[7] With the recent large increase in extra heavy oil production in Canada and in Venezuela, the share of imports from these two countries are now higher and increasing. Transportation costs encourage importing oil from the closest sources. However, transportation costs are still low enough that refineries in the United States are looking for bargain-priced crude oils from all over the world. Therefore, although less than 20% of the petroleum imported by the United States comes from the Persian Gulf, the production rates and prices charged by this region greatly affect the price of all petroleum imported into the United States. It is truly a world petroleum market with active negotiations between buyers and sellers.

1.2.4 When Will Petroleum Run Out?

In 1977, when the author joined Exxon Corporate Research and three years after the Arab oil embargo, many people, including the petroleum companies, believed that we were running out of a reliable, secure supply of petroleum. At that time, OPEC refused to lower the price of petroleum. Shortages caused long lines at gas stations. Exxon and other petroleum companies were doing large-scale research and development on the conversion of coal and shale oil to hydrocarbon liquids. In the 1980s the threat went away. In 1981 the average annual price of oil peaked at over $30 per barrel, and by 1986 the average annual price was close to $12 per barrel with short-term prices below $10 per barrel. There were no more gas lines, and Exxon ceased construction of a shale oil conversion plant in Colorado. What happened? The sharp increase in the price of gasoline caused a large increase in conservation in the United States. Four-cylinder cars became popular, the U.S. government mandated an increase in automobile gas mileage over time from the automobile manufactures, and speed limits were reduced on highways.

The higher price of petroleum spurred the exploration and discovery of more petroleum all over the world, including in deeper seas. In June 1977, the first oil was pumped down the Trans-Alaska Pipeline that opened the Prudhoe Bay Field, the largest field of petroleum within the United States, over 13 billion barrels.[7] The demand for petroleum decreased by 5 million barrels per day by 1986, and the supply of petroleum by non-OPEC countries increased by 14 million barrels per day, causing the price to drop. In an attempt to recover OPEC control, Saudi Arabia flooded the market with petroleum. The world was far from running out of petroleum, but the laws of economics eventually worked their magic. Unfortunately, people proceeded to stick their heads in the sand and forgot there ever was a problem.

Now, in 2007, six-cylinder engines and SUVs are popular, the mandate on automobile gas mileage has been repealed, speed limits are back up, the production of petroleum from the Prudhoe Bay Field has peaked, and the price of petroleum has reached above $90 per barrel. Although the response of the U.S. government has been slow, economic forces are responding. Active oil exploration has discovered oil fields in offshore West Africa, offshore Brazil, and in still deeper water of the Gulf of Mexico. The production of extra heavy oil has ramped up in Canada and Venezuela, biodiesel and gasoline-containing ethanol is being marketed, and hybrid cars have become popular. However, the production of petroleum peaked in United States in 1971, and the discovery of new conventional petroleum reserves in the world has not been keeping up with consumption. Is the supply of petroleum about to run out?

Paul Roberts,[8] using well-balanced reporting, provides his reader with enough information to make guesses as to the future supply of petroleum. He arrives at the possibility that the production of non-OPEC petroleum could peak in 2015, and OPEC petroleum could peak in 2025. Of course, no one really knows. Even this guess is for conventional, light crude oil. As the price of petroleum increases, the search will move into still deeper oceans and into the Artic. More oil will be extracted from existing wells where past recoveries have been only 25 to 33%. Of course, the huge resources of extra heavy oil in Alberta, Canada, and in Venezuela are only beginning to be tapped. Liquid hydrocarbons are already being manufactured from natural gas in a few locations. Although fuels from agricultural products will never replace the huge consumption volumes of petroleum, they can make a significant contribution. Ethanol from cellulose, such as switchgrass, is much more promising than from corn. Butanol may be preferred over ethanol because of its lower solubility in water and higher solubility in gasoline. Biodiesel from canola oil produced from rapeseed (the third most produced vegetable oil in the world) is more promising than that from soybean oil because of the former's higher production per acre. Of course, the technologies for obtaining liquid hydrocarbons from the conversion of shale oil and coal are on the shelf, waiting to be reactivated. Therefore, although we are not likely to run out of liquid hydrocarbon fuels for a very long time, cheap, light petroleum is a very limited natural resource. Petroleum price is already high enough to activate alternatives, but the prospects are that the price will increase further in the future to maintain a high supply of liquid hydrocarbon fuels, which include even more expensive alternatives. Clearly, the time for greater conservation is already here.

What about hydrogen? Unfortunately, there are no hydrogen reservoirs that can be tapped by drilling a well. It is most economically produced from natural gas or from hydrocarbon liquids of petroleum. Hydrogen from water is far from energy efficient unless it involves reactions with carbonaceous materials (gasification). It is the preferred fuel for fuel cells if their technical obstacles can be overcome to achieve their high theoretical efficiencies. Even in this case, for vehicles the hydrogen will likely be made onboard by reforming liquid hydrocarbons to carry safely enough fuel in a limited volume. Although this method promises to conserve hydrocarbons by the much higher fuel efficiency, fuel cells operating on hydrogen do not completely free us from our dependence on petroleum.

1.2.5 REFINERY CAPACITY

For many years there was an overcapacity for petroleum refining in Europe and North America. As a result, refining was not a very profitable business. The large integrated petroleum companies made most of their profits from petroleum production. They maintained refineries to assure a market for the petroleum they produced and to provide cheap and plentiful supplies of feedstocks for petrochemicals. Smaller refineries do little more than distill the petroleum into boiling-point fractions. With increasing environmental restrictions on petroleum transportation fuels, such as very low amounts of sulfur, the smaller refineries could not afford to add the processes to meet environmental restrictions. As a result, many of the smaller refineries in North America were closed, whereas the capacities of larger refineries were increased. Just before the recession that started in 2001, refineries in North America were operating near full capacity for the first time in decades. Because refineries are periodically shut down for maintenance, they cannot be operated at 100% capacity. During these maintenance periods, units are repaired, improved, and cleaned of foulants. Nevertheless, just before the 2001 recession and after the recovery of the economy, refining has become profitable as it is approaching full capacity. In 2005, the United States had 144 operating refineries with a total capacity of 17 million barrels per day[9] or an average capacity of 118,000 barrels per day. The largest two refineries in the United States are Exxon Mobil's refineries in Baytown, Texas (563,000 barrels per day) and Baton Rouge, Louisiana (501,000 barrels per day). Why hasn't a new refinery been built in the United States since 1976? One reason is that overcapacity and low profitability did not justify it until recently. A second reason is that because a small refinery is not viable, a new refinery would cost many billions of dollars. A third reason is that it would take about 10 years, if ever, to receive environmental permits for a new refinery. Instead, refinery capacity has been increasing in the United States by about 2% per year by adding capacity to existing refineries. For instance, Motiva, equally owned by Shell and Saudi Refining Co., announced plans to increase the capacity of their Port Arthur, Texas, refinery from 285,000 to 600,000 barrels per day at a cost of $7 billion.[10] This will make it the largest refinery in the United States.

With light crude oils becoming less available and more costly with time, does it make sense to increase the capacity of conventional refining based on light crude oils? No, the United States is not limited by the refining capacity of light crude oils, but the refining capacity of heavy and extra heavy oils. The close distance to the large sources of heavy and extra heavy oils in Canada, Venezuela, and Mexico assure a plentiful, convenient supply. As a result, the capacity limitation in the United States is on resid conversion unit capacity rather than light crude conversion capacity. Because the rest of the world is well behind the United States in resid conversion capacity, they have even farther to catch up. However, by understanding petroleum macromolecules and their chemistry, resid conversion processes can be greatly improved over their present state. With the past emphasis on refining of light crude oils, petroleum macromolecules have not received the attention of the smaller molecules in petroleum. This book is directed at fulfilling this need.

Because the consumption of petroleum in the United States is at 20.6 million barrels per day whereas the refining capacity is at 17 million barrels per day, it should be clear that the United States already imports refinery products. Canada, as for crude oil, is the largest exporter of refined products to the United States. In the discussion of OPEC, it already has been mentioned that OPEC, especially Saudi Arabia, is building and expanding refineries to export refined products (additional 5.9 million barrels per day by 2012). India with government-controlled ceilings on local prices of refined products is also a net exporter of refined products (17% of total refined) to other countries in Asia, with plans to greatly expand. India is conveniently located across the Arabian Sea from the Persian Gulf and has low-paid, but highly technically qualified, workers. The Reliance Refinery in India, already one of the largest in the world, is being doubled to be the first refinery with a capacity over 1 million barrels per day at a cost of $6.1 billion.[11] Reliance expects to export 90% of their petroleum products and extend the market to Europe, the United States, and Africa. Even government-owned refineries in India are increasing refining capacity with the same objective. Although India is not finding the expected increase in demand for refinery fuel products within India, they are finding refining to be a good international business and project it to remain so in the future. Because India's refineries can command much better prices internationally than locally, exporting makes good business sense. The danger to the United States, of course, is that in the future it could become dependent on the importation of refined products as it is presently dependent on the importation of petroleum. In the near term, this does not seem likely, but it might be a concern in the longer-range future, especially if refineries become overregulated in the United States.

1.2.6 Effect of Threat of Global Warming

There is now consensus among scientists that the temperature of the Earth is increasing and that carbon dioxide in the atmosphere is increasing. There is suspicion that the increase in carbon dioxide is from the burning of fossil fuels and that this is promoting the increase in temperature, along with other greenhouse gases, such as methane, nitrous oxide, and gaseous fluorocarbons. However, others argue that this could be part of a normal cycle of the Earth, not affected much by man. Certainly, the warming of oceans promotes the frequency of hurricanes and typhoons as has taken their toll in 2004 and 2005. How much should we reduce the burning of fossil fuels, including petroleum, because of the fear of causing global warming? Too tight a restriction on the use of fossil fuels would cause the economy to contract, but in 2005 the United States observed the strong negative impact on their economy and human lives caused by hurricanes. We cannot ignore the possibility that unless the use of fossil fuels is reduced, the Earth will continue to get warmer and hurricanes will continue to be more frequent and more severe.

The 1997 Kyoto Protocol[12] was an attempt to reach an agreement among nations to reduce the emissions of greenhouse gases. Upon ratification by Russia (of the agreement reached for reduction by 55% of the carbon dioxide emissions in 1990 by countries included in Annex I) on November 18, 2004, this agreement came in force on February 16, 2005. By 2006, a total of 162 nations have ratified the agreement, with the notable exceptions of the United States and Australia. Under the agreement, by 2010, industrialized nations will reduce the total emissions of greenhouse gases by 5.2% of that in 1990. The reduction requirements vary with country. For example, the reductions required by the European Union is 8%, by the United States is 7%, by Japan is 6%, and by Russia is 0%. Australia is permitted an 8% increase. Because China and India are not included among the Annex I countries, they are exempt from any restrictions, despite China being the country with the second largest emission of greenhouse gases. However, on a per-person basis, the emissions of greenhouse gases by China and India are among the smallest in the world. Although many claim that the Kyoto Protocol does not go far enough, the United States, the largest emitter of greenhouse gases, refuses to sign because it feels the degree of reduction is unfair relative to other countries and that such a reduction will harm its economy.

Whatever the status of the agreement on the Kyoto Protocol, it is clear that the United States needs to move in the direction of reducing carbon dioxide emissions as opposed to the continued increase that we have come to expect. The first reductions should be those that are positive for both our economy and the environment. We have already discussed the need to conserve the use of petroleum-derived fuels for vehicles. This has the positive side effect of reducing carbon dioxide emissions. The manufacture of biofuels, such as ethanol, butanol, and biodiesel, promises to consume as much carbon dioxide by agricultural growth as produced from combustion as long as fossil fuels are not used for refining them. Electrical power generation is the industry with by far the greatest carbon dioxide emissions. We already need to burn more natural gas than coal to reduce emissions of gases that cause acid rain, and sulfur and nitrogen oxides. This switch has the added feature of a reduction in carbon dioxide emissions because methane combustion produces more water vapor relative to carbon dioxide than the combustion of coal. An increase in electrical power generation from nuclear, water power, solar, and wind sources is even better as they produce no carbon dioxide or acid rain.

Petroleum refining is the industry with the second largest emission of carbon dioxide, and energy is second only to crude oil among direct costs to refineries. Therefore, refineries have the opportunity to reduce substantially their operating costs and the emission of carbon dioxide at the same time. One method is to optimize the heat exchanger network in the refinery. Another method is to mitigate the fouling of heat exchangers and fired heaters. Foulants cause energy loss by forming insulating layers on heat exchange surfaces. Fouling mitigation has the extra benefit of extending the period between refinery shutdowns for maintenance, thereby increasing the effective refinery capacity. Methods to accomplish fouling mitigation are discussed in chapter 6.

Carbon dioxide can be absorbed out of power plants or refinery smokestacks, liquefied, and pumped into oil wells to increase the recovery of oil (secondary oil recovery). About half the carbon dioxide injected remains within the formation. Therefore, carbon dioxide flooding, with increased petroleum production, is the most economical way to sequester carbon dioxide. Unfortunately, power plants are normally not located near petroleum production sites. The alternatives are pipelining carbon dioxide to petroleum production sites or sequestering the carbon dioxide as a supercritical gas in deep saline reservoirs (CCS, carbon capture and sequestration). In addition, there is not enough petroleum reservoir volume in the United States to contain all the carbon dioxide from the burning of fossil fuels at power plants and refineries. The good news is that there are no technical hurdles to widespread CCS.[13,14] Of course, the consumer will have to pay much more for energy, but if introduced gradually, the economy will not collapse.[13,14] Technological advances should even cushion the economic impact, taking advantage of the advanced tools of petroleum engineers and geologists. Canada is already moving to CCS to reduce the impact of exploiting the tar sands in Alberta on greenhouse gas emissions.[15] Meanwhile, the United States is waiting for new public policy and legislation by a more enlightened government and public.

1.3 ORIGIN OF PETROLEUM

The origin of petroleum is still not completely understood, the details of which are still being argued at scientific meetings. Nevertheless, for the purposes of this book, a simplified version of the generally accepted view will be given. Those wishing more details should consult the book by Tissot and Welte.[16] Petroleum primarily was formed by the decay of plants and animals in an environment nearly free of oxygen. However, the decay of microscopic plankton organisms is believed to be largely responsible for most petroleum. These organisms usually accumulated in either river deltas or at the bottom of oceans and were covered by silt, sand, or sedimentary rocks (formed from fossils of shells and bones). Under high pressures and temperatures caused by eventually being buried at great depths, the organic matter initially formed solid kerogen, insoluble in water and common organic solvents.

Over long periods of geological time, the kerogens partially broke down to form liquids that were capable of flowing or migrating. If the liquids flowing through porous rock became contained by basins of nonporous rock, they might have accumulated in pools large enough to form a reservoir that can be economically produced through wells. Sulfur largely became incorporated into organic compounds by reaction with inorganic sulfur compounds in the rocks. Likewise, the magnesium complexed to the porphyrin, chlorophyll, in plants, and the iron complexed to the porphrin, hemoglobin, in animals, become exchanged with the more stable, vanadium oxide and nickel. Because petroleum was formed under water, the final petroleum reservoir often contains water, usually seawater containing dissolved sodium, calcium, and magnesium chlorides. Reservoirs also often contain gases that were a by-product of the petroleum maturation process. These gases are usually mostly methane, but can include ethane, propane, butanes, carbon dioxide, and hydrogen sulfide. The gases are dissolved into the petroleum up to their solubility limit to form what is termed "live" oil. Insoluble gases can also form a gas cap above the reservoir if the formation is capable of containing it.

Oil composition is mainly caused by the nature and evolution of the source. Marine organic matter usually formed paraffinic–naphthenic (containing both linear alkanes and cycloalkanes) or aromatic (less than 50% saturated hydrocarbons) crude oils. Terrestrial organic matter from plants usually generated paraffinic (containing mostly normal and isoalkanes) crude oils. However, at higher temperatures caused by higher depths of burial, and at longer maturation times, the oils in the reservoirs can be thermally cracked to form lighter (lower-density) oils containing higher fractions of low-boiling-point liquids. The thermal reactions can also convert paraffinic–naphthenic oils into aromatic oils. On the other hand, reservoir oils can become lighter by the precipitation of larger, more aromatic molecules caused by the solubility of light paraffinic hydrocarbon gases into the oil (called *deasphalting*).

If the petroleum migrates close enough to the surface to be exposed to flowing water containing oxygen, it likely will be biologically degraded by microorganisms. Unfortunately, the appetite of microorganisms is similar to that of man for the more paraffinic and lower boiling portion of the petroleum. As a result, heavily biodegraded oil has little oil boiling in the range of gasoline and diesel fuel. Such oil[17] is called heavy oil (viscosity of 100 to 10,000 centipoises at reservoir conditions and specific gravity of 0.93 to 1.00 g/mL), or extra heavy oil and bitumen (viscosity of greater than 10,000 centipoises at reservoir conditions and specific gravity greater than 1.00 g/mL). Examples of extra heavy oil are the tar sand deposits in northern Alberta, Canada, and the Cero Negro region of Venezuela. Each is reported to have similar amounts of petroleum as exists in Saudi Arabia, albeit much heavier (higher density) and less valuable. Nevertheless, with the increase in price of light crudes, production in both of these regions has recently greatly expanded by upgrading the extra heavy oil to synthetic crude oil near the production sites. Most of the upgraded heavy oil in Venezuela and Canada is sold to refineries in the United States. In addition, heavy oil without upgrading from Venezuela, Mexico, and Canada is often diluted with gas condensate or light crude for lower viscosity for transportation and sold to refineries in the United States. This is particularly convenient by pipeline from Alberta, Canada, to the midwest and northwest United States. Recently, a pipeline to the midwest from the Gulf of Mexico was reversed to allow oil from Alberta to flow to the refineries along the Gulf of Mexico and compete with oil from Mexico and Venezuela. As a result, the United States imports more petroleum from Canada than any other country, and most refineries in the United States have processes to convert the higher boiling fraction of petroleum into transportation fuels (usually delayed coking). These heavy and extra heavy oils contain the high proportions of petroleum macromolecules that are emphasized in this book. Because they were formed by the removal of the smaller molecules, it makes sense that even light crude oils usually contain petroleum macromolecules, but not the high proportions as in those that were biologically degraded. In addition, most heavy and extra heavy crude oils contain significant levels of naphthenic acids, a by-product of the biological degradation process. Naphthenic acids are discussed in section 2.6.2.

1.4 PRODUCTION AND TRANSPORT OF PETROLEUM

The emphasis of this book is on the upgrading and refining of heavy oils. Therefore, only a simplified description of the production and transport of petroleum will be given to provide background to the reader. Conventional oil wells are drilled by rotary drilling rigs. Offshore rigs are built on ships or barges. The derrick is the support structure that holds the drilling equipment, with a pulley system that allows for raising and lowering the drill pipe and casing. The drill pipe is formed by screwing together sections with a drill bit at the end. The casing is a larger-diameter concrete pipe that lines the drill hole and prevents the hole from collapsing. The drill is rotated by its connection to a rotating turntable at the surface. Mud is pumped down the drill pipe through the drill bit and up through the space between the drill pipe and the casing. This removes rock cuttings from the drill bit to the surface. High-pressure valves under the rig are set to open and release pressure above a preset value to prevent an uncontrolled gush of oil and gas to the surface that could blow out the well or cause a fire.

Once the final depth is reached, the well is completed to permit the oil to flow into the casing. A perforating gun is lowered into the well to the production depth, and explosive charges are set off to create perforating holes in the casing. After the drill is removed, tubing is placed in the hole to transport the oil, water, and gas to the surface. A packer is run down the outside of the pipe and expanded just above the production level to form a seal. Often the well needs to be stimulated to get sufficient flow of oil into the well. If the formation is mainly limestone, hydrochloric acid is pumped down the hole to form channels in the rock. If the formation is mainly sandstone, sand suspended in water is pumped down the well to form fractures in the formation.

If the pressure is high enough and the viscosity of the oil is low enough, the oil flows to the surface naturally through a series of valves at the surface called a Christmas tree. Otherwise, a pump is required. Often, a reciprocating, horsehead pump is used. This pushes and pulls a rod down the center of the tubing and attached to the pump in the production zone. If the viscosity of the oil is too high for pumping, a second hole may be drilled into the production zone and steam injected. The higher temperature lowers the viscosity of the oil, and the additional pressure helps push the oil to the well so that it can be pumped to the surface. Another method is to drill the well vertically to the production zone and horizontally in the production zone. This greatly increases the surface area of the perforated casing. Thus, even if the rate of flow of oil is low, the accumulated flow of oil through the large surface area can be high. For extra heavy oil (bitumen), a combination of steam injection into a horizontal well above a second horizontal well for production, called steam-assisted gravity drainage (SAGD)[18] may be used. Because the Athabasca tar sands is not in porous rock, the deposits within 100 meters of the surface can be mined by giant shovels and dumped into huge trucks or pumped as a slurry in water through pipes. The bitumen is separated from clay and sand by mixing with caustic and hot water. More recently, a hydrocarbon liquid similar to *n*-hexane has been piped to the site to dissolve the bitumen off the sand and clay. This method has the additional advantages of leaving the least soluble asphaltenes on the inorganic material and reducing the use of water that is in scarce supply.[19]

Typically, only about 25 to 33% of the oil is removed from reservoirs. However, when the oil price is high as at present, more expensive methods have been developed to remove more oil from reservoirs, called secondary or enhanced oil recovery. In secondary recovery, water or gases are injected into the reservoir to increase the pressure and displace the oil. Although water is better at displacement, gases, such as methane, ethane, or carbon dioxide, can dissolve into the oil and reduce the viscosity. In enhanced oil recovery, such methods are used as flooding with water-soluble polymer mixtures or with mixtures of water and surfactants.[20]

Petroleum coming out of a well often contains seawater, clay, sand, and dissolved gases. Most of these are usually removed at a central processing site and where crude oils from many wells are blended to obtain more uniform properties.

1.5 REFINERY PROCESSING

There are probably no two refineries that are alike. They differ according to the type of crude oils they are designed to run and the range of products they are designed to produce. Different countries have different specifications on products, especially those related to the environment. A minor number of refineries produce lubricating oils or are connected to petrochemicals production. Of course, the volume of crude oil processed can vary from only about 6000 to one million barrels per day (under construction in India). The objective of this book is not to describe refineries in detail, but refinery processing is summarized in this section to provide the reader enough background to understand the context of processing heavy oils and resids. For more detail about refinery processing, the reader should consult Gary, Handwerk, and Kaiser,[21] and Self et al.[22] and about resid processing, the book by Gray[23] is recommended.

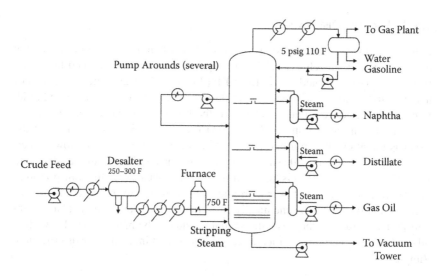

FIGURE 1.1 Schematic of a crude-processing unit. (From F Self, E Ekholm, and K Bowers, *Refining Overview: Petroleum Processes and Products*, AIChE, 2000. With permission.)

1.5.1 CRUDE PROCESSING UNIT

Nearly all refineries have a least one crude processing unit with atmospheric distillation as shown in the schematic of Figure 1.1. Crude oil is pumped from large tanks through heat exchangers that heat it to 250 to 300°F under pressure to a desalter. Desalters are described in more detail in chapter 7, section 7.1, but basically the crude oil is mixed with fresh water and then separated into two phases with the help of an electric field to coalesce water drops. The water containing soluble salts flows out of the bottom to a water treatment plant. Meanwhile, the oil flowing out the top of the desalter is further heated by heat exchangers and by a furnace to 650 to 750°F (343 to 399°C) and then separated into various streams in an atmospheric fractionator where part of the oil flashes above as a vapor and part flows down as a liquid. Steam strips the gas oil out of this liquid, and it flows out the bottom as the atmospheric resid. The fractionator contains trays for contacting liquid and vapor. At various points the liquid is withdrawn, cooled by heat exchangers, and pumped to a lower tray in the tower (pump arounds) to make sure there is sufficient vapor–liquid contact even though the liquid is withdrawn. Side streams are drawn off at various points: atmospheric gas oil (650 to 725°F or 343 to 385°C), heavy distillate (525 to 650°F or 274 to 343°C), light distillate (365 to 525°F or 185 to 274°C), and naphtha (200 to 365°F or 93 to 185°C). Light straight-run gasoline (90 to 200°F or 32 to 93°C) is taken overhead and cooled where gas and water are separated. Steam or reboiler stripping is used to remove the light tail off each of the side streams. The smallest refineries have only the crude processing unit with just enough hydrotreating capacity to meet product specifications, such as sulfur content. The atmospheric resid is sold for heavy fuel oil to electrical power plants or as bunker fuel for ships.

1.5.2 VACUUM DISTILLATION

Figure 1.2 shows a schematic of a complex refinery that includes more processing units than nearly all refineries. The atmospheric bottoms are pumped through a furnace where it is heated to 730 to 850°F (388 to 454°C) and into a fractionator under 25 to 50 mm Hg where the vapor flashes off and cools the remaining liquid. Because the residence time is greatest at the bottom, it needs to be kept below thermal cracking temperatures (650 to 700°F or 343 to 371°C). Vacuum on the tower is maintained with a barometric steam ejector system on the tower over-head. One or two gas oil side streams are withdrawn and the vacuum resid (950 to 1050°F+ or 510 to 566°C+) is taken off the bottom. Because all the transportation fuels have boiling points lower than atmospheric resid, all the products of vac-uum distillation need to be chemically converted to form lower-molecular-weight transportation fuels. Otherwise, the vacuum resid can be blended into heavy fuel oil or, if meeting certain specifications, used as asphalt in making roads or roofing shingles.

1.5.3 FLUID CATALYTIC CRACKING

The fluid catalytic cracking (FCC) process is the greatest refinery moneymaker. It takes heavy oil, vacuum gas oil and some atmospheric resids, that can only be used for heavy fuel oil and selectively converts them into lower-boiling trans-portation fuels, primarily gasoline, which have much greater value. An extra bonus is that cracking and isomerization during FCC lowers the density and thus increases the volume of liquids. Because refinery products are sold by the volume, this further increases the profits earned by FCC. All this is achieved today by combining very active acidic catalyst and oil at a high-temperature, short-contact-time riser reactor. The oil is separated from the solid catalyst using cyclones and stripping steam. The coke is burned off the catalyst in a fluidized bed regenerator to provide the heat for the process where the equilibrium cata-lyst is withdrawn and fresh catalyst is added. The oil from the reactor is fed to a fractionator where it is separated by distillation into light fuel gas, C_3/C_4 gas for alkylation, gasoline, heavy cat naphtha, light cat cycle oil, heavy cat cycle oil, and clarified slurry oil (also called decant oil or cat fractionator bottoms). As the name implies, part of the cycle oils may be recycled to the reactor for further conversion, whereas the rest is withdrawn as product. The greatest advantage of acidic cracking over thermal cracking is much lower yield of hydrocarbon gases relative to gasoline.

FCC feed may contain atmospheric resids, but it requires much higher qual-ity feeds than coking or hydroconversion. This is because of expensive zeolite catalysts; intolerance to sodium, nickel, vanadium, and basic nitrogen; as well as limitations on the amount of coke that can be burned in the regeneration step by cooling capacity. As a result, the feed in resid catalytic cracking is at worst an excellent quality atmospheric resid, but mixtures of vacuum gas oil and atmo-spheric resids are more common.

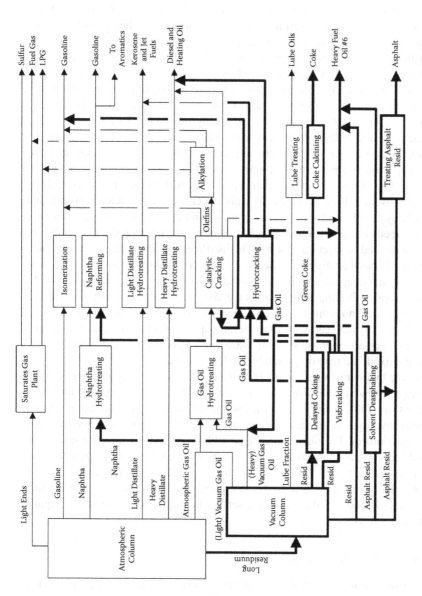

FIGURE 1.2 Schematic of a complex refinery. (From F Self, E Ekholm, and K Bowers, *Refining Overview: Petroleum Processes and Products,* AIChE, 2000. With permission.)

1.5.4 GAS OIL HYDROCRACKING

Because compounds containing polynuclear aromatics have little value except as a heavy fuel oil, there is a need for a process to reduce especially two- and three-ring aromatics into one-ring aromatics and paraffins. Gas oil hydrocracking is the one process with ring reduction as the principle objective. This is accomplished by hydrogenating the aromatic rings and cracking the resulting naphthenes. The acid catalysts used for the cracking are often poisoned by sulfur and basic nitrogen. Therefore, it is usually necessary to desulfurize and denitrogenate them nearly completely in the hydrogenation step. In addition, any vanadium, nickel, and olefins are removed. This type of process is only worthwhile for a refinery having sufficient streams with high concentrations of two- and three-ring aromatics. As a result, on average gas oil hydrocracking capacity is only 8% of the crude distillation capacity. Whereas most refineries have a cat cracker, a minor fraction of refineries have a gas oil hydrocracker. Typical feeds that contain high concentrations of two- and three-ring aromatics are coker gas oil, visbreaker gas oil, and the FCC cycle oils, such as light cat cycle oil and heavy cat cycle oil. These are all good feeds to a gas–oil hydrocracker. Because the main product is paraffinic, the most desired products are the middle distillates: kerosene, jet fuel, diesel fuel, and home furnace heating oil. Conventional hydrocracking is done at high pressures and high temperatures in a fixed-bed reactor to reach high conversions. As a result, naphtha boiling-range liquids are also produced. Although this naphtha is low in olefins and aromatics, the octane for gasoline is still medium because of isomerization by acid catalysis. However, the quality for solvent applications and reforming feed are quite good. Mild hydrocracking is only carried to low conversions in a fixed-bed reactor, producing little naphtha.

Severe hydrocracking is typically done in a two-stage hydrocracker with the first stage at moderate hydrogen pressure, about 600 to 1000 psig, and moderate temperature, 700°F (370°C), with a hydrogenation catalyst. The gas products, including hydrogen sulfide and ammonia, are flashed off, and the liquids are sent to the fractionator. The unconverted gas–oil boiling-range liquids are sent to the second reactor that contains an acid-cracking catalyst. This second reactor is at high hydrogen pressure, 1500 to 2500 psig, and high temperature, about 800°F (427°C). The gases are flashed off and a purge liquid stream is sent to the fuel oil, whereas the bulk of the liquids are sent to the same fractionator as the liquids from the first reactor.

1.5.5 NAPHTHA REFORMING

After hydrotreating to remove sulfur, the process is reversed in reforming to dehydrogenate naphthenes to form aromatics. This is done because aromatics have much higher octane than naphthenes in gasoline, the principal product of naphtha. In addition, linear paraffins are isomerized to increase octane. A side benefit is that hydrogen is produced, to be used for hydrotreating. This is the lowest-cost hydrogen available to a refinery. Finally, valuable petrochemical feedstocks, such as benzene, toluene, and xylenes, can be separated from reformed naphtha. It is no wonder that reforming is second only to fluid catalytic cracking as a refinery moneymaker.

1.5.6 RESID PROCESSING

Because atmospheric and vacuum resids are petroleum macromolecules, resid processing is the emphasis of this book and is discussed in detail and, thus, only briefly reviewed here. As North America has plentiful supplies of coal and natural gases to fuel electrical power plants, there has not been a large heavy fuel oil market in North America for some time as opposed to the rest of the world. Therefore, many of the refineries in North America, particularly in the United States, have at least one resid conversion unit, usually delayed coking. Today, with the increased demand for transportation fuels, and increased use of heavy and extra heavy crude oils containing larger fractions of vacuum resids, there is an even greater need to increase the conversion of vacuum resids to transportation fuels both in North America and the rest of the world.

1.5.6.1 Visbreaking

This is a process that initially was used to thermally crack a heavy fuel oil to reduce its viscosity. Today, it is usually used to thermally crack an atmospheric resid to convert 20 to 30% to transportation fuels and the remaining can still be blended into heavy fuel oil. Because it does not convert enough heavy fuel oil and because the vacuum gas oil portion of atmospheric resids is a valuable feed to fluid catalytic cracking, there are very few visbreakers in North America. They are more common in smaller refineries of Europe and Asia that do not have fluid catalytic crackers. A visbreaker is either a furnace coil followed by a fractionator or a furnace coil, an up-flow soaker drum, and a fractionator. They are designed to be at short enough residence time at a given reaction temperature (450 to 475°C) not to produce significant solid coke. In chapter 4, the mechanism and kinetics of coke formation in visbreaking will be introduced, and visbreaking processes will be discussed in chapter 9.

1.5.6.2 Coking

This is by far the most common resid conversion process. Coking processes are thermal conversion processes that completely convert vacuum resids to lighter products and solid coke. Because solid coke is an expected product, a method for its removal is required. Although part of the liquid product is in the vacuum resid boiling range, this is usually recycled back to the coker after removing the lighter liquids in the coker fractionator. By far the most common coker is delayed coking, which consists of a furnace coil followed by two tall insulated drums. The hot resid from the furnace coil is pumped into the bottom of one of the drums where the coke accumulates while volatile liquids and gases flow out of the top of the drum to the coker fractionator. When coke fills up the drum, the hot resid is fed to the bottom of the second drum. Meanwhile, the coke in the first drum is cooled and drilled out to form a solid-fuel by-product. If the coke meets certain specifications, it might also be used for making electrodes for aluminum or steel manufacture. There are also coking processes in which the resid is sprayed on a fluidized bed of hot coke. In Fluid Coking, part of the coke is burned to provide heat for the process, with the remainder removed as a fuel coke by-product. In Flexicoking, all but a purge stream of coke is gasified with air and steam to provide the heat for the process and, after absorbing out the hydrogen sulfide, to form a fuel gas (hydrogen, carbon monoxide, and nitrogen) for the refinery or a nearby electrical power plant. Coking processes are discussed in more detail in chapter 8.

1.5.6.3   Resid Hydrotreating and Hydroconversion

These are catalytic processes with hydrogen under pressure designed to add hydrogen to atmospheric and vacuum resids. Hydrotreating is done primarily to remove sulfur, nitrogen, vanadium, and nickel and to reduce Conradson carbon residue (coke-forming tendency) in fixed-bed reactors. Hydroconversion has the additional objective to convert resids to lower-boiling liquids. This may be done also in packed beds, but conversion of vacuum resids is limited to about 50%. By using ebullated-bed reactors (LC-Fining or H-Oil), the spent catalyst can be removed and fresh catalyst added continuously, enabling conversions of the order of 65%. Although not currently commercially practiced, several hydroconversion processes are under development that use dispersed catalysts and promise to achieve 95% conversion of vacuum resids. Hydroconversion is discussed in more detail in chapter 10.

1.6   CHALLENGES FOR PROCESSING PETROLEUM MACROMOLECULES

1.6.1   Aromatic By-Product

As we will learn in chapters 2 and 3, petroleum macromolecules in vacuum resids are not just higher in molecular weight than lower-boiling liquids in petroleum, but they contain higher amounts of heteroatoms: sulfur, nitrogen, vanadium, and nickel. Moreover, the greatest difference is that petroleum macromolecules have more and larger polynuclear aromatics that cause thermal stability and insolubility. As a result, all economical processes that convert petroleum macromolecules to lower-boiling liquids require an aromatic by-product. However, we will show that the existing processes for converting petroleum macromolecules have considerable room for improvement, and will point the direction for the improvement. For instance, even for coking processes that are considered to be mature processes, we will show that lab experiments can achieve 76% yield of nearly equal quality liquid products compared to the highest commercial coking yield of 61%. Part of this is achieved by learning that many polynuclear aromatics are formed during coking and discovering how their formation can be reduced. Thus, although an aromatic by-product of resid conversion cannot be prevented, its yield certainly can be reduced from current commercial practice.

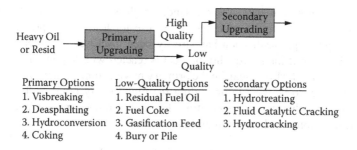

Primary Options	Low-Quality Options	Secondary Options
1. Visbreaking	1. Residual Fuel Oil	1. Hydrotreating
2. Deasphalting	2. Fuel Coke	2. Fluid Catalytic Cracking
3. Hydroconversion	3. Gasification Feed	3. Hydrocracking
4. Coking	4. Bury or Pile	

FIGURE 1.3 Conversion of heavy oils and resids involves two steps.

1.6.2 Produce Feed for Secondary Upgrading

As is shown in Figure 1.3, the objective of the processing of resid at the refinery and the upgrading of heavy and extra heavy oils near the source are to produce, in a primary upgrading step, a higher-quality feed for secondary catalytic processes. Of course, it is desired to maximize the yield of feed for secondary upgrading at a given quality. The advantage of this approach is to remove as much of the catalyst poisons as possible in the primary upgrading step with minimum hydrogen consumption to protect the more expensive catalysts and use hydrogen more effectively in the secondary upgrading step. The secondary upgrading options that combine hydrotreating with either fluid catalytic cracking or hydrocracking are much more selective for the valuable transportation fuels than the primary upgrading options. However, the primary upgrading options are more tolerant for catalyst poisons: sulfur, nitrogen, vanadium, nickel, and polynuclear aromatics. These catalyst poisons are mostly rejected with the necessary aromatic by-product in the low-quality stream. Another advantage of this approach is that the quality of the feed to the primary upgrading processes can vary greatly, but can produce a feed to secondary upgrading of uniform quality.

The primary upgrading options of visbreaking, hydroconversion, and coking have already been discussed with the corresponding low-quality product options of heavy (or residual) fuel oil and fuel coke. The feed to fluid catalytic cracking may have boiling points in the vacuum resid range as long as it meets certain quality specifications. Therefore, a process that separates the vacuum resid into high- and low-quality fractions is a primary upgrading option. Solvent deasphalting, a process used more often in producing heavy lubricating oils, is a possibility that will be discussed in detail in chapter 7. Deasphalting also can be used in combination with visbreaking, hydroconversion, and coking.

There are some locations where the low-quality product from primary upgrading has no market value, as it is near the source of extra heavy oils. In the past, some have just left coke in pits, waiting to find a use. However, with a greater regard for the environment, this is becoming unacceptable. Some then bury the coke underground in the reservoir after removing the extra heavy oil. At least one upgrader partially extracts the extra heavy oil from tar sands with an organic solvent, leaving asphaltenes with the sand, which is returned to the resource.[16] Nevertheless, gasification feed is the future use for low-quality products from primary upgrading. Gasification has already been discussed as a step in Flexicoking. However, if the low-quality product is reacted with oxygen instead of air along with steam, hydrogen can be the primary product (unfortunately along with carbon dioxide) to be used in upgrading the high-quality fraction. The energy released can supply steam for generating electricity, production of extra heavy oil from the reservoir, and heat for upgrading.

1.6.3 MINIMIZE HYDROCARBON GAS BY-PRODUCT

The chemical equilibrium products of thermal conversion of hydrocarbons are coke and hydrocarbon gases, such as methane, ethane, propane, and butanes. Instead, the upgrader of petroleum resids is rewarded for maximizing the yield of the intermediate, volatile liquids, and minimizing the yield of coke and hydrocarbon gases. Especially, the production of hydrocarbon gases is unfortunate because they contain such high concentrations of hydrogen that is sorely needed for high-quality volatile liquids. Therefore, producing hydrocarbon gases in upgrading processes can be equated to wasting hydrogen, a valuable reagent.

1.6.4 MITIGATE FOULING

The processing of petroleum is done continuously on a huge scale. Especially today, when upgraders and refineries are operating at full capacity, any cause for one or more units to be shut down is extremely costly. However, polynuclear aromatics in petroleum macromolecules have a strong tendency to become insoluble in the oil and to foul and form coke on the heat exchanger, furnace tube, and reactor walls as well as on catalysts. This can cause process units to be shut down frequently for scheduled and unscheduled cleaning. In addition, these foulants insulate surfaces designed for heat transfer and result in greatly increased energy cost, second only to crude oil as the direct cost to a refinery. As a result, there is a strong emphasis in this book on fouling mitigation. Chapter 4 on thermal conversion kinetics will show that coke formation is triggered by the phase separation of converted asphaltenes. Chapter 5 on phase behavior will provide the oil compatibility model, a tool for predicting how to keep asphaltenes in solution. Finally, Chapter 6 will describe a systematic procedure for mitigating fouling from any possible source, including asphaltenes and coke.

1.7 CHEMICAL ENGINEERING APPROACH

Research on petroleum macromolecules greatly benefits from an interdisciplinary approach involving chemists, physicists, chemical engineers, material scientists, mathematicians, geologists, biologists, etc. With the exception of geology, chemical engineers usually have taken courses in the other disciplines and understand their approach to research. However, those in these other disciplines rarely take a course in chemical engineering. Therefore, if they are to learn the chemical engineering approach, they must do it during years of industrial research. Meanwhile, chemical engineers traditionally have the role of scaling up research in the lab to the pilot-plant scale in development, and then from the pilot-plant scale to the commercial scale to design industrial scale processes and to oversee operations in the refinery. Therefore, petroleum companies find that, having a minor number of chemical engineers in basic research, is useful to ease the movement of ideas in the lab to development where chemical engineers are in a majority. Although this is true, research chemical engineers have a different approach to research than the other scientists, the value of which is often not recognized. This book offers examples of the research approach of one research chemical engineer, but this section is meant to introduce the reader to this approach in an exaggerated, humorous way.

Physicists are mostly concerned with measurements and developing or perfecting instruments to make the best measurement. Therefore, they are defined as those who make very accurate measurements on very impure materials. Yes, they are not very concerned about the materials, only that they make accurate measurements. On the other hand, chemists are extremely concerned about materials and want to know the chemical structure of every molecule in the material. Therefore, chemists are defined as those who make very inaccurate measurements on very pure materials. Yes, they are not much concerned about the quality of their instruments, only that they have very pure materials of known chemical structure.

Of all the engineering disciplines, only chemical engineering applies both physics and chemistry. The author may be biased, but he thinks that chemical engineering combines the best of both fields. Therefore, chemical engineers are defined as those who make very inaccurate measurements on very impure materials. One might wonder why this is an attribute. In the time it takes the physicist to perfect his instrument, and the chemist to purify his material, the chemical engineer has scaled up, developed, and commercialized a process to manufacture this material. How is this possible? Engineers have learned to deal with uncertainty. They have learned to draw a black box around part of a system they do not understand, and by probing the black box and seeing the response, understanding it sufficiently to move on. They do not worry about the details in the black box that are not necessary to accomplish their goal. They tell people that they invented a field called "fuzzy logic" where they measure variables only qualitatively, like hot and cold. Actually, this is a step-out improvement in accuracy over where they used to be called "hairy logic." They use very simple mathematical models. When these models do not correctly predict the data, they do not throw away the models. Instead, they calculate the actual to predicted ratio and correlate these ratios on dimensionless plots. They actually give names to these ratios: friction factor, activity coefficient, fugacity, etc. To account for this uncertainty, they insert large safety factors in their process design (lagniappe or lan-yap) and in their cost estimates (contingencies). Thus, when the process comes in under budget and delivers at 120% of design, everyone is happy.

In this book the reader should find that the author applied this chemical engineering approach to his research. If one does not understand the details, the approach is to back off to a level one can deal with and move on. Petroleum is an ultracomplex material because it is made up of over a million different compounds. Rather than try to isolate and identify each molecule, the author backed off and used solubility and adsorption to separate petroleum resids into classes and developed chemical kinetics and phase equilibria using these classes instead of compounds. When he needed to describe the molecules a little closer, he represented the molecules in these classes by different combinations of only two building blocks that he called pendants and cores. Although he did not know the detailed chemical structures within the building blocks, he found that this representation could describe much that has been observed about the process chemistry of petroleum macromolecules. He used hydrogen content and hydrogen balance extensively to evaluate constants in chemical kinetics, measure the attraction energy among petroleum macromolecules, characterize the building blocks, and determine the maximum yield from a coking process. The outcome is that petroleum is actually not so complex because it is composed of mostly carbon, hydrogen, and sulfur atoms. As a result, in many cases, simple models describe petroleum much better than most pure compounds. The author is not satisfied with these limited successes of his models but expands the range of data to test each model and determine where each model fails. This provides the direction to improve the model. Meanwhile, the real objective is not the model or even knowledge, but how to apply the model and the knowledge to develop a new or improved process for heavy oils. Such innovations are not accidental. One needs to work continuously at innovation by determining what the barriers are to improvement. Are these barriers real or a false interpretation of the data? The barriers determined to be false give one the opportunity for step-out improvements in the technology. The barriers determined to be true enable one to turn one's attention to more fruitful areas without wasting time and effort. For example, in chapter 4, the reader will learn that one false barrier was that coke was believed to be a direct-reaction product of resid thermolysis. Instead, the author discovered that coke results from a phase separation that provided the means to greatly delay its formation. However, throughout the book, the reader will learn that polynuclear aromatic units in resids are real thermal reaction limits. Even when it is possible to convert these polynuclear aromatics by hydroconversion, one cannot do it economically. Thus, one must accept that a heavy by-product to resid conversion is a requirement. There is potential in minimizing polynuclear aromatics from being formed during resid conversion, but those that are already present in the resid provide a real barrier.

The danger of the chemical engineering approach to research is to become infatuated with the success of models that conceal the lack of detailed knowledge. Yes, they enable one to move on to devise new or improved processes. However, they remain to be placeholders until the ability for greater detailed knowledge is reached. Once the physicist perfects his instrument and obtains unique data, and once the chemist produces a new understanding of the molecules in heavy oil, the models need to be improved to reflect the new insight. Instead of backing off, one might actually be able to move in and produce a more detailed model. For example, in chapter 2, it is described how high-performance liquid chromatography provided a more detailed view of heavy oils than solubility classes. In the near future, ultrahigh-resolution Fourier transform ion cyclotron resonance mass spectrometry may actually enable one to identify and measure the concentration of nearly every compound in heavy oils. This will enable much more sophisticated and more complex models while providing the means to break down more barriers to process improvements.

REFERENCES

1. IA Wiehe, KS Liang. Asphaltenes, resins, and other petroleum macromolecules. *Fluid Phase Equilibria*, 117: 201–210, 1996.
2. EB Sirota, D Weitz, T Witten, J Isaelachvili. *Complex Fluids*, Vol. 248, Pittsburg: Materials Research Society, 1992.
3. OPEC Web site, www.opec.org, 2006.
4. TC Fishman. *China Inc.* New York: Scribner, 2005.
5. TL Friedman, *The World is Flat.* New York: Farrar, Straus, and Giroux, 2005.
6. Infoplease Web site, www.infoplease.com, 2004 Statistics.
7. Gilson Consulting Web site, www.gravmag.com, 2006 Statistics.
8. P Roberts, *The End of Oil.* Boston: Houghton Mifflin, 2004.
9. U.S. Department of Energy Web site, www.eia.doe.gov, 2005.
10. Reuters. Shell to Begin $7 Billion Port Arthur Refinery Expansion. London/New York, and www.shell.com, September 21, 2007.
11. International Herald Tribune Web site, Bloomberg News, www.iht.com, March 23, 2006.
12. Wikpedia Website, www.en.wikpedia.org, 2006.
13. E Rubin, Overview and Economics of CO_2 Capture. SPE/AIChE Workshop on Practical Strategies for Managing CO_2 Emissions—Today Not Tomorrow, Sonoma, CA, 2008.
14. JR Katzer, The future of coal-based power generation. *Chem Eng Progress* 104(3): S15–S22, 2008.
15. P Dittrick, Oil sands and sustainability. *Oil & Gas J* 106.9: 15, March 3, 2008.
16. BP Tissot, DH Welte. *Petroleum Formation and Occurrence.* 2nd ed. Berlin: Springer-Verlag, 1984.
17. BP Tissot, DH Welte. *Petroleum Formation and Occurrence.* 2nd ed. Berlin: Springer-Verlag, 1984, p. 471.
18. RM Butler. Gravity drainage to horizontal wells. *J Can Pet Tech.* 31: 31–37, 1992.
19. WJ Power. The Athabasca oil sands project: The commercial application of innovations in technology. Proceedings of the 6th International Conference on Petroleum Phase Behavior and Fouling, Amsterdam. Paper No. 9, 2005.

20. DO Shah, RS Schechter. *Improved Oil Recovery by Surfactant and Polymer Flooding.* New York: Academic Press, 1977.

21. JH Gary, GE Handwerk, MJ Kaiser. *Petroleum Refining Technology and Economics.* 5th ed. Boca Raton: CRC Press, 2007.

22. F Self, E Ekholm, K Bowers. *Refining Overview—Petroleum, Processes and Products.* CD-Rom, AIChE—South Texas Section, 2000.

23. MR Gray. *Upgrading Petroleum Residues and Heavy Oils.* New York: Marcel Dekker, 1994.

2 Characterization of Petroleum Macromolecules

2.1 CLASS SEPARATION[1]

Petroleum contains in the order of a million different molecules without a repeating molecular unit. Because the number of possible isomers rapidly expands with increasing molecular weight, most of this diversity exists in the vacuum resid boiling range, the petroleum macromolecules. Thus, it makes little sense to try to identify each petroleum macromolecule (section 2.6.3, discusses progress in doing just this). Instead, petroleum macromolecules are commonly separated into classes in which macromolecules within a class are more similar than those between classes. It is then hoped that different resids or a resid before and after processing can be described in terms of the relative amounts in each of these classes.

2.1.1 VOLATILES AND COKE

Although it is common to separate petroleum macromolecules into saturates, aromatics, resins, and asphaltenes, the procedures used vary widely. Figure 2.1 outlines the general separation procedure used by the author. As this procedure will be applied to vacuum resids before and after processing, it also includes volatiles and coke that are not present in the unprocessed vacuum resid. The volatile liquids are removed after processing by vacuum distillation. The nonvolatile product was mixed in a flask with 15 parts toluene and let sit for at least 16 h at room temperature. This mixture was poured through a fine (4 to 5.5 μm pores) fritted-glass filter. The solids on the filter were washed with at least 25 parts additional solvent, and washing continued until the solvent passed through the filter without color. The toluene insolubles were vacuum dried on the filter at 100°C for at least 16 h to form the class, called coke. The toluene was removed from the toluene solubles by rotary evaporation at 50°C, followed by vacuum drying for 16 h at 50°C.

2.1.2 ASPHALTENES AND RESINS

Asphaltenes are the insolubles separated in the next step that was performed on the toluene solubles for the processed resid or on the unprocessed resid. This followed the same procedure as the toluene separation, except that *n*-heptane was

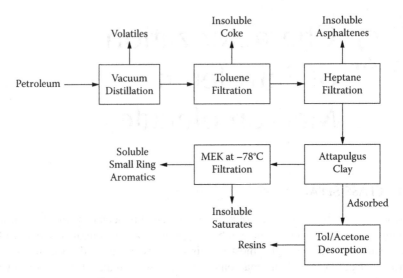

FIGURE 2.1 General separation procedure of unreacted and reacted petroleum oils. (From IA Wiehe. *Ind Eng Chem Res*, 31: 530–536, 1992. Reprinted with permission. Copyright 1992. American Chemical Society.)

used instead of toluene, and *n*-heptane was not removed from the heptane solubles. Instead, the heptane solubles were mixed in a flask with Attapulgus clay, 30 times the weight of heptane solubles, for a minimum of 16 h. The Attapulgus clay used was National Bureau of Standards qualified for ASTM D2007-86, a clay–gel chromographic separation test. This clay contains a controlled amount of water. Not only is this required for reproducible adsorption, but only with water on the clay can one completely desorb the hydrocarbons off the clay. The mixture of clay, *n*-heptane, and oil was filtered with a fine, fritted-glass filter. The clay on the filter was washed with at least 200 parts of *n*-heptane per part of solubles, and washing was continued until the heptane ran through the filter without color. The *n*-heptane was removed from the solution so produced by rotary evaporation followed by vacuum drying at 50°C.

The fraction remaining on the clay was desorbed by mixing with a mixture of 50% toluene and 50% acetone on the filter. At least 100 parts of this mixture per heavy oil was used to wash the clay. This was followed by washing with at least the same amount of 90% toluene and 10% methanol, which was continued until the solvent mixture passed through the clay without color. The solvent was removed by rotary evaporation and vacuum drying at 50°C, leaving the fraction called resins. It is significant that although the resins are quite toluene soluble, toluene alone cannot remove much of the resins from a polar surface. Acetone and methanol need to be present to interact with the polar functionality on the clay surface and weaken the interaction between the clay surface and the resins. This is because acetone and methanol form stronger complexes with the functionality on the clay surface than do the resins. Only then can the resins be desorbed and dissolved into toluene. This is also important when one wishes to clean resins and asphaltenes from process surfaces such as heat exchangers. One needs to select a cleaning agent that both desorbs and dissolves the foulant.

2.1.3 SATURATES AND AROMATICS

The heptane-soluble oil that was not adsorbed on clay was mixed with 30 parts methyl ethyl ketone (MEK) in a flask and cooled to −78°C in a dry ice–isopropyl alcohol bath. After 4 h, this cold mixture was filtered through a fine, fritted-glass filter by adding dry ice directly on the filter. The filtration was done under a flow of dry nitrogen to prevent water from the air condensing and adding to the MEK solution. The waxy solids remaining on the filter were dissolved off with room-temperature n-heptane, and the heptane was removed by rotary evaporation and vacuum drying at 50°C, leaving the oil fraction that was called saturates. The MEK was removed from the solution that passed through the filter by rotary evaporation and vacuum drying at 50°C, leaving the oil fraction that was called aromatics. As we will see, this fraction is more properly called small-ring aromatics because it is not as aromatic as either the resin or the asphaltene fraction, and it has less coke-forming tendency.

2.1.4 REPRODUCIBILITY OF CLASS SEPARATION

It is very difficult to reproduce class separation between laboratories. The procedure has to be followed very closely. For instance, the asphaltene precipitation requires waiting at least 4 h in n-heptane before filtering, or a lower yield of asphaltene will be obtained. This is because one needs to wait for asphaltene flocs to grow in size by agglomeration, or the individual submicroscopic, insoluble asphaltene particles will pass through the filter. Thus, the yields are affected by the filter material as well as the size of the filter pores. The effect of other changes in the asphaltene separation has been reported.[2] However, if the procedure is kept completely the same, the results of the class separation can be very reproducible within the same laboratory.

2.1.5 THE CONCEPT OF ASPHALTENES

As the highest-molecular-weight, most-aromatic fraction usually separated from unprocessed crude oil or resids, asphaltenes have attracted much attention and controversy. This strong attention has encouraged counter arguments from Bunger and Cogswell[3] and from Boduszynski et al.,[4] and Altgelt.[5] These authors correctly point out that asphaltenes are merely the insolubles precipitated from petroleum or resid to produce a slice out of the continuum of molecules. The yield and quality of the asphaltenes so produced depend on the variables of the separation as well as the feed selected. No single property distinguishes asphaltenes from the range of properties of other fractions.[3] When asphaltenes are partially converted, the remaining asphaltenes have different properties than the starting asphaltenes.[3]

Therefore, asphaltenes are not a unique compound group or compound class[4] and it is misleading to treat them as a reactant in chemical kinetics.[3]

Although the author agrees with the data behind the preceding arguments, he does not agree with the conclusions. As will be shown later in this chapter, asphaltenes, as other classes, can be distinguished uniquely in terms of two properties. Instead of denying that asphaltenes cannot correctly be considered a reactant in chemical kinetics, the author in chapters 3 and 4 accepts the challenge of how to change the chemical kinetics so that they can be used as a reactant. Surprisingly, asphaltenes will be found to react thermally according to first-order kinetics, just how one would expect a pure compound to react. As it turns out, the insolubility nature of asphaltenes is important in the thermal kinetics of resid conversion. Therefore, it is fortunate that an insolubility class was selected as a reactant rather than a class based on acidic or basic functionality. Instead, a theme in this book will be that the functionality most responsible for insolubility and thermal stability in asphaltenes are the polynuclear aromatics. Therefore, insolubility in n-heptane and adsorption on clay are good methods to separate classes with similar concentrations of polynuclear aromatics. Yes, one should always realize that each of the classes include a very large distribution of molecules without sharp borders between molecules in different classes. However, one should still try to develop the simplest model that describes the chemistry and physics of this complex system rather than claim that it is impossible. As will be demonstrated many times in this book, the important features can be captured in simple models of these complex systems to produce very useful and predictive approximations.

2.1.6 Unreacted Resid Fractions

Cold Lake 1050°F+ resid was separated by the preceding separation procedure with yields and analyses of the fractions shown in Table 2.1. Typically, the mass and elements are well balanced except that, because of air oxidation, the fractions total higher oxygen and lower hydrogen than the starting resid. This does not seem to affect the coke-forming tendency, reported as Conradson carbon residue (CCR) and measured by the microcarbon method, because the sum of the CCR of the fractions equals the CCR of the starting resid. The terms, saturates and aromatics, have only relative meaning and should not be taken on an absolute basis. Thus, the saturates are more saturated than the other fractions, but still contain aromatic carbon and sulfur. A long paraffinic chain on a single aromatic ring will cause this molecule to separate with the saturates fraction. One should expect petroleum macromolecules to have multiple functionalities in the same molecule. Thus, the best the class separation can do is to separate according to the dominant functionality of the particular macromolecule. Nevertheless, the saturates fraction contains little or no oxygen, nitrogen, vanadium, and nickel, and exhibits little coke-forming tendency. The aromatics fraction, less aromatic than the resins and asphaltenes, but more aromatic than the saturates, is also relatively free of heteroatoms other than sulfur.

TABLE 2.1
Analysis of Cold Lake Vacuum Resid

Fraction	Yield (wt%)	C (wt%)	H (wt%)	H/C (atomic)	S (wt%)	O[a] (wt%)	N (wt%)	V (ppm)	Ni (ppm)	VPO[b] MW	CCR[c] (wt%)	% Arom C
Saturates	18	84.54	12.31	1.73	2.74	0.0	0.03	0	0	920	1.8	15
Aromatics	17	81.87	10.00	1.46	5.56	0.0	0.12	0	0	613	8.7	37
Resins	40	82.08	9.50	1.38	6.09	1.72	0.77	260	110	986	24.6	45
Asphaltenes	25	81.93	7.94	1.15	7.50	1.55	1.15	740	230	2980	49.6	50
Total	100	82.45	9.70	1.40	5.75	1.08	0.62	289	102	1040	24.2	40
Starting	100	82.95	9.92	1.42	5.73	0.77	0.77	300	100	1183	24.0	41

[a] Oxygen determined by neutron activation.

[b] VPO number average molecular weight measured in o-dichlorobenzene at 130°C.

[c] Conradson carbon residue measured by microcarbon residue technique (ASTM Test D4530).

Source: IA Wiehe. *Ind Eng Chem Res.* 31: 530–536, 1992. Reprinted with permission. Copyright 1992. American Chemical Society.

The resins fraction is sometimes called "polar aromatics" because of its high oxygen and nitrogen content relative to saturates and aromatics. However, the average resin molecule has only one oxygen or one nitrogen atom out of a molecular weight of 986. Thus, it is not as polar as many common low-molecular-weight compounds that contain oxygen or nitrogen, such as acetone or pyridine. As will be discussed later, the present evidence indicates that resins adsorb on clay more because they contain large-ring aromatics than because of oxygen and nitrogen functionality. Nevertheless, there is a sharp increase in the concentration of vanadium, nickel, molecular weight, and CCR in going from aromatics to resins. On the other hand, the asphaltenes, as is well known,[3] have still higher concentration of each of the undesirable components and properties: sulfur, nitrogen, vanadium, nickel, molecular weight, CCR, and aromatic carbon. It is no wonder that problems in converting resids usually force one to focus on asphaltenes. However, because the yield of resins is much higher than the yield of asphaltenes, as a percentage of the resid, the resins have comparable amounts of undesirable components and properties as the asphaltenes.

One of the odd features of the separation of resids is that the fractions have nearly constant carbon content. For this separation, all the carbon contents are within 83 ± 1.5 wt%. The lower hydrogen content in the more aromatic fractions is compensated by the higher heteroatom content, keeping the carbon content nearly constant. As we shall see, this feature also carries over to resid thermal conversion products. Thus, the hydrogen-to-carbon atomic ratio that tends to measure aromaticity (lower value for more aromatic) is really equivalent to a constant times the hydrogen content for these fractions.

2.1.7 FRACTIONS OF THERMALLY CONVERTED RESID

The Cold Lake 1050°F⁺ resid was thermally reacted for 60 min at 400°C in small batch-tubing bomb reactors under 7 MPa of nitrogen. After the gases were vented, the resid product was separated according to the separation scheme, but without vacuum distillation. The yields and analysis of each fraction are shown in Table 2.2. The gases and the fraction that volatilized during solvent removal were not analyzed, and the molecular weight of the coke could not be measured because of its insolubility in o-dichlorobenzene.

Again, the carbon content is relatively constant, with the carbon content of saturates being slightly higher than that of coke. The average molecular weight of each of the fractions, except coke, is decreased by thermal conversion. Although the hydrogen content of saturates and aromatics are not significantly changed by thermal conversion, the hydrogen content of resins and asphaltenes are significantly decreased. This demonstrates one of the objections of Bunger and Cogswell[3] in treating asphaltenes as a reactant. Asphaltenes change properties during conversion, as well as converting to other fractions. The conversion is more of a continuum than a discrete transformation. In chapter 4, kinetic models

TABLE 2.2
Analysis of Fractions of Cold Lake Vacuum Resid Reacted at 400°C for 60 Min

Fraction	Yield (wt%)	Carbon (wt%)	Hydrogen (wt%)	H/C (atomic)	Sulfur (wt%)	Nitrogen (wt%)	VPO[a] mol. wt
Gases	1.5						
Lt. volatil.	12.3						
Saturates	12.9	85.25	12.53	1.75	2.18	0.08	690
Aromatics	25.8	83.77	10.10	1.44	5.04	0.11	470
Resins	20.5	81.72	8.27	1.21	5.26	1.15	899
Asphaltenes	18.3	83.16	6.34	0.91	7.23	1.50	2009
Coke	8.7	82.19	5.54	0.80	7.63	1.78	

[a] VPO number average molecular weight measured in *o*-dichlorobenzene at 130°C.

Source: IA Wiehe. *Ind Eng Chem Res.* 31: 530–536, 1992. Reprinted with permission. Copyright 1992. American Chemical Society.

will be developed by assuming two asphaltene species, unreacted and converted, to account for reactions that change asphaltene properties without completely converting it to other fractions. Nevertheless, toluene-insoluble coke, not present in the resid, forms and exceeds the asphaltenes in concentration of aromatics (measured by H/C [hydrogen-to-carbon]), sulfur, and nitrogen.

2.1.8 FRACTIONS OF THE THERMAL CONVERSION OF RESID FRACTIONS

A more direct way to track resid thermal conversion is to react the resid fractions individually and then to separate the products according to the separation scheme, recognizing that reactions and intermolecular interactions among the fractions may alter the chemistry when reacted together in the resid. This was again performed for 60 min at 400°C in tubing bomb reactors. A summary of the result is shown in Table 2.3. To have sufficient quantity for reaction, the saturates and aromatics were reacted together in the proportion found in the resid.

In each case the thermal conversion of a fraction formed the next more aromatic and higher-molecular-weight fraction and the whole series of less aromatic and lower-molecular-weight fractions. Thus, saturates plus aromatics formed the more aromatic and higher-molecular-weight resins with a by-product of gas and light volatiles that evaporate with the solvents. The resins formed more aromatic and higher-molecular-weight asphaltenes with a by-product of lower-molecular-weight and less aromatic saturates and aromatics. The asphaltenes formed more aromatic and higher-molecular-weight coke with by-products of lower-molecular-weight and less aromatic resins, aromatics, and saturates. Reactants of increasing

TABLE 2.3

Analysis of Fractions of Cold Lake Vacuum Resid Fractions Reacted at 400°C for 60 Min

Reactant	Product	Yield (wt%)	C (wt%)	H (wt%)	H/C (atomic)	S (wt%)	N (wt%)	VPO[a] mol wt
Saturates + aromatics	Volatiles	29.5						
	Saturates	32.1	85.59	12.50	1.74	1.79	0.0	694
	Aromatics	38.4	83.95	10.59	1.50	4.17	0.0	345
	Resins	19.4	83.06	9.01	1.29	5.32	0.0	839
Resins	Volatiles	10.8						
	Saturates	5.7	84.36	12.03	1.70	2.77	0.17	670
	Aromatics	30.8	81.60	9.99	1.46	4.78	0.12	442
	Resins	30.6	82.01	8.44	1.23	4.60	2.18	804
	Asphaltenes	22.1	83.17	6.66	0.95	6.76	1.98	1841
Asphaltenes	Volatiles	10.4						
	Saturates	2.6	83.87	12.63	1.79	2.54	0.0	
	Aromatics	14.2	81.79	10.36	1.51	4.92	0.0	422
	Resins	12.4	81.60	8.05	1.18	6.68	0.90	622
	Asphaltenes	21.0	82.00	6.18	0.90	7.96	2.18	1557
	Coke	39.4	82.42	5.50	0.80	7.92	1.73	7525

[a] VPO number average molecular weight measured in *o*-dichlorobenzene at 130°C.

Source: IA Wiehe. *Ind Eng Chem Res.* 31: 530–536, 1992. Reprinted with permission. Copyright 1992. American Chemical Society.

aromaticity and molecular weight produce an increasing yield of higher-molecular-weight, more aromatic by-product at the same thermal reaction conditions.

As was seen with the whole resid, the saturates and aromatic fractions maintain similar aromaticity (H/C atomic ratio) during thermal conversion, but decrease molecular weight. This is even the case when saturates and aromatics are formed from resins and asphaltenes. On the other hand, the resins and asphaltenes become more aromatic and decrease in molecular weight prior to being converted to another fraction. Again, the situation is no different if the resin or asphaltene is the reactant, a product of another fraction, or a product of the entire resid. One difference is that resins formed from saturates and aromatics contain below-detectable nitrogen level. Thus, resins need not contain nitrogen to adsorb on Attapulgus clay, showing that the separation is made on the basis of aromaticity or the size of aromatic rings.

2.2 SOLVENT–RESID PHASE DIAGRAM[1]

In this section, the type of data just described on class separation of resids and resid thermal conversion products will be organized on a compositional map. This will enable one to picture how the distribution of types of petroleum macromolecules exist in resids and how they change with thermal processing.

2.2.1 THE CONCEPT OF A COMPOSITIONAL MAP

Long[6] was the first to suggest that the distribution of asphaltenes could be displayed on a conceptual heavy-oil map of molecular weight versus polarity. This was later extended by Long and Speight[7] to a real map of the composition of heavy oil in terms of molecular weight versus solubility parameter. They used desorption off Attapulgus clay with various solvents to separate Cold Lake crude and Arab Heavy vacuum resid into eight fractions. Using the solubility parameter of the desorbing solvent and the 95% and 5% points in the molecular-weight ranges determined by gel permeation chromatography (GPC), they found that these two feeds gave similar molecular-weight ranges for fractions eluded by the same solvent. However, the relative amounts of the fractions were different for the two feeds.

In agreement with Bunger and Cogswell,[3] this author has found that no property uniquely differentiates one petroleum fraction from another. Therefore, possibly more important than the exact selection of GPC molecular-weight and solubility parameter as the independent variables in a compositional map is the concept of Long and Speight that petroleum fractions may be distinguished by combinations of two properties: one that measures molecular attraction and one that measures molecular size. The solubility parameter of the eluding solvent is neither a direct or revealing measure of molecular attraction. Although measuring molecular-weight distribution is preferred to measuring average molecular weight, GPC suffers from adsorption effects with petroleum fractions, which become more severe after partial conversion. Therefore, different independent variables for a compositional map were further investigated.

2.2.2 MEASUREMENT OF MOLECULAR ATTRACTION

The feature that causes petroleum fractions to adsorb strongly on Attapulgus clay and to be insoluble in paraffinic liquids is their aromaticity. Because the more aromatic fractions have higher oxygen and nitrogen content, it is not easy to discount completely the influence of these heteroatoms. However, as already pointed out, resins formed from saturates and aromatics had no nitrogen, showing that it was not necessarily a part of resins. In addition, the average resin of molecular weight 600 to 1000, and the average asphaltene of molecular weight 1500 to 3000, have one or fewer oxygen or nitrogen atoms, yet the resins have at least 45% aromatic carbons and asphaltenes have at least 50%. Thus, while oxygen and nitrogen functionality might play a small role, aromaticity is the dominant cause of the adsorption on solids and insolubility in paraffinic liquids. This conclusion contradicts the current view,[8] but more evidence will be given in chapter 5. However, if one wanted to extend this approach to coal liquefaction products where oxygen functionality plays a major role, a measure of oxygen functionality would be required in addition to aromaticity and molecular weight. This was the conclusion of Snape and Bartle[9,10] who found that three independent variables, such as number average molecular weight, the proportion of internal aromatic carbon to total carbon, and the percent of acidic OH, were needed to distinguish among benzene insolubles, asphaltenes, and n-pentane solubles for coal and petroleum-derived liquids. Thus, their map is three dimensional.

The simplest measure of aromaticity is the hydrogen-to-carbon (H/C) atomic ratio. Ouchi[11] has shown that H/C is a linear function of the fraction of aromatic carbons of petroleum fractions, as measured by ^{13}C-NMR (carbon nuclear magnetic resonance). However, because we have shown that the carbon content of petroleum fractions from saturates to coke is nearly the same (81.6 to 86.5 wt%), without much loss in accuracy, hydrogen content can be used to measure aromaticity and, thus, molecular attraction.

2.2.3 Measurement of Molecular Weight

The petroleum scientific community cannot agree on the correct method for measuring asphaltene molecular weight. This topic has resulted in many heated discussions at scientific meetings. Nothing says more about the lack of knowledge of asphaltenes, forcing one to reach the conclusion that asphaltene molecular-weight measurement is not an exact science.[12] The value one measures depends on the technique,[13] the solvent, and the temperature.[14] As will be discussed later in this chapter, the problem is that asphaltenes fall in that gray area between being in solution and being a colloid because of their tendency to self associate and form aggregates that can be detected by small-angle x-ray and neutron scattering. As a result, techniques that measure molecular weight in solution tend to give values that are too high. On the other hand, the low volatility of asphaltenes interferes with mass spectrometry techniques and produce molecular-weight measurements that tend to be low. Also, the strong tendency of asphaltenes to adsorb causes GPC techniques to produce low-molecular-weight measurements because the higher-molecular-weight, more adsorbing asphaltenes are supposed to pass through the gel column first. The contrary view is that, if the absolute average molecular weight of asphaltenes were really important for a property, then we could use that property to measure average molecular weight. Vapor pressure or boiling point would be one such property, except that even high-vacuum distillation volatilizes little or no asphaltenes. This would seem to require asphaltene molecular weights to be higher than 1000. In section 2.6.3 more will be discussed on this topic, including evidence that the average asphaltene molecular weight might be less than 1000.

In chapter 5, it is shown that using two-dimensional solubility parameters, the best solvents for carbonaceous solids, such as asphaltenes, were determined to be quinoline, trichlorobenzene, and o-dichlorobenzene. Therefore, the author

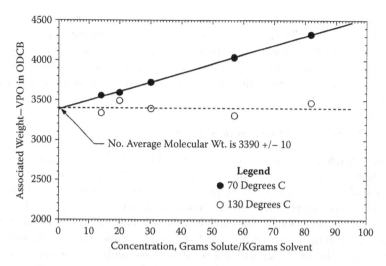

FIGURE 2.2 Vapor pressure osmometry data on Arabian Heavy asphaltenes at two temperatures. (From IA Wiehe. *Ind Eng Chem Res*, 31: 530–536, 1992. Reprinted with permission. Copyright 1992. American Chemical Society.)

considered these for solvents in vapor pressure osmometry. Quinoline was rejected because it tends to oxidize when exposed to air. Trichlorobenzene was rejected because the effect of temperature was desired to be studied, and trichlorobenzene's vapor pressure is very low at low temperatures. Therefore, o-dichlorobenzene was used, with results shown in Figure 2.2 for Arabian Heavy asphaltenes. Typically, one measures the associated molecular weight as a function of concentration and extrapolates to infinite dilution to obtain what one hopes is the unassociated molecular weight. Yet, even with this extrapolation, when using a poor solvent for asphaltenes, such as toluene at 50°C, one measures an infinite dilution molecular weight for the same asphaltenes that is definitely too high (4900). However, using o-dichlorobenzene at 70°C, the extrapolation to infinite dilution measures the molecular weight to be 3380. By using the maximum temperature of the instrument at 130°C and o-dichlorobenzene, the molecular weight is found to be independent of asphaltene concentration with an average, 3400, which is within experimental error of the extrapolated value at 70°C. Thus, it is concluded that the best solvent or the highest temperature can be used to dissociate asphaltenes to measure the average molecular weight of asphaltenes with vapor pressure osmometry. It is recommended that both the best solvent, o-dichlorobenzene, and the highest temperature, 130°C, be used, particularly when measuring converted asphaltenes or coke, which associate even more than unconverted asphaltenes.

Nitrobenzene and pyridine have been proposed to be superior solvents for vapor pressure osmometry of petroleum asphaltenes because they yield low-molecular-weight measurements.[14,15] However, observation of mixtures of these two liquids with asphaltenes between a glass slide and a cover slip with an optical microscope at 600X reveals that neither nitrobenzene nor pyridine completely dissolve asphaltenes. Thus, the low-molecular-weight measurements were a result of the higher-molecular-weight fraction not being in solution. In this study, all mixtures of asphaltenes and coke with o-dichlorobenzene were checked with an optical microscope to ensure that they were in solution to the limit of resolution (0.5 μm).

2.2.4 Recommended Compositional Map

The recommended measure of molecular attraction is hydrogen content, and that of molecular size is the vapor pressure osmometry (VPO) molecular weight using o-dichlorobenzene at 130°C. Therefore, the recommended compositional map is these two properties plotted against each other as shown in Figure 2.3 for the class separation of eight different resids and their thermal reaction products. One of the significant features is that each of the five classes occupies unique areas of the graphs. Thus, the combination of number average molecular weight and hydrogen content is capable of differentiating between petroleum fractions, whether or not the fraction has been partially converted. For convenience, solid curves are drawn to show that all species with molecular weight and hydrogen content that lie to the right of one curve are n-heptane soluble, and to the right of the other curve are

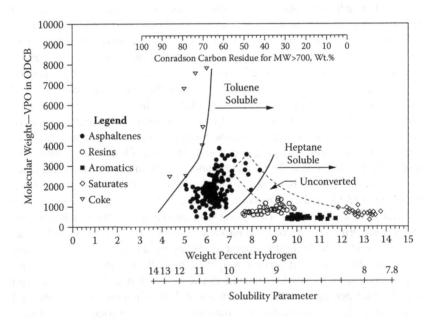

FIGURE 2.3 The solvent–resid phase diagram of eight different resids and their thermal reaction products displays each of the five classes in unique areas. (From IA Wiehe. *Ind Eng Chem Res*, 31: 530–536, 1992. Reprinted with permission. Copyright 1992. American Chemical Society.)

toluene soluble. This, plus the unique area for each of the petroleum fractions, suggests that the diagram is similar to a phase diagram, but here the different "phases" are due to their solubility behavior in particular solvents or due to their adsorption from solution. This is why Figure 2.3 is called a "solvent–resid phase diagram."

The dashed curve in Figure 2.3 encloses all the points representing fractions from unconverted resids. The area representing converted resids is much larger than that occupied by unconverted resids. Actually, the aromatics and saturates occupy nearly the same area for converted and unconverted resids. However, the resins and asphaltenes, upon conversion, tend to move to lower hydrogen content (more aromatic) and equal or lower molecular weight. Because a pure compound would be a point on this diagram, the smaller the area for a class, the better is its approximation as a pure component. The areas for saturates, aromatics, and resins are about the same size, but the resins from converted resid can be more clearly differentiated. The area for asphaltenes is much larger with clear separation between converted and unconverted asphaltenes. This will be justification in chapter 4 to have separate converted and unconverted asphaltene pseudocomponents in kinetic models. The area for coke, that is, everything to the left of the toluene line, is much larger than the area for asphaltenes. Of course, a minority of the toluene-insoluble coke samples were soluble enough in o-dichlorobenzene to measure the molecular weight in order to appear on the diagram. These were only the coke samples formed by reacting asphaltenes at short reaction times.

Figure 2.3 shows two alternative x-axes to hydrogen content. This is because the same large-ring aromatics that cause petroleum macromolecules to have low hydrogen content also cause them to have low solubility in solvents and other oils and to form coke. Therefore, in chapter 3, it will be shown that Conradson Carbon is a linear function of hydrogen content for oil fractions over 700 in average molecular weight, and in chapter 5 it will be shown that the solubility parameter of oil fractions is a linear function of the reciprocal of hydrogen content.

2.2.5 TRANSFORMATIONS BETWEEN CLASSES

The solvent–resid phase diagram provides insight on the transformation of one class to another. Resins can be formed from aromatics by either molecular-weight growth, or by decreasing hydrogen content by cracking off more saturated fragments, or by aromization of naphthenoaromatics. On average, both mechanisms occur, as was found with resins formed from saturates and aromatics. Resins typically become more aromatic (lower hydrogen content) during thermal conversion by cracking off more saturated fragments and by aromization of naphthenoaromatics. If two or three of these more aromatic resin molecules combine, they form an asphaltene with properties similar to those of an unreacted asphaltene. As will be seen in chapter 4, this was part of the evidence to propose such a reaction in a resid conversion kinetic model. Likewise, asphaltenes typically form lower-molecular-weight and more aromatic asphaltenes during thermal conversion. In chapters 3 and 4, the limit of this more aromatic asphaltene will be called the asphaltene core. If two or more of these more aromatic asphaltenes combine, they form toluene-insoluble coke. Of course, as we saw in reacting the individual classes, each of the thermal reactions form the desired range of volatile liquid by-products of low-molecular-weight and higher hydrogen content that do not appear on the map of nonvolatile fractions.

2.2.6 CONCLUSIONS AND IMPLICATIONS

The solvent–resid phase diagram points out some of the advantages and over-comes the greatest disadvantage in using solubility and adsorption fractions for tracking resid chemical changes. It shows that classes characterized by solubil-ity and adsorption can be distinguished from each other by areas on a plot of molecular weight versus hydrogen content. By being areas rather than points, it also clearly shows that each class represents a collection of a large number of dif-ferent macromolecules and that chemical changes can occur without completely changing a member of one class into others. However, even when this occurs, by measuring the hydrogen content and the molecular weight, one can continue to track the path of the chemical change using the solvent–resid phase diagram.

A mechanism that can rationalize much of the data in this section is that the aromatic moieties that impart insolubility in solvents, adsorption on clay, and low volatility also impart a thermal reaction limit to the resid. As is represented in Figure 2.4, the saturates can nearly be completely converted thermally to vola-tile products. However, the aromatics, resins, and asphaltenes tend to have low-volatility fragments that are more aromatic and lower in molecular weight after cracking off fragments that are less aromatic and lower in molecular weight. Even-tually, these more aromatic fragments combine to convert to a less soluble class by molecular-weight growth. Naturally, as the aromaticity of a macromolecule is increased, it produces a greater fraction of more aromatic products (toward the left in Figure 2.4).

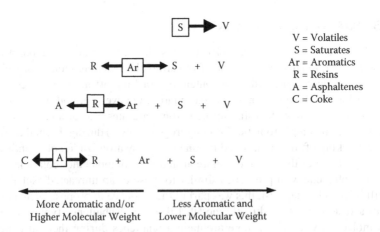

V = Volatiles
S = Saturates
Ar = Aromatics
R = Resins
A = Asphaltenes
C = Coke

More Aromatic and/or   Less Aromatic and
Higher Molecular Weight   Lower Molecular Weight

FIGURE 2.4 Thermal conversion reaction path. (From IA Wiehe. *Ind Eng Chem Res*, 31: 530–536, 1992. Reprinted with permission. Copyright 1992. American Chemical Society.)

2.3 HIGH PERFORMANCE LIQUID CHROMATOGRAPHY

Although the class separation method has contributed much to the understanding of the diversity of molecules in unconverted and converted resids, it is too labor intensive and time consuming for a routine analytical tool. Therefore, in this section, high performance liquid chromatography (HPLC) is discussed as an analytical tool that is capable of providing even better information on molecular diversity and on a routine basis with small sample sizes. It even includes aromatic size information over one to four aromatic rings. Although the HPLC technique was developed for vacuum gas oils, it can be applied to the deasphalted portion of resids.

2.3.1 HPLC Instrument

The HPLC method that will be described here is one developed by the author's colleagues, Bob Overfield, Joel Haberman, and Winston Robbins,[16–18] at Exxon Corporate Research. The basic parts of the HPLC are two columns that perform the separation by selective adsorption, followed by selective desorption using programmed changes in solvents. This enables separation of the oil into saturates, one-ring aromatics, two-ring aromatics, three-ring aromatics, four-ring aromatics, and polars, which include any aromatic rings larger than four as well as nitrogen- and oxygen-containing molecules. An ultraviolet diode array detector is used to measure the concentration of aromatic carbons, whereas an evaporative mass detector measures the total mass.

The HPLC system is made up of purchased components with details given by Robbins.[18] A computer controlled the system, collected the ultraviolet (UV) spectra, made calculations, and stored the data on a disk drive. The ternary solvent pump is controlled by the computer through cables. The autosampler allowed automatically running a tray of samples. A flow cell was used to help match the sensitivities of the two detectors because the UV detector is 10 times more sensitive than the mass detector. The evaporative mass detector sprays the oil solution into hot nitrogen. This evaporates the solvent, and the remaining mass of oil droplets are measured by the decrease in light intensity caused by the scattering from the droplets.

2.3.2 Separation

Two different columns were used for the oil separation. A Whatman propylaminocyano (PAC) column was used for separating saturates and one-ring aromatics. A dinitroanilinopropyl (DNAP) column was used to separate the larger aromatic rings, including polars. Because each DNAP column was a little different, model compound standards were used for calibration. The retention times for the six fractions are first set from the separation of a hexane solution of model compounds: hexadecane for saturates, nonadecylbenzene for one-ring aromatics, naphthalene for two-ring aromatics, dibenzothiophene for three-ring aromatics, and pyrene for four-ring aromatics. Also, fluorene and phenanthrene are used that elude later among the two- and three-ring aromatics, respectively. An oil quality-control sample is used to make the final adjustments to the valve timing that dictates the start of two-ring aromatics and to adjust the solvent program to separate four-ring aromatics from polars. On the other hand, for the differentiation between saturates and one-ring aromatics, the software located and used the valley between peaks to account among sample types. The details on the calibration of the two detectors can be found elsewhere.[16–18]

2.3.3 Aromaticity Measurement Using the Diode Array Detector

It was demonstrated[16,17] that the aromaticity of one-ring or higher aromatic-ring compounds can be measured from UV spectra based on first principles and more than 80 model compounds. There is a nearly constant response per aromatic carbon π–π atomic transition when spectra are converted to an energy basis for each wavelength, and the absorbance integrated over the range corresponding to 204 to 430 nm. These integrated absorption energies, oscillator strengths, include all transitions between a ground state of a molecule and its excited states. Thus, they count the aromatic carbons. Preparative HPLC was done to collect fractions large enough to measure percent aromatic carbon by ^{13}C-NMR over 0.2 to 70% aromaticity. Good agreement was obtained between the aromaticity measured by HPLC with the diode array detector and the aromaticity measured by ^{13}C-NMR with only a slight bias for higher values with HPLC. The aromaticity measured by the diode array detector is as a weight percent of aromatic carbons, whereas the aromaticity measured by ^{13}C-NMR is as a percent of the carbons. Therefore, the diode array aromaticity needs to be divided by the weight fraction of the sample that is carbon or the ^{13}C-NMR aromaticity needs to be multiplied by the weight fraction of the sample that is carbon to compare these aromaticities in the same units (Table 2.4).

TABLE 2.4
Comparison of ^{13}C-NMR and Diode Array Aromaticities for Preparatory HPLC Fractions

	Heavy Vacuum	Gas Oil	Cat-Cracking	Fract. Bot.
Fraction	Diode Array (wt% Arom. C)	^{13}C-NMR[a] (wt% Arom. C)	Diode Array (wt% Arom. C)	^{13}C-NMR[a] (wt% Arom. C)
1-Ring aromatics	12.6	15.9	36.2	34.2
2-Ring aromatics	29.7	29.0	44.0	NA
3-Ring aromatics	41.6	37.0	68.9	70.4
4-Ring aromatics	52.6	51.8	73.3	75.1

[a] Calculated from mol% aromatic carbon multiplied by wt% carbon.

Source: WK Robbins. *J Chrom Sci.* 36: 457–466, 1998. With permission from Preston Publications.

2.3.4 MASS MEASUREMENT USING THE EVAPORATIVE MASS DETECTOR

The operating conditions are individually optimized for each evaporative mass detector in its HPLC system. The point-by-point calibration accommodates most peak shapes and allows the system to handle the nonlinear behavior of the evaporative mass detector over an 800-fold concentration range. There can be errors when low-boiling fractions evaporate with the solvent, polar fractions irreversibly adsorb to the column, the droplets are dark in color, the sample contains cyclohexane insolubles, and crystalline compounds (wax or model compounds) form "snow flakes" rather than droplets. The first four cause low mass balances, whereas the last one causes high mass balances. Nevertheless, mass balances of $100 \pm 20\%$ are considered normal. If a mass balance falls outside this range, the cause is determined and changes are made to accommodate the sample.

The evaporative mass detector was compared with preparative HPLC samples where the solvent was evaporated and the mass of each fraction was determined by weighing. As is shown in Table 2.5, the two methods compare favorably in most cases, but with greater variation in the polars. This is understandable given the polars' dark color and their strong adsorption on the columns.

2.3.5 HPLC ANALYSIS

The evaporative mass detector limits this HPLC technique to vacuum gas oils or heavier (>650°F boiling point) because lighter oils evaporate with the solvent. These oil samples, 150 mg, are dissolved in 5 mL of cyclohexane, placed in vials,

---

TABLE 2.5
Comparison of Evaporative Mass Detector (EMD) with Gravimetric Mass Measurements (wt%)

Fraction	Extract[a] EMD	Extract[a] Grav.	HVGO[b] EMD	HVGO[b] Grav.	HCCO[c] EMD	HCCO[c] Grav.
Saturates	24.3	26.6	52.1	53.8	17.9	17.6
1-Ring aromatics	18.3	15.1	17.3	17.8	11.3	12.9
2-Ring aromatics	20.8	20.6	11.9	15.0	15.6	12.3
3-Ring aromatics	13.1	14.8	6.6	8.8	17.7	19.0
4-Ring aromatics	14.0	11.7	5.6	6.2	26.7	23.8
Polars	9.4	11.0	6.2	1.3	10.9	14.4

[a] Lube extract.

[b] Heavy vacuum gas oil.

[c] Heavy cat-cycle oil.

Source: WK Robbins. *J Chrom Sci.* 36: 457–466, 1998. With permission from Preston Publications.

and loaded into the autosampler. Samples are automatically injected into n-hexane and passed through the DNAP column, followed by the PAC column. Saturates do not adsorb on either column, and one-ring aromatics are eluded off the PAC column with n-hexane. The valve is switched to the DNAP column, and two-ring aromatics are eluded with n-hexane. The gradient is started with 2 to 5% methylene chloride in n-hexane, eluding the three-ring aromatics. As the gradient increases to 90% methylene chloride, the four-ring aromatics elute. Isopropanol is initiated and gradually increased to 10% in methylene chloride while eliminating n-hexane to elute the polars. Both columns are flushed with 100% methylene chloride and equilibrated with 100% n-hexane before injecting a new sample. This solvent program is shown in Table 2.6.

The two detectors acquire data at 3-s intervals after a dead volume of 3 min and continuing until the end of the 36-min run. The absorption energies from the diode array detector are integrated separately from 204 to 230 nm, and from 232 to 430 nm. The diode array data are used to calculate the percent aromatic carbon for each of the six fractions, using response factors, injection volume, flow rate,

TABLE 2.6
HPLC Solvent Program

Time (min)	n-Hexane (vol%)	Methylene Chloride (%)	Isopropyl Alcohol (%)	Flow Rate (mL/min)	Comment
0	100	0	0	1.5	Injection
3	100	0	0	1.5	S and 1-ring elute from PAC
9.5	100	0	0	1.5	2-ring elute from DNAP
11	100	0	0	1.5	End of 2-ring elution
11.05	95	5	0	1.5	Start gradient for 3-ring
18	98	2	0	1.5	End 3-ring gradient
20	75	25	0	1.5	Steep gradient for 4-ring
23	80	20	0	1.5	End 4-ring gradient
26	10	90	0	1.5	Gradient to elute polars
29	0	90	10	1.5	IPA to elute strong polars
37	0	75	25	1.5	Clean column
37.5	0	100	0	2.0	Regenerate DNAP
41	0	100	0	2.0	Regenerate both columns
42	0	100	0	2.0	Methylene chloride on PAC
44	100	0	0	2.0	Flush methylene chloride
50	100	0	0	2.0	Flush DNAP bypass loop
52	100	0	0	2.0	Flush PAC bypass loop
54	100	0	0	2.0	Equilibrate to n-hexane
75 (end)	100	0	0	1.5	Equilibrate to n-hexane

Source: WK Robbins. *J Chrom Sci.* 36: 457–466, 1998. With permission from Preston Publications.

follicle path length, data acquisition rate, and sample concentration. The mass of each fraction was determined from the evaporative mass detector. Although the mass is reported as a percent of the mass measured, the percent mass recovered is calculated based on the mass injected. The percent aromatic carbons are reported as a weight percent of the total injected sample. Thus, they can be added over all the fractions to determine the weight percent aromatic carbons in the total injected sample. They also can be subtracted from the mass percent for each fraction to determine the weight percent that is not aromatic. In chapter 3, the representation of petroleum macromolecules as pendants and aromatic cores will be discussed. This reporting of HPLC analysis fits well into the pendant-core format. However, if one wishes to determine the weight percent aromatic carbon in a cut, such as two-ring aromatics, the percent aromatic carbon of that cut as a weight percent of injected sample needs to be divided by the mass fraction of that cut (7.14/.221 = 32.3% for two-ring aromatics).

If one decides to inject a fraction separated by prep HPLC or by class separation, one needs to lower the concentration to that expected for that fraction in a full-range sample. Otherwise, one will overload the capacity of the columns for that fraction because the HPLC system has been optimized for full-range samples of oil.

2.3.6 EXAMPLE APPLICATIONS OF ANALYTICAL HPLC

2.3.6.1 Estimate of Average Carbon Number and Molecular Weight[18]

Table 2.7 shows HPLC results for nine boiling-point cuts from distillation of a sample of unconverted vacuum gas oil. Because for this fraction, it is a good assumption that there is at most one aromatic structure per molecule, the average carbon number of each fraction can be calculated. For instance, the 850 to 900°F cut in Table 2.7 has 2.5 wt% of the cut as carbons in one-ring aromatics. Subtracting this from the wt% of the injected sample, that is, one-ring aromatics (16.5 − 2.5 = 14.0%), one obtains the weight percent that is not aromatics. Assuming that the nonaromatics are either saturated rings or paraffinic side chains, each of these carbons will carry a mass of 14 because they will be $-CH_2-$ (ignoring the $-CH_3$ end groups). Thus, the ratio of the number of pendant carbons to aromatic core carbons is 14.0:14 divided by 2.5:12, which equals 4.80. Multiplying this by the number of aromatic carbons in a single aromatic ring, which is 6, gives the number of pendant carbons to be 28.8. Adding the number of aromatic core carbons, which is 6, gives the average carbon number to be 34.8. In general, the following equation can be used:

$$\text{Carbon No.} = N + 12 N [\text{mass \%} - \text{wt\% aromatic C}]/[(\text{wt\% aromatic C})(14)]$$

where N = 6, 10, 14, 18, and 18 for one-ring, two-ring, three-ring, four-ring, and polars, respectively. Polars adsorb more strongly than four-ring aromatics because they are either five-ring aromatics (unsubstituted six-ring aromatics

TABLE 2.7
HPLC Analyses of Distillation Cuts of Heavy Vacuum Gas Oil

	650–700 (°F)	700–750 (°F)	750–800 (°F)	800–850 (°F)	850–900 (°F)	900–950 (°F)	950–1000 (°F)	1000–1050 (°F)	>1050 (°F)	Wt. Sum All Cuts	Initial HVGO[a]
Wt% Cut	3.35	1.84	7.32	14.44	24.42	20.31	17.97	6.56	4.19		
Wt% Arom. C											
Sats	0.0	0.0	0.0	0.0	0.0	0.0	0.1	0.1	0.1	0.0	0.0
1-ring arom.	2.4	2.8	2.8	2.7	2.5	2.2	2.1	1.8	1.4	2.3	2.8
2-ring arom.	4.9	5.2	4.8	4.4	3.8	3.3	3.2	2.8	2.4	3.7	4.3
3-ring arom.	2.5	3.1	3.0	2.9	2.5	2.3	2.4	2.2	2.1	2.5	2.9
4-ring arom.	2.1	2.4	2.6	3.0	3.3	3.5	3.8	4.2	5.0	3.4	3.6
Polars	3.0	2.5	2.6	2.7	3.0	4.3	3.8	4.4	5.8	3.4	3.3
Total Arom. C	14.9	16.0	15.8	15.7	15.1	15.6	15.4	15.5	16.8	15.4	17.0
Norm. wt%											
Sats	68.7	65.9	60.1	61.9	57.0	51.8	48.4	43.4	35.0	54.3	52.3
1-ring arom.	10.7	14.4	14.7	16.4	16.5	16.0	16.4	15.8	14.2	15.9	17.1
2-ring arom.	5.3	8.3	8.9	9.7	10.2	10.5	11.5	11.6	11.8	10.3	11.8
3-ring arom.	3.2	4.7	6.6	4.5	5.2	6.5	7.1	7.4	8.7	6.0	6.8
4-ring arom.	4.4	3.0	4.6	3.4	4.5	6.2	7.1	9.1	13.0	5.8	6.7
Polars	7.7	3.7	5.0	4.0	6.6	9.0	9.5	12.6	17.3	7.9	5.2
Recovery	61.3	77.4	89.3	94.8	101.9	106.6	104.2	104.8	103.4	98.8	97.0

	Est. C No.										
1-ring arom.	23.8	27.3	27.9	32.1	34.8	38.3	41.0	46.0	53.0	35.9	32.8
2-ring arom.	10.7	15.1	17.3	20.3	24.4	28.7	32.2	36.9	43.6	25.4	24.9
3-ring arom.	17.4	20.2	28.4	20.6	27.0	35.9	37.5	42.4	51.7	30.8	30.2
4-ring arom.	34.9	21.9	29.9	20.1	23.6	29.9	31.4	36.0	42.7	28.8	31.1
Polars	42.2	25.4	32.2	25.4	36.5	34.9	41.1	46.8	48.6	38.3	26.9

[a] Heavy vacuum gas oil.

Source: WK Robbins. *J Chrom Sci.* 36: 457–466, 1998. With permission from Preston Publications.

have boiling points higher than 1050°F) or smaller aromatics with oxygen or nitrogen functionality. Thus, the approximation of polars as four-ring aromatics is a compromise.

An estimate can also be obtained for the molecular weight by assuming each aromatic carbon is bonded to one hydrogen, and each pendant carbon is bonded to two hydrogens. This overestimates the number of hydrogens, but does not include the atomic weight of heteroatoms. The estimated molecular weight is given by

$$MW = 13\ N + (14)(12)\ N\ [mass\ \% - wt\%\ aromatic\ C]/[(wt\%\ aromatic\ C)(14)]$$

$$MW = N\ \{13 + 12\ [mass\ \% - wt\%\ aromatic\ C]/[wt\%\ aromatic\ C]\}$$

For the example one-ring aromatic in the 850 to 900°F boiling point cut,

$$MW = 6\{13 + 12\ [16.5 - 2.5]/2.5\} = 481$$

2.3.6.2 Distillation of Heavy Vacuum Gas Oil[18]

The HPLC of distillation cuts of a heavy vacuum gas oil (HVGO) in Table 2.7 shows that, with increasing boiling point, the saturates continuously decrease in weight percent, the one-ring aromatics increase and then decrease, and the two- to four-ring aromatics and polars continuously increase. However, the saturates have the greatest weight percent for all the boiling point cuts, and one-ring aromatics have the second greatest weight percent for all boiling point cuts except the highest. The lightest two cuts have less than 80% recovery because of evaporation with the solvent where the saturates and one-ring aromatics would be even larger weight percent. The two-ring aromatics fraction has the largest weight percent of aromatic carbons for cuts with boiling points below 900°F, and the polars fraction has the largest percent of aromatic carbons for cuts with boiling points above 900°F. As expected, there is a trend of higher carbon number with increasing boiling point for each of the fractions, but there are some exceptions.

2.3.6.3 Hydrotreating[18]

Hydrotreating of vacuum gas oils is done to remove heteroatoms, sulfur, nitrogen, nickel, and vanadium, and to hydrogenate the larger aromatic rings. HPLC does not measure heteroatom removal except indirectly in its role in converting polars to aromatics. However, Table 2.8 shows that hydrotreating of a vacuum gas oil reduces the aromaticity from 30.9 to 25%. The mass of three rings, four rings, and polars are reduced because they are converted to one ring and saturates that increase in mass. In addition, the aromaticity of the remaining three rings, four rings, and polars are decreased. Thus, HPLC provides a good measure of the effectiveness of hydrotreating processes in hydrogenating large-ring aromatics.

TABLE 2.8
HPLC Analysis of Hydrotreating and Fluid Catalytic Cracking Feeds and Products

	Hydrotreating HVGO Feed	Hydrotreating Product	Cat-Cracking Refinery A HVGO Feed	Cat-Cracking Refinery A H. Cycle Oil	Cat-Cracking Refinery B HVGO Feed	Cat-Cracking Refinery B Cat Bottoms
			Wt% Arom. C			
Sats	0.04	0.08	0.08	0.04	0.03	0.1
1-ring arom.	2.6	4.9	2.9	1.2	2.6	0.3
2-ring arom.	6.7	7.2	4.5	14.8	4.1	2.1
3-ring arom.	8.4	5.8	2.7	31.6	2.8	7.4
4-ring arom.	10.7	5.7	2.7	21.6	2.6	32.5
Polars	2.5	1.3	1.7	2.9	1.2	24.8
Total arom. C	30.9	25.0	14.6	72.0	13.4	67.2
			Norm. wt%			
Sats	40.3	46.7	55.5	14.6	57.3	10.8
1-ring arom.	13.5	19.5	17.9	2.2	16.4	1.7
2-ring arom.	12.4	11.7	10.9	8.0	10.4	2.0
3-ring arom.	12.4	9.6	6.2	39.2	5.7	10.1
4-ring arom.	18.0	9.9	6.4	32.4	5.4	45.1
Polars	3.4	2.6	3.2	3.6	4.6	30.2
			Est. C No.			
1-ring arom.	28	21	33	10	33	30
2-ring arom.	17	15	22	—	23	10
3-ring arom.	20	22	30	17	26	18
4-ring arom.	29	29	39	26	35	24
Polars	24	33	32	22	—	~21

Source: WK Robbins. *J Chrom Sci. 36:* 457–466, 1998. With permission from Preston Publications.

2.3.6.4 Fluid Catalytic Cracking[18]

Ideally, fluid catalytic cracking cracks all the pendants off the aromatic cores (leaving methyls on one-ring cores) and cracks the saturates selectively into the gasoline boiling range (~100 to 430°F) while minimizing the formation of hydrocarbon gases. At the same time, fluid catalytic cracking is expected to isomerize paraffins and aromatize single-ring naphthenes for greater octane gasoline. Although HPLC cannot analyze the gasoline product, it excels in characterizing the vacuum gas oil feed for quality as a fluid catalytic cracking feedstock. In addition, HPLC can analyze the heavy by-products, heavy cycle oil, and cat fractionator bottoms and determine how much potential conversion remains. For refinery A in Table 2.8, in comparing the feed with the heavy cycle oil, one can see that a significant fraction of the saturates, one- and two-ring aromatics, were converted, leaving mostly three- and four-ring aromatics remaining in the heavy cycle oil. The remaining three- and four-ring aromatics actually have more pendants (difference between weight percent and percent aromatic carbon) in the heavy cycle oil than in the feed. This and the 14.6 wt% saturates in the heavy cycle oil show that greater conversion is possible despite the heavy cycle oil containing 72 wt% aromatic carbon. The two-ring aromatics in the heavy cycle oil appear to have greater weight percent aromatic carbon than total weight percent. This is a sure sign that a large fraction of the two-ring aromatics evaporated with the solvent in the evaporative mass detector. These evaporated two-ring aromatics most likely belong in the light cycle oil fraction.

For refinery B, the vacuum gas oil feed to the fluid catalytic cracker is compared with the cat fractionator bottoms. One needs to make sure that the cat fractionator bottoms are soluble in cyclohexane before injecting into an HPLC as this stream can even be partially insoluble in toluene.[19] Even the cat fractionator bottoms in Table 2.8, which are cyclohexane soluble, are greater than 75 wt% four-ring aromatics and polars. However, 10.8 wt% saturates remain even though there are little one- and two-ring aromatics. This is because saturates react slower in fluid catalytic cracking than one- and two-ring aromatics. By going to higher severity, these saturates can be greatly reduced in cat fractionator bottoms, but the polars would become insoluble in cyclohexane.

2.3.7 Preparative HPLC of Heavy Coker Liquids

2.3.7.1 Heavy Coker Gas Oil

The HPLC can also be used on a larger scale to separate gram quantities of fractions of heavy oils to better characterize the fractions. This was done at Exxon by the author, with Larry Kaplan and Win Robbins, using larger columns. Preparative HPLC was achieved by repeatedly and automatically injecting 300-mg samples of a heavy oil dissolved in cyclohexane from the same vessel, warmed to keep waxes from precipitating. The UV detector with a thinner cell was used to monitor the UV response with higher concentrations. In addition, a refractive

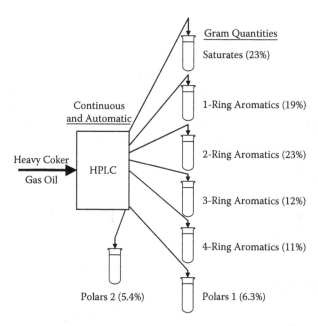

FIGURE 2.5 A schematic of the prep HPLC procedure for a heavy coker vacuum gas oil from a Flexicoker.

index detector was used to locate the separation between saturates and one-ring aromatics. Although the larger scale has different elution times than the analytical scale, the two nondestructive detectors enabled monitoring the location of the peaks and performance. Instead of the product going into an evaporative mass detector, it was sent to a fractional collector that dropped each of the fractions into separate glass tubes under nitrogen purge in a heated aluminum block and within a laboratory hood. In this way, solvents were evaporated off each of the fractions as they were collected. Although it took 5 to 7 days at 24 hours per day, gram quantities of HPLC fractions were collected that could be characterized by elemental analysis, GC-simulated distillation, and proton and ^{13}C-NMR. Figure 2.5 shows a schematic of the procedure for a heavy coker gas oil from a Flexicoker. Because the polars showed two peaks on the UV detector, they were collected separately and called polars 1 and polars 2.

The concept of representing petroleum fractions by average chemical structures has correctly been criticized[20] because it fails to characterize the large diversity of molecules present. However, by representing each of the HPLC fractions by a few chemical structures, one obtains a better view of the diversity of molecular types. For the heavy Flexicoker gas oil, this is shown in Figure 2.6 to Figure 2.8 for saturates, one- to four-ring aromatics, polars 1, and polars 2. Table 2.9 to Table 2.15 reveals how well the structure or composite of structures match the analytical data on each of the prep HPLC fractions. Although these 17 chemical

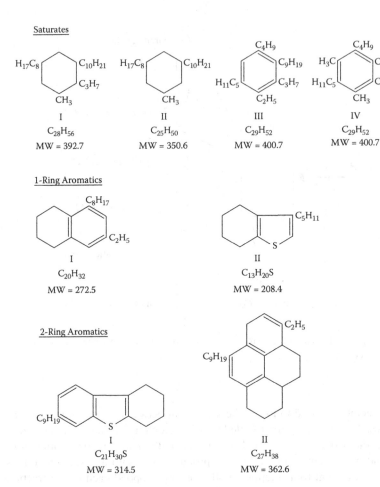

FIGURE 2.6 Representative chemical structures of saturates, one-ring aromatics, and two-ring aromatics of heavy Flexicoker vacuum gas oil separated by prep HPLC.

structures do show a lot of diversity, they still cannot adequately characterize the diversity of many thousand of molecules in heavy coker gas oil. Nevertheless, being able to picture a relatively small number of chemical structures provides more insight than staring at tables of analytical data. Thus, we now see that highly substituted one-ring aromatics are separated with the saturates, and that polars 2 tend to be three-ring aromatics with nitrogen or oxygen functionality, whereas polars 1 are similar to four-ring aromatics, but with fewer paraffinic side chains. It is also easy to see why heavy coker gas oil does not make a very good feed to fluid catalytic cracking unless it is hydrotreated to saturate aromatic rings and remove sulfur, nitrogen, and oxygen. As discussed in the previous section, in fluid

3-Ring Aromatics

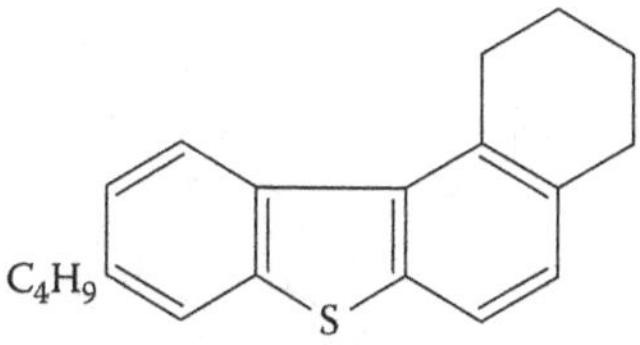

$C_{20}H_{22}S$

MW = 294.4

4-Ring Aromatics

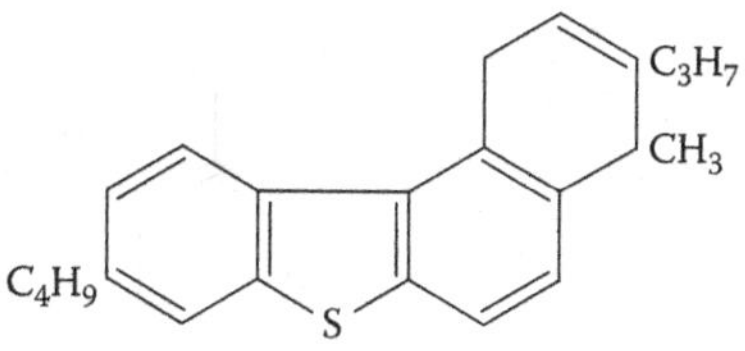

$C_{24}H_{26}S$

MW = 346.5

Polars 1

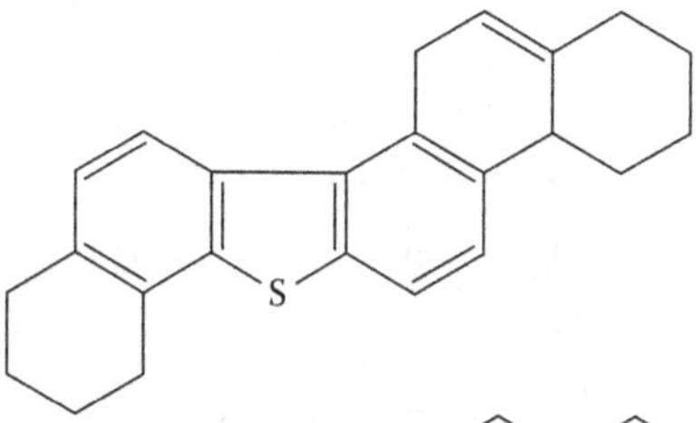

I

$C_{24}H_{22}S$

MW = 392.5

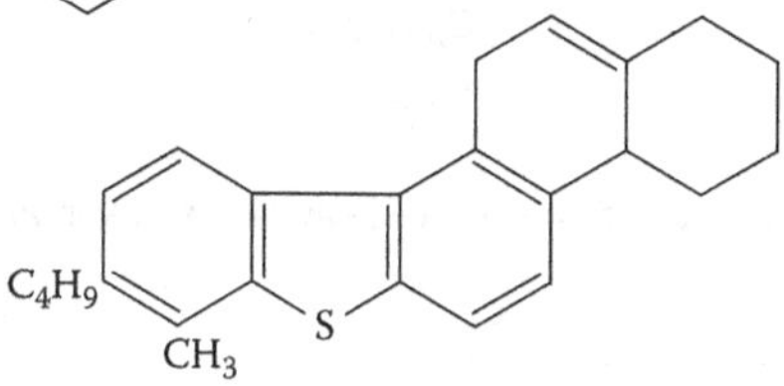

II

$C_{24}H_{24}S$

MW = 344.5

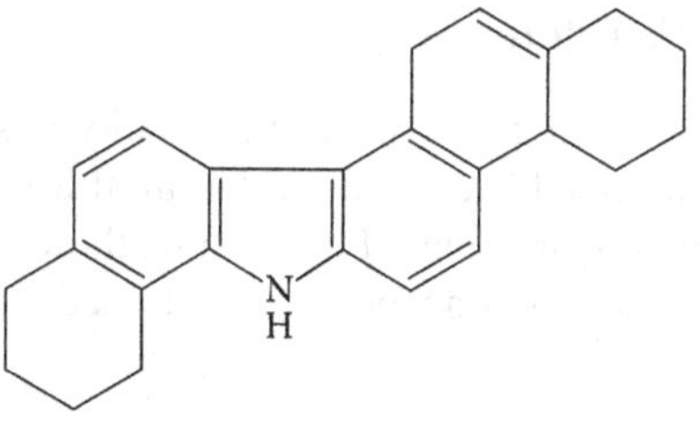

III

$C_{24}H_{23}N$

MW = 325.4

FIGURE 2.7 Representative chemical structures of three-ring aromatics, four-ring aromatics, and polars 1 of heavy Flexicoker vacuum gas oil separated by prep HPLC.

catalytic cracking, one would like to produce alkyl-substituted one-ring aromatics, but it cannot crack aromatic rings. Thus, saturating the aromatic rings down to at least one-ring aromatics better gives fluid catalytic cracking a chance to meet its objective. Removing aromatic sulfur, nitrogen, and oxygen also greatly opens up the structures for further cracking. Many refineries prefer to hydrocrack heavy coker gas oil because hydrocracking does better at aromatic ring reduction than hydrotreating followed by fluid catalytic cracking.

Polars 2

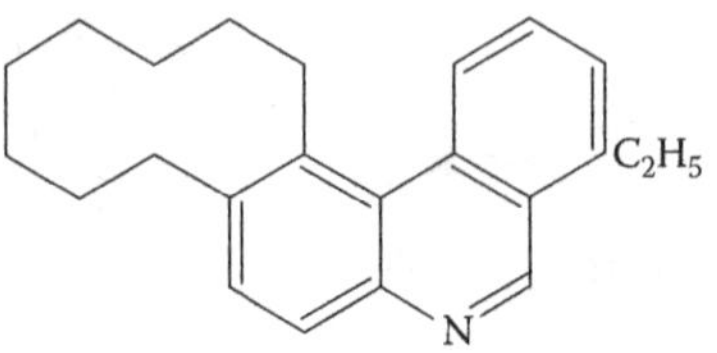

I

$C_{24}H_{27}N$

MW = 329.5

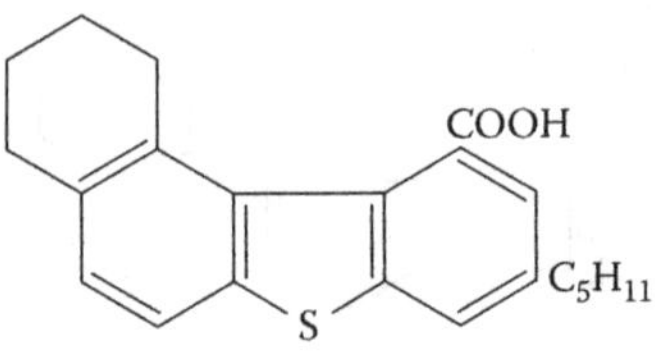

II

$C_{22}H_{24}O_2S$

MW = 352.5

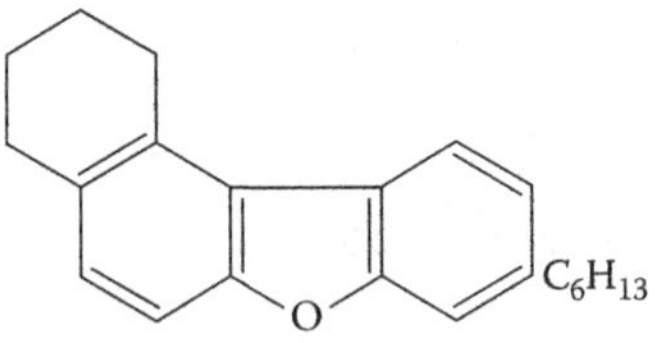

III

$C_{22}H_{26}O$

MW = 306.4

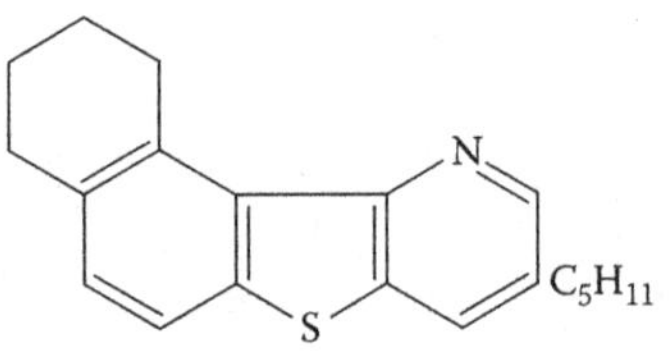

IV

$C_{20}H_{23}SN$

MW = 311.4

FIGURE 2.8 Representative chemical structures of polars 2 of heavy Flexicoker vacuum gas oil separated by prep HPLC.

2.3.7.2 Once-Through Coker Scrubber Bottoms

Not all of the liquids that evaporate out of a coker are of the quality desired. Therefore, usually a heavy fraction is recycled back to the coker until it is converted to lighter liquids and coke. In Fluid Coking and Flexicoking, these heavy (approximately 975 to 1200°F atmospheric boiling points) condensed liquids

TABLE 2.9
Structural Representation of Saturates from Heavy Coker Gas Oil as Separated by Prep HPLC

Property	Structure I	Structure II	Structure III	Structure IV	.3I + .25II + .25III + .2IV Composite	Measured
% Arom. H	0.0	0.0	1.92	0.0	0.48	0.5
% Arom. C	0.0	8.0	20.7	20.7	11.3	11.4
H, wt%	14.37	13.88	13.08	13.08	13.67	13.63
C, wt%	85.63	86.12	86.92	86.92	86.33	85.66
H/C atomic	2.00	1.92	1.79	1.79	1.88	1.90

TABLE 2.10
Structural Representation of One-Ring Aromatics from Heavy Coker Gas Oil as Separated by Prep HPLC

Property	Structure I	Structure II	.75I + .25II Composite	Measured
% Arom. H	6.25	5.0	6.0	6.2
% Arom. C	30.0	30.8	30.2	30.6
H, wt%	11.84	9.68	11.30	11.27
C, wt%	88.15	74.94	84.84	85.04
H/C atomic	1.60	1.46	1.58	1.58
S, wt%	0.00	15.39	3.85	4.17

TABLE 2.11
Structural Representation of Two-Ring Aromatics from Heavy Coker Gas Oil as Separated by Prep HPLC

Property	Structure I	Structure II	I + II Composite	Measured
% Arom. H	10.0	7.89	8.9	9.6
% Arom. C	38.1	37.0	37.6	37.4
H, wt%	9.62	10.56	10.09	10.06
C, wt%	80.19	89.43	84.81	84.51
H/C atomic	1.42	1.41	1.42	1.42
S, wt%	10.19	0.0	5.10	4.67

TABLE 2.12
Structural Representation of Three-Ring Aromatics from Heavy Coker Gas Oil as Separated by Prep HPLC

Property	Structure	Measured
% Arom. H	23.5	22.7
% Arom. C	60.0	60.9
H, wt%	7.53	7.80
C, wt%	81.58	84.04
H/C atomic	1.10	1.11
S, wt%	10.89	8.23

TABLE 2.13
Structural Representation of Four-Ring Aromatics from Heavy Coker Gas Oil as Separated by Prep HPLC

Property	Structure	Measured
% Arom. H	7.56	7.50
% Arom. C	66.7	66.6
H, wt%	7.56	7.50
C, wt%	83.19	84.44
H/C atomic	1.08	1.06
S, wt%	9.25	8.06

TABLE 2.14
Structural Representation of Polars 1 from Heavy Coker Gas Oil as Separated by Prep HPLC

Property	Structure I	Structure II	Structure III	Composite Average	Measured
% Arom. H	27.3	25.0	26.1	26.1	26.8
% Arom. C	66.7	66.7	66.7	66.7	67.1
H, wt%	6.47	7.02	7.12	6.87	6.75
C, wt%	84.16	83.69	88.58	85.48	83.70
H/C atomic	0.917	1.00	0.958	0.958	0.961
S, wt%	9.36	9.31	0.0	6.22	5.71
N, wt%	0.0	0.0	4.31	1.44	1.13

TABLE 2.15
Structural Representation of Polars 2 from Heavy Coker Gas Oil as Separated by Prep HPLC

Property	Structure I	Structure II	Structure III	Structure IV	Composite Average	Measured
% Arom. H	22.2	16.7	19.2	17.4	18.9	18.2
% Arom. C	54.1	54.5	54.5	55.0	54.5	53.1
H, wt%	8.26	6.86	8.55	7.45	7.78	7.80
C, wt%	87.48	74.96	86.23	77.14	81.45	81.80
H/C atomic	1.12	1.09	1.18	1.15	1.14	1.14
S, wt%	0.00	9.10	0.00	10.30	4.85	4.52
N, wt%	4.25	0.00	0.00	4.50	2.19	2.80
O, wt%	0.00	9.08	5.22	0.00	3.58	3.80

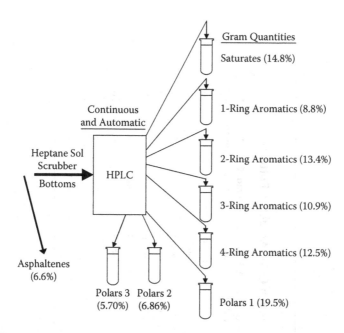

FIGURE 2.9 A schematic of the prep HPLC procedure for once-through Flexicoker scrubber bottoms.

are recycled to a scrubber where they remove the small coke particles from the volatile products in the coker reactor. Thus, they are called scrubber bottoms. However, in one case, a Flexicoker was operated without recycling the scrubber bottoms to determine the potential value of once-through scrubber bottoms. The recycled scrubber bottoms would be of lower quality because aromatic fragments would be recycled many times before converting to coke. Figure 2.9 is a schematic that shows how the once-through coker scrubber bottoms were separated. Because it contains asphaltenes, they need to be separated by n-heptane precipitation before preparative (prep) HPLC. Otherwise, the asphaltenes tend to be irreversibly adsorbed to the HPLC column. Therefore, the asphaltenes become one fraction, and prep HPLC is used to separate the n-heptane solubles into saturates, one- to four-ring aromatics, and polars 1 to 3. A third polars peak was discovered in scrubber bottoms, which was not present in coker gas oil. Analytical data on these nine fractions are shown in Table 2.16.

The surprising feature of once-through scrubber bottoms is that nearly half (47.9 wt%) the saturates and one- to three-ring aromatics is of good quality, which should be added to the heavy coker gas oil rather being recycled back to the coker. As will be shown in the next section, much of this good-quality portion is resid of lower boiling points that evaporated out of the coker with little reaction. In chapter 7 on separation and in chapter 8 on coking, methods for separating and for avoiding coking the higher-quality part of resids will be discussed. Meanwhile, the

TABLE 2.16
Analytical Data on Preparative HPLC Fractions of Once-Through Coker Scrubber Bottoms

Fraction	Yield (wt%)	Carbon (wt%)	Hydrogen (wt%)	H/C (atomic)	Sulfur (wt%)	Nitrogen (wt%)	Σ Elements (wt%)	Conrad. C (wt%)	V (ppm)	Ni (ppm)
Saturates	14.8	86.85	13.33	1.83	0.27	0.0	100.45	0.0	0	0
1-Ring arom.	8.8	85.49	11.27	1.57	2.55	0.0	99.31	0.0	0	0
2-Ring arom.	13.4	86.00	9.84	1.36	3.77	0.34	99.95	1.4	0	0
3-Ring arom.	10.9	85.92	9.22	1.28	4.51	0.34	99.99	5.4	0	0
4-Ring arom.	12.5	81.29	7.82	1.15	4.56	0.89	97.16	19.6	0	0
Polars 1	19.5	82.43	8.64	1.25	5.40	0.57	99.34	36.3	74	0
Polars 2	6.9	84.04	7.31	1.04	4.61	1.37	98.83	45.3	284	30
Polars 3	5.8	81.69	8.51	1.24	3.26	0.07	96.28	24.8	78	54
Asphaltenes	6.6	87.06	6.56	0.897	4.01	1.94	99.57	61.3	489	40
Total	99.2	83.67	9.37	1.33	3.64	0.53	98.29	18.9	71	10

remainder of the scrubber bottoms is of low quality and, thus, needs to be processed in a coker. In the next chapter it will be shown that the Conradson carbon residue of these fractions with molecular weights below 700 amu (atomic mass unit) underestimate the coking potential. The analytical data on polars 3 indicate that this fraction is probably an artifact. By difference, this fraction appears to have almost 4 wt% oxygen. However, the resid feed does not contain enough oxygen to warrant this high level in the scrubber bottoms. Still, the scrubber bottoms should contain considerable olefins that are one carbon from aromatics (conjugated olefins), resulting from the thermal cracking. These olefins are easily oxidized on exposure to air and, therefore, these oxygen-containing functionalities probably formed after the sample was taken. As a result, polars 3 may not really be a fraction of coker scrubber bottoms during the coking operation.

2.4 SHORT-PATH DISTILLATION (DISTACT DISTILLATION)

By far the most common separation in refining is by distillation. The highest atmospheric boiling point separation with vacuum distillation is at 730 to 850°F with a pressure of 25 to 50 mmHg, obtaining atmospheric boiling points up to 950 to 1050°F. However, vacuum resids can be distilled further in the laboratory by using higher vacuum (0.001 mmHg) while using short residence times at high temperatures to minimize thermal reactions. In this way, vacuum resids can be distilled up to about 1300°F (700°C) atmospheric boiling points to obtain distillation fractions for analytical characterization. These high cut-points without thermal cracking are achieved by a diffusion pump that pulls a high vacuum (0.001 mmHg) and a short path length before vapors are condensed on a cold finger and flowed into a receiver. It is often called Distact distillation after the commercial name of the equipment. This unit is actually a thin-film evaporator (Figure 2.10 and Figure 2.11) that splits the resid into one light and one heavy stream per pass. By running multiple passes, a resid may be cut into multiple cuts. Altgelt and Boduszynski[21] have used short-path distillation and atmospheric equivalent boiling point extensively to characterize petroleum resids.

The data on topped Athabasca bitumen[22] in Table 2.17 are for increasingly deeper cuts on the same feed to show how the increasing yield affects the quality of the lighter product, and are compared with a cut using a standard vacuum distillation (ASTM D1160). Increasing the cut-point increases the yield of lighter liquids, but decreases the quality. For the data on a Cold Lake vacuum resid (975°F+) in Table 2.18, the feed was first cut at 1300°F and the lighter product was rerun at a cut-point of 1093°F to obtain three cuts. Thus, the quality varies sharply for these three cuts with good mass and elemental balances. As a result, the 1302°F+ fraction that is 56% of the resid contains 66% of the sulfur, 77% of the nitrogen, 88% of the nickel, 85% of the vanadium, and 85% of the Conradson carbon residue. Likewise, the short-path distillation data on Arabian Heavy vacuum resid (975°F+) in Table 2.19 has a 1305°F+ fraction, that is, 55% of the resid, that contains 66% of

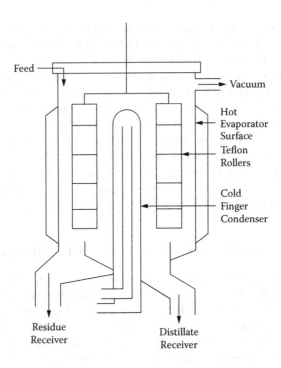

FIGURE 2.10 A schematic of short-path (Distact) distillation showing Teflon® rollers and the cold finger condenser. (From CJ Domansky. *Aostra J Res.* 2: 191–196, 1986. With permission.)

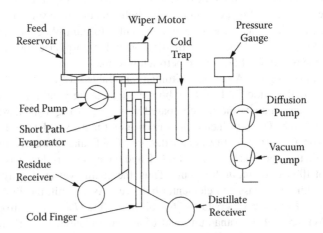

FIGURE 2.11 A schematic of entire short-path (Distact) distillation equipment. (From CJ Domansky. *Aostra J Res.* 2: 191–196, 1986. With permission.)

TABLE 2.17
Properties of the Lighter Cut with Increasing Cut-Point for Athabasca Bitumen 343°C+

Property	ASTM D1160	Short Path Distillation					
		Cut #1	Cut #2	Cut #3	Cut #4	Cut #5	Cut #6
Cut range, °C	343–524	343–475	343–500	343–540	343–580	343–620	343–660
Yield, wt%	37.9	26.5	31.3	39.6	47.4	54.9	60.4
Density, g/mL	0.9630	0.9541	0.9602	0.9664	0.9727	0.9811	0.9819
C_7 Asphaltenes, wt%	0.04	0.00	0.19	0.07	0.17	1.05	0.52
Conrad. carbon, wt%	0.1	0.1	0.1	0.3	1.1	3.4	3.5
Sulfur, wt%	3.16	3.08	3.08	3.25	3.42	3.67	3.82
Nickel, ppm	0	0	0	0	1.8	13.8	15.4
Vanadium, ppm	0.6	1.2	1.1	1.3	5.4	39.1	45.5
Iron, ppm	2.0	5.5	1.8	3.8	4.5	28.1	14.8

Source: From CJ Domansky. *Aostra J Res.* 2: 191–196, 1986. With permission.

the sulfur, 75% of the nitrogen, 97% of the nickel, 82% of the vanadium, and 90% of the Conradson carbon residue. On the other hand, this shows that the fraction of resids with boiling points below 1300°F is of much better quality than the entire resid, and the 975–1120°F fraction is still much better. In chapter 9, this will be used to advantage to show how coking can be improved.

TABLE 2.18
Short-Path Distillation Cuts of Cold Lake Vacuum Resid

Property	Lightest Cut	Middle Cut	Heaviest Cut	Weighted Average	Feed
Cut range, GCSD, °F	975–1119	1119–1302	1302+		975+
Yield, wt% resid	27.1	17.1	55.8	100	100
Carbon, wt%	84.40	83.03	82.76	83.25	83.81
Hydrogen, wt%	11.09	10.61	9.12	9.91	9.97
Sulfur, wt%	4.04	4.65	6.55	5.54	5.44
Oxygen, wt%	0.46	0.57	0.88	0.71	0.61
Nitrogen, wt%	0.31	0.35	0.90	0.65	0.54
Total	100.30	99.21	100.21	100.06	100.37
H/C atomic	1.56	1.52	1.31	1.42	1.42
Nickel, ppm	11	60	180	114	117
Vanadium, ppm	34	193	444	290	309
Conrad. Carbon, wt%	4.8	11.8	32.9	21.7	21.8

TABLE 2.19
Short-Path Distillation Cuts of Arabian Heavy Vacuum Resid

Property	Lightest Cut	Middle Cut	Heaviest Cut	Weighted Average	Feed
Cut range,GCSD, °F	975–1120	1120–1305	1305+		975+
Yield, wt% resid	25.4	20.0	54.6	100	100
Carbon, wt%	84.48	84.15	83.20	83.72	83.67
Hydrogen, wt%	11.24	10.95	9.19	10.06	10.18
Sulfur, wt%	3.99	4.28	6.23	5.27	5.13
Oxygen, wt%	0.35	0.34	0.61	0.49	0.54
Nitrogen, wt%	0.22	0.25	0.58	0.42	0.42
Total	100.28	99.97	99.81	99.96	99.94
H/C Atomic	1.59	1.55	1.32	1.43	1.45
Nickel, ppm	2	13	98	57	55
Vanadium, ppm	18	80	285	176	190
Conrad. Carbon, wt%	5.0	11.6	36.9	23.7	22.1

2.5   COMBINING SHORT-PATH DISTILLATION AND PREPARATIVE HPLC

Short-path distillation and HPLC are laboratory separation tools that enable the diverse petroleum macromolecules in vacuum resids to be divided in different fractions so that we can use analytical methods to characterize the fractions and, thus, the diversity of the vacuum resid. The combination of short-path distillation and preparative HPLC is particularly strong because we obtain a view of how the compound classes vary with atmospheric boiling point. Table 2.20 shows how this is achieved on the 1305°F+ fraction of Arabian Heavy vacuum resid. Because this fraction contains the asphaltenes, only the heptane solubles are injected in the preparative HPLC, and the asphaltenes become one of the compound classes. However, more than half the 1305°F+ fraction are asphaltenes and only 12.5 wt% (saturates and one- to three-ring aromatics) is of good quality. Although thermal conversion of this fraction can still produce volatile liquids of good quality, one should avoid contacting this fraction with expensive catalysts.

2.6   OTHER METHODS TO CHARACTERIZE PETROLEUM MACROMOLECULES

There are many other methods to characterize the chemical properties of petroleum macromolecules in more detail than covered in this chapter. However, because they are not applied to the process chemistry in this book, they are not discussed. Instead, the reader is referred to the books by Altgelt and Boduszynski,[21] and by Speight.[23] As will be seen, much can be understood, modeled, and used without knowing the detailed chemical structures present in petroleum macromolecules.

TABLE 2.20
Elemental Analysis of Preparative HPLC Fractions of Arabian Heavy 1305°F+

Fraction	Yield (wt%)	C (wt%)	H (wt%)	H/C (Atomic)	S (wt%)	N (wt%)	Σ Elmts (wt%)
Saturates	1.8	85.78	13.59	1.89	0.46	NA	99.83
1-Ring arom.	1.5	84.96	12.54	1.76	2.08	NA	99.58
2-Ring arom.	4.3	84.04	11.97	1.70	2.86	NA	98.87
3-Ring arom.	4.9	83.52	11.07	1.58	4.79	NA	99.38
4-Ring arom.	14.2	82.10	10.12	1.47	6.30	NA	98.52
Polars 1	12.2	82.41	8.83	1.28	6.88	0.87	98.99
Polars 2	1.6	81.26	8.45	1.24	6.72	0.54	96.97
Polars 3	2.9	81.77	8.58	1.25	6.42	0.81	92.87
Asphaltenes	56.6	82.18	8.15	1.18	8.07	0.92	99.32
Total	100.0	82.42	9.00	1.30	5.95	0.66	98.94
1305°F+ feed	100.0	83.20	9.19	1.32	6.23	0.58	99.20

On a first level, the important concept is that the petroleum macromolecules contain polynuclear aromatics that impart thermal stability and insolubility, which will be discussed in the next chapter on the pendant-core building-block model. However, with continued advancement in characterization tools, the future challenge will be to incorporate the new knowledge in advanced models of the process chemistry. An example is discussed in section 2.6.3. Also, a few additional areas of importance to the characterization of petroleum macromolecules will be summarized.

2.6.1 METALS

Although vanadium and nickel are present in petroleum in trace amounts, they are important for processing because they accumulate on catalysts to plug pores, to attack the catalyst structure, or to promote undesirable side reactions. They can also accumulate in gasifiers to form a slag. Although iron can also be a naturally occurring metal in petroleum, most is a result of corrosion during production and processing. For a long time it was thought that only a portion of vanadium and nickel existed in porphyrin structures, those that show absorption in visible light between 400 and 570 nm, called the Soret band. However, more recent data with EXAFS (extended x-ray absorption) and XANES (x-ray absoption near-edge structure) show that all the vanadium (as an oxide) and nickel in petroleum typically are coordinated with nitrogen as exists in porphyrin structures.[24] An example porphyrin containing vanadium oxide[25] is shown in Figure 2.12. As will be discussed later, it is difficult to physically separate porphyrins from other polynuclear aromatics. Therefore, vanadium and nickel are typically removed chemically by hydrotreating, although oxidation methods are also possible.[26]

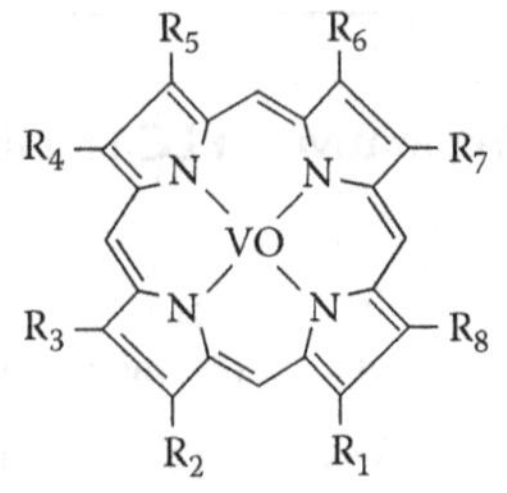

FIGURE 2.12 Chemical structure of vanadyl porphyrins.

2.6.2 NAPHTHENIC ACIDS

This is another trace component of petroleum that can significantly influence its processing chemistry. The term naphthenic acid implies a saturated ring structure with a carboxylic acid attached. However, it is commonly measured by total acid number (TAN), using ASTM D664 or D974. These tests measure the number of milligrams of potassium hydroxide required to neutralize 1 g of oil. Thus, these tests measure the concentration of all acids in the oil, whether or not they are attached to a saturated ring. TAN values of crude oils from 0 to 8.0 have been reported.[27] The primary processing problem with naphthenic acids is corrosion, but they have also been implicated in causing stable water emulsion formation in desalting. A very minor amount of naphthenic acids are even caustic extracted from crude oils, recovered, and sold as a commercial product.[28] Metal naphthenates are soluble in organics and are used as antifungal agents, pesticides, paint drying catalysts, and fuel- and lubricating-oil additives. Naphthenic acids are also used to make low-cost surfactants, and are known to be by-products of biodegradation of crude oils.[29] Typically, naphthenic acids are C_{10}–C_{50} compounds with 0 to 6 fused saturated rings and with the carboxylic acid attached to a ring by a short side chain.[30] Example naphthenic acid chemical structures are shown in Figure 2.13.

2.6.3 PETROLEOMICS

The assertion that there are too many different molecules in heavy oils to measure the quantity of individual compounds has been recently challenged by a research group at the National High Magnetic Field Laboratory at Florida State University. Rogers, Marshall, and coauthors[31,32,33] have taken advantage of the uniquely high mass accuracy and high mass resolving power of ultrahigh-resolution Fourier transform ion cyclotron resonance (FT-ICR) mass spectrometry to attempt the analysis of individual compounds of petroleum. As a result, they coined the term "petroleomics" for the relationship between the chemical composition of petroleum-derived materials and their physical properties. They have been able to resolve over 17,000 compounds in a South American crude oil.[31]

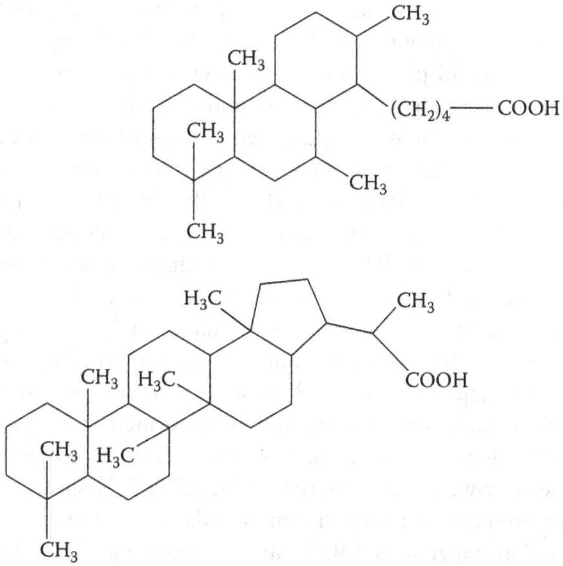

FIGURE 2.13 Example naphthenic acid chemical structures.

Although this is a significant achievement, it is likely that this is still a minor fraction of the compounds actually present. However, they continue to improve their mass spectrometry methods and instruments to resolve an increasing number of compounds. Although they measure the molecular weight of asphaltenes to be less than 1000, mass spectrometry typically obtains low molecular weight because only the low-molecular-weight asphaltenes can be volatized. However, they also measure higher-molecular-weight species that break down to low-molecular-weight species with mild thermal energy.[33] One is tempted to conclude that the higher-molecular-weight species are associated asphaltenes. Mullins and coauthors[34,35] also concluded from time-resolved florescence depolarization and molecular diffusion measurements that the average molecular weight of asphaltenes is of the order of 700. In addition, Ruiz-Morales and Mullins[36] combined optical absorption and fluorescence emission measurements with molecular orbital calculations to conclude that the mean number of fused rings for unconverted asphaltenes is 7, with a range from 4 to 10. When this is combined with the low molecular weight, Ruiz-Morales and Mullins conclude that the average asphaltene molecule contains only one polynuclear aromatic in contrast with the archipelago model of asphaltenes with several polynuclear aromatics in the average asphaltene molecule.

The main feature of the method for measuring molecular weight in this book is its consistency with a lot of different types of data.[37] In section 2.2.3, the evidence was presented for using vapor pressure osmometry (VPO) at 130°C with o-dichlorobenzene as a solvent for measuring molecular weight and obtaining asphaltene average molecular weights of the order of 3000. In section 2.2.4 it was shown that this molecular weight measurement helps provide the solvent–resid phase diagram, showing consistency with measurement for fractions other than asphaltenes and thermally converted products. In Table 2.1 we saw that the number average molecular weight of the fractions gave close to the measured molecular weight of the total Cold Lake vacuum resid. Both the measured molecular weights of the saturates and resin fractions are higher than 700. Because these fractions do not associate like asphaltenes, there is no explanation why VPO would give much higher values. Table 2.2 shows as expected that thermal conversion decreases the VPO molecular weight of fractions. This change for asphaltenes in VPO molecular-weight measurement with thermal reaction time will be quantitatively described by a kinetic model in chapter 3. Table 2.3 shows that resins formed from aromatics, asphaltenes formed from resins, and coke formed from asphaltenes are all higher VPO molecular weight than their reactant. As expected, these are formed by molecular-weight growth reactions. Table 2.21 shows VPO molecular-weight measurements on known compounds containing polynuclear aromatics in o-dichlorobenzene at 130°C, and in toluene at 50°C. Because these two methods agree, these model compounds do not associate in contrast with asphaltenes. When compared with the actual molecular-weight values, the VPO measurements tend to be slightly high. However, the measurements are not wrong by the orders of magnitude required to obtain 3000, rather than 700. Finally, in Table 2.22, the properties of asphaltenes from the slurry oil of fluid catalytic cracking (FCC) is compared with resid asphaltenes. Because the FCC asphaltenes in slurry oil volatilized off the catalyst in a riser reactor at about 500°C, it is not surprising that their average molecular weight is of the order of 700. As the two VPO techniques agree for FCC asphaltenes, no association is occurring, unlike for resid asphaltenes, even though the FCC asphaltenes are more aromatic (lower hydrogen-to-carbon ratio). If resid asphaltenes have a molecular weight of 700, why don't they volatilize out of cokers (also operated at about 500°C) or distill over in Distact distillation? Actually, in Table 2.16,

TABLE 2.21
VPO Molecular-Weight Measurements on Pure Compounds

Compound	Actual mol. wt.	Av VPO mol. wt. (130°C, o-DCB)	Av VPO mol. wt. (50°C, Toluene)
Triphenyl methane	244	247	255
o-Terphenyl	230	240	240
Pyrene	202	218	225
Benzene-hexa-n-heptanoate	846	803	1075
2,3,6,7,10,11-Hexaoctoxy-triphenylene	998	1153	1020

TABLE 2.22
Comparison of FCC and Resid Asphaltenes

Asphaltenes	H/C (atomic)	Av VPO mol. wt. (130°C, o-DCB)	Av VPO mol. wt. (50°C, toluene)	Conradson Carbon res. (Wt%)
FCC	0.820	679	682	49.4
Resid	1.15	2980	5600	50.0

we showed that 6.6% thermally converted asphaltenes were found in the scrubber bottoms of a Flexicoker and, thus, volatilized out of the reactor. Therefore, some of the asphaltenes may be less than 1000 in molecular weight after thermal cracking, but this hardly represents the average because the resid feed contained 25% asphaltenes.

Because petroleum macromolecules are so complex, it is difficult to be absolutely sure about any structural detail, such as even molecular weight. This is both the challenge and the fun in working with petroleum macromolecules. One needs to observe many different types of data and determine if your view is consistent. This has been done for VPO molecular weight measurements. Are the newer methods correct that the average molecular weights of asphaltenes are about 700? We need to be open to that possibility. However, measurements on one asphaltene fraction is not sufficient. It needs to be tested with different types of data as the author has done for VPO molecular weight. On the other hand, if the average molecular weight of asphaltenes is actually 3000, the structure would have to be of the archipelago type with multiple polynuclear aromatic structures in the same molecule.

The other challenge and fun is to make progress without knowing the detailed structure. In chapter 3, the Pendant-Core Building Block Model will show that much can be understood about petroleum macromolecules and their processing by only knowing that they contain at least one thermally stable polynuclear aromatic, something we can all agree. In chapter 4, this will be extended to developing a model for resid thermal kinetics. Such progress does not diminish the need to understand the detailed structure of petroleum macromolecules. With such understanding, much greater detailed predictive models could be made, and much better processes could be devised. However, we need not wait for the scientific community to agree on the detailed structures of petroleum macromolecules to make predictions and improve processes based on our present level of understanding. This is an evolving procedure that always should enable incorporating new information when it is obtained. All knowledge, especially about petroleum macromolecules, is an approximation. We should not build barriers and refuse to change our models in the face of contrary experimental data. On the other hand, we should not accept new models that are not shown to be consistent with all the experimental data. Our models and understanding need to be tested in as many ways as we can conceive so that we can recognize the gaps in these approximations in need of improvement. Only by filling these gaps can we create new knowledge.

2.7 PHYSICAL STRUCTURE

2.7.1 SCATTERING DATA

Petroleum has a unique physical structure that is revealed by x-ray diffraction, small-angle x-ray, and neutron scattering. The author collaborated with Ken Liang[38] at Exxon Corporate Research (also previously at Xerox) who obtained the x-ray diffraction data in terms of intensity versus two times the scattering angle in Figure 2.14 for class fractions of Arabian Heavy vacuum resid: asphaltenes, resins, and saturates plus aromatics. The saturates plus aromatics have a broad peak in the region expected for paraffins. The asphaltenes have overlapping peaks from

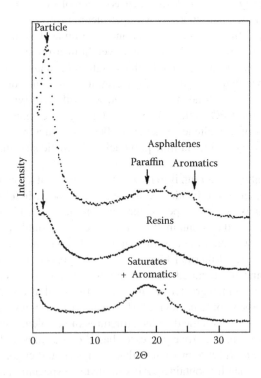

FIGURE 2.14 X-ray diffraction data of intensity versus twice the scattering angle for class fractions of Arabian Heavy vacuum resid: asphaltenes, resins, and saturates plus aromatics. (From IA Wiehe, KS Liang. *Fluid Phase Equilibria*. 117: 201–210, 1996. With permission.)

the paraffins and the large-ring aromatics. Approximate analysis from the location and width of the aromatic peak indicates that the distance between aromatics is about 3.55 A with a correlation length of about 11.35 A. This means that there are on average about 4.2 polynuclear aromatics in a stack. In addition, there is a low-angle peak that is representative of the distance between agglomerates of 39 A, or approximately the diameter of the agglomerates. The x-ray diffraction on the resins has a broad peak for the paraffins with an aromatic shoulder and a slight particle peak at low scattering angle. Thus, resins have some polynuclear aromatics with a slight tendency to associate, but much less than for the asphaltenes.

Data were collected by a combination of x-ray scattering and small-angle scattering with high-intensity x-rays from a synchrotron on asphaltenes and coke separated after thermally converting Arabian Heavy vacuum resid. Figure 2.15 is the resulting plot of the logarithm of the intensity versus the logarithm of the scattering vector, Q, 4 pi times the sine of the scattering angle and divided by the wavelength of the x-ray. A straight line on this plot indicates fractal behavior as is shown for the asphaltenes from a Q of 0.002 to 0.06 with a slope of −3.45. This might imply that agglomerates across these scales will look alike or possess self similarity, but, in this case, it is probably a result of surface roughness. The

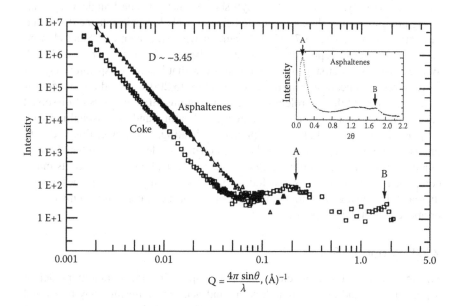

$$Q = \frac{4\pi \sin\theta}{\lambda}, \, (\text{\AA})^{-1}$$

FIGURE 2.15 The combination of x-ray scattering data and small-angle scattering data with high-intensity x-rays from a synchrotron in terms of intensity versus scattering angle, Q, on asphaltenes and coke separated after thermally converting Arabian Heavy vacuum resid. Insert of x-ray diffraction of unreacted asphaltenes from Figure 2.14 included for comparison. (From IA Wiehe, KS Liang. *Fluid Phase Equilibria*. 117: 201–210, 1996. With permission.)

scattering from coke shows parallel fractal behavior from a Q of 0.0015 to 0.01. The coke also shows an aromatic particle peak (A) and an aromatic diffraction peak (B), like asphaltenes. However, unlike asphaltenes, coke shows little or no diffraction from paraffins (between A and B). Thus, coke contains similar aromatic structures as the asphaltenes, but with fewer paraffins, and is consistent with other evidence that coke is formed from asphaltenes after thermally cracking off paraffinic side chains.

2.7.2 EFFECT OF TEMPERATURE

The effect of temperature on asphaltene cluster size and solubility is somewhat controversial in the literature. Storm, Barresi, and Sheu[39] report that asphaltenes cluster more, but do not phase separate, at elevated temperatures above 200°C, based on viscosity and small-angle x-ray scattering data. However, the viscosity data are not definitive for structure, and the small-angle x-ray data were done at room temperature after heating and cooling. In addition, their bombs containing resid were heated to 350 to 400°C, temperatures known to cause thermal cracking. In contrast, Winans and Hunt[40] used small-angle neutron scattering data on 5 wt% asphaltenes in d_{10}-1-methyl naphthalene measured directly at temperature and showed decreasing asphaltene aggregate size with increasing temperature to 340°C. Previously, Espinat and Ravey[41] showed that, by small-angle x-ray scattering at temperature, the agglomerated size of asphaltenes in toluene decreased in heating from −27°C to 77°C. This agreed with Andersen and Stenby[42] who observed increased asphaltene solubility on heating from 24 to 80°, and with Wiehe[43] who used hot-stage microscopy to show that insoluble asphaltenes redissolve in resid on heating from room temperature to 200°C. To add to the confusion, it is known[44] that petroleum crude that contains large amounts of dissolved gases (live oils) can precipitate asphaltenes on heating, and in deasphalting with propane or butane, asphalt is precipitated on heating.[45] However, this is due to the well-known phenomenon[46] of the rapid change of density and other thermodynamic properties as a mixture approaches the critical point of one of its major components. Therefore, it is concluded that the solubility of asphaltenes always increases by raising the temperature unless one of the major components is near or above its critical temperature.

2.7.3 MODEL OF SIROTA

Sirota[47] points out that much of the data supporting the colloidal model of asphaltenes, including small-angle x-ray and neutron scattering, may be caused by the solution behavior of associating molecules (asphaltenes) near the solution critical point and below the glass transition temperature of asphaltenes. He draws the analogy of polystyrene (also soluble in toluene and insoluble in n-heptane) with a glass transition temperature about 100°C, which, when rapidly precipitated in isooctane at 0°C, exhibits a "morphological behavior similar to asphaltenes." This reminds us that one should keep an open mind about how good an approximation any model is, particularly for a complex system such as petroleum, and continue to test it to determine its limits. However, there are also serious questions about the Sirota model. If the scattering from asphaltenes is a result of concentration fluctuations because the asphaltene solution is near a solution-critical point, one would expect the scattering to disappear at a little higher temperature and insolubility of asphaltenes to occur at a little lower temperature. Because Winans and Hunt[40] observed asphaltene particles, albeit smaller ones, at 340°C in d_{10}-1-methyl naphthalene by small-angle, and Espinat and Ravey[41] found asphaltene particles, albeit larger ones, but not insolubility, at −27°C in toluene, it is difficult to accept the Sirota critical solution point explanation near room temperature. Nevertheless, Sirota also mentions another explanation that might be more plausible. He compares asphaltenes to block copolymers in which one block is insoluble in a solvent, a second block is soluble in the same solvent, but the two blocks are connected by a covalent bond. The polynuclear aromatic potions of asphaltenes could act as an insoluble block, and the paraffinic side chain and the small-ring aromatic portions could act as the soluble block. This would produce the observed particles in solution that increase in size with lower temperatures and poorer solvents, and decrease in size with higher temperatures and better solvents. This explanation also fits well with the pendant-core model described in the next chapter.

REFERENCES

1. IA Wiehe. A solvent-resid phase diagram for tracking resid conversion. *I & EC Res.* 31: 530–536, 1992.
2. JG Speight, RB Long, TD Trowbridge. Factors influencing the separation of asphaltenes from heavy petroleum feedstocks. *Fuel* 63: 616–620, 1984.
3. JW Bunger, DE Cogswell. Characteristics of tar sand bitumen asphaltenes as studied by conversion of bitumen by hydropyrolysis. In: JW Bunger, NC Li, Ed. *Advances in Chemistry*, Series 195. Washington, D.C.: American Chemical Society, 1981, pp. 219–236.
4. MM Boduszynski, JF McKay, DR Latham. Asphaltenes, where are you? *Asphalt Paving Technol.* 49: 123–143.
5. KH Altgelt, MM Boduszynski. *Composition and Analysis of Heavy Petroleum Fractions.* New York: Marcel Dekker, 1994, pp. 463–474.
6. RB Long. The concept of asphaltenes. In: JW Bunger, NC Li, Ed. *Advances in Chemistry*, Series 195. Washington, D.C.: American Chemical Society, 1981, pp. 17–27.
7. RB Long, JG Speight. Studies in petroleum composition. Development of a compositional "map" for various feedstocks. *Rev Inst Fr Pet.* 44: 205–217, 1989.
8. JG Speight. *The Chemistry and Technology of Petroleum.* 3rd ed. New York: Marcel Dekker, 1999, pp. 412–498.
9. CE Snape, KD Bartle. Definition of fossil fuel derived asphaltenes in terms of average structural properties. *Fuel.* 63: 883–887, 1984.
10. CE Snape, KD Bartle. Further information on defining asphaltenes in terms of average structural properties. *Fuel.* 64: 427–429, 1985.
11. K Ouchi. Correlation of aromaticity and molecular weight of oil, asphaltene and preasphaltene. *Fuel.* 64: 426–427, 1985.

12. JG Speight, DL Wernick, KA Gould, RE Overfield, BML Rao, DW Savage. Molecular weight and association of asphaltenes: A critical review. *Rev Inst Fr Pet.* 40: 51–61, 1985.

13. JP Dickie, TF Yen. Macrostructures of the asphaltic fractions by various instrumental methods. *Anal Chem.* 39: 1857–1867, 1967.

14. SE Moschopedis, JF Fryer, JG Speight. Investigation of asphaltene molecular weights. *Fuel.* 55: 227–232, 1976.

15. MM Al-Jarrah, AH Al-Dujaili. Characterization of some Iraqi asphalts. II. New findings on the physical nature of asphalts. *Fuel Sci Technol Int.* 7: 69–88, 1989.

16. JI Haberman, RE Overfield, WK Robbins. Method for spectroscopic analysis of hydrocarbons. U.S. Patent 4,988,446, 1991.

17. RE Overfield, WK Robbins, JI Haberman. Method for refining or upgrading hydrocarbons with analysis. U.S. Patent 5,076,909, 1991.

18. WK Robbins. Quantitative measurement of mass and aromaticity distributions for heavy distillates 1. Capabilities of the HPLC-2 system. *J Chrom Sci.* 36: 457–466, 1998.

19. IA Wiehe, RJ Kennedy. Application of the oil compatibility model to refinery streams. *Energy Fuels.* 14: 60–63, 2000.

20. MM Boduszynski. Limitations of average structure determination for heavy ends in fossil fuels. Prep Paper—*Div Pet Chem.* 28: 1376–1390, 1983.

21. KH Altgelt, MM Boduszynski. *Composition and Analysis of Heavy Petroleum Fractions.* New York: Marcel Dekker, 1994, pp. 9–73.

22. CJ Domansky. Short path distillation of Athabasca bitumen. *Aostra J Res.* 2: 191–196, 1986.

23. JG Speight. *The Chemistry and Technology of Petroleum,* 3rd ed. New York: Marcel Dekker, 1998, pp. 215–495.

24. KH Altgelt, MM Boduszynski. *Composition and Analysis of Heavy Petroleum Fractions.* New York: Marcel Dekker, 1994, pp. 440–444.

25. AJG Barvise, EV Whitehead. Characterization of vanadium porphyrins in petroleum residues. Preprints. ACS, *Div Pet Chem.* 25: 268–279, 1980.

26. KA Gould. Oxidative demetalization of petroleum asphaltenes and residua. *Fuel.* 59: 733–736, 1980.

27. E Babaian-Kibala, PR Petersen, MJ Humphries. Corrosion by naphthenic acids in crude oils. Preprints. ACS, *Div Pet Chem.* 43: 106–110, 1998.

28. JA Brient. Commercial utility of naphthenic acids recovered from petroleum distillates. Preprints. ACS, *Div Pet Chem.* 43: 131–133, 1998.

29. SD Olsen. The relationship between biodegradation, total acid number and metals in oils. Preprints. ACS, *Div Pet Chem.* 43: 142–145, 1998.

30. WK Robbins. Challenges in the characterization of naphthenic acids in petroleum. Preprints. ACS, *Div Pet Chem.* 43: 137–140, 1998.

31. RP Rogers, GC Klein, LA Stanford, AG Marshall. Petroleomics: FT-ICR high resolution mass spectral analysis of petroleum derived materials. 5th International Conference on Petroleum Phase Behavior and Fouling, Banff, Canada, June 13–17, 2004.

32. AG Marshall, RP Rogers. Petroleomics: The next grand challenge for chemical analysis. *Accounts Chem Res.* 37(1): 53–59, 2004.

33. AG Marshall, CL Hendrickson, GC Klein, JM Purcell, TM Schaub, DF Smith, LA Stanford, RP Rogers. Fourier transform ion cyclotron resonance mass spectrometry: The platform for petroleomics. 7th International Conference on Petroleum Phase Behavior and Fouling, Asheville, NC. June 25–29, 2006.

34. H. Groenzin, OC Mullins. Petroleum asphaltene size and structure. Preprints. ACS, *Div Fuel Chem* 44: 728–732, 1999.
35. G Andreatta, CC Goncalves, G Buffin, N. Bostrom, CM Quintella, F Artega-Larios, E Perez, OC Mullins. Nonaggregates and structure-function relations in asphaltenes. *Energy Fuels.* 19: 1282–1289, 2005.
36. Y Ruiz-Morales, OC Mullins. Characterization of asphaltene fused ring systems comparing molecular orbital calculations with optical spectroscopy. 7th International Conference on Petroleum Phase Behavior and Fouling, Asheville, NC. June 25–29, 2006.
37. IA Wiehe. In defense of vapor pressure osmometry for measuring molecular weight. *J Dispersion Sci Technol.* 28: 431–435, 2007.
38. IA Wiehe, KS Liang. Asphaltenes, resins, and other petroleum macromolecules. *Fluid Phase Equilibria.* 117: 201–210, 1996.
39. DA Storm, RJ Barresi, EY Sheu. Flocculation of asphaltenes in heavy oils at elevated temperatures, *Fuel Sci Technol Int.* 14: 243–260, 1996.
40. RE Winans, JE Hunt. An overview of resid characterization by mass spectrometry and small angle scattering techniques. PrepPaper—ACS *Div Fuel Chem.* 44: 725–732, 1999.
41. D Espinat, JC Ravey. *Soc Pet Eng.* 25187: 365–373, 1993.
42. SI Andersen, EH Stenby. Thermodynamics of asphaltene precipitation and dissolution investigation of temperature and solvent effects. *Fuel Sci Technol Int.* 14: 261–288, 1996.
43. IA Wiehe. Thermal reactivity of heavy oils. ACS Tutorial, *Div Pet Chem*, San Francisco. 1997.
44. KJ Leontaritis. The asphaltene and wax deposition envelope. *Fuel Sci Technol Int.* 14: 13–39, 1996.
45. JY Low, RL Hood, KZ Lynch. Valuable products from the bottom of the barrel using rose technology. PrepPaper—ACS *Div Pet Chem.* 40: 780–784, 1995.
46. JS Rowlinson. *Liquids and Liquid Mixtures.* London: Butterworth Scientific Publications, 1959, pp. 231–235.
47. EB Sirota. The physical structure of asphaltenes. *Energy Fuels.* 19: 1290–1296, 2005.

3 Pendant-Core Building Block Model of Petroleum Resids

3.1 APPROACH TO DEVELOP APPROXIMATIONS

In the previous chapter we learned that analytical tools do not yet exist to identify and measure the concentration of each molecular species in heavy oils. Therefore, our description of these petroleum macromolecules has to be an approximation. The choice then is either to determine the best approximation to petroleum macromolecules that can be made with present instruments, or the simplest approximation that describes our phenomenological data. In this book our choice will be the latter. The approach will be to try the simplest approximation that incorporates concepts, previously determined, to compare the predictions with experimental data and adjust the approximation, when needed, to better describe the data. Finally, the approximation will be tested by comparing the predictions with either a different type of data or the same type, but at significantly different conditions. As a result, if these data comparisons are successful, we are more confident at predicting results not directly measured. However, more likely, the approximation will be verified at most, but not all, conditions. This provides the direction to further improve the approximation. The result is that one continuously learns more about the system through these constant combinations of approximations and experimental challenges. It is this creation of knowledge that we call science, and the utilization of the knowledge for commercial benefit that we call innovation. The advantage of the simplest approximation approach is that it leads more quickly to new innovations that provide the motivation for our research. In this chapter and those that follow, examples will be provided where this approach to research has led to new innovations in the process chemistry of petroleum macromolecules. In addition, at the end of each chapter, the reader will be given recommendations of how both the approximations and innovations can be further improved.

One may instead wish to take the alternate approach of obtaining the best approximation that can be made with the best instruments available, rather than the simplest approximation. This is a more purely scientific approach. Although this allows for a better approximation to the truth, one often spends more time worrying about the trees and forgetting about the forest. It is difficult to select in advance, with this approach, what are the most important problems to work on to advance the field. As a result, one tends to work on all of the problems, and progress is thorough but slow. In the end, both are valid approaches and are often made based on whether the motivation is more to obtain innovation or the best approximation to the truth.

3.2 THE SIMPLEST APPROXIMATION TO THE DISTRIBUTION OF PETROLEUM MACROMOLECULES

In the previous chapter we learned that each fraction of heavy oil can be defined by a region on a plot of molecular weight versus hydrogen content, the solvent-resid phase diagram (Figure 2.3). Thus, a natural starting point to approximate each petroleum macromolecule is by a method that can be defined by a combination of molecular weight and hydrogen content. We also learned in the previous chapter that the greatest intermolecular attraction between petroleum macromolecules is between polynuclear aromatics. This will be emphasized in chapter 5 as the primary cause of the insolubility of petroleum macromolecules. In addition, aromatics are thermally stable, and polynuclear aromatics (PNAs) of five or greater fused rings do not volatize appreciably at typical conversion conditions. A bare 5-ring aromatic, benzopyrene has an atmospheric boiling point of 495°C (923°F). With the expected addition of even methyl and ethyl groups, the boiling point would put it in the vacuum resid range (565°C+ or 1050°F+). As a result, these large PNAs within petroleum macromolecules remain when heavy oils are completely thermally cracked in an open reactor to combine and form the carbonaceous by-product we call coke. Thus, these large PNAs are a thermal reaction limit that prevent heavy oils from being completely converted to volatile liquid products. Our approximation for petroleum macromolecules must allow for part of the macromolecule forming distillable liquids, and part of the macromolecule forming coke when thermally cracked.

The simplest approximation of petroleum macromolecules that provides for a molecular-weight distribution would be as a polymer. This would have an identical repeating chemical group but varying number of these groups per molecule. However, because each repeating group would have the same hydrogen content, the hydrogen content of the macromolecules would be the same. If thermal cracking would break the bonds between repeating groups, at complete conversion, either all distillable liquids or all coke would be formed, depending on whether the repeating group was volatile or nonvolatile. Thus, the polymer approximation clearly does not describe our experimental data.

The next simplest approximation of petroleum macromolecules would be as a copolymer. Thus, each macromolecule would be made up of combinations of two repeating chemical groups. Because the repeating groups would have different molecular weights and hydrogen content, these would vary according to the number of groups of each type in each macromolecule. If one group, the cores, represents the nonvolatile PNAs that become coke and the other group, the pendants, represents the volatile fragments that become volatile liquids when cracked

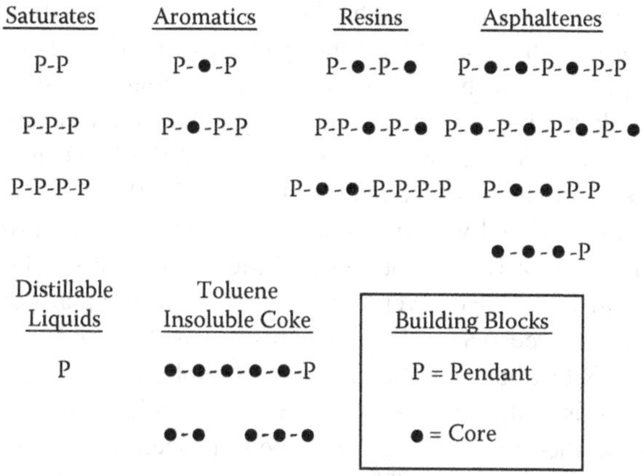

Saturates	Aromatics	Resins	Asphaltenes
P-P	P-●-P	P-●-P-●	P-●-●-P-●-P-P
P-P-P	P-●-P-P	P-P-●-P-●	P-●-P-●-P-●-P-●
P-P-P-P		P-●-●-P-P-P-P	P-●-●-P-P
			●-●-●-P

Distillable Liquids	Toluene Insoluble Coke	Building Blocks
P	●-●-●-●-●-P	P = Pendant
	●-● ●-●-●	● = Core

FIGURE 3.1 The Pendant-Core Building Block Model allows for a wide distribution of species. (From IA Wiehe. *Energy and Fuels* 8: 536–544, 1994. Reprinted with permission. Copyright 1994. American Chemical Society.)

off the petroleum macromolecule, it would meet our second requirement for an approximation. This approximation is called the Pendant-Core Building Block Model. The term *building block* is used instead of *chemical group* to emphasize that the groups are assumed to be identical. Of course, we believe that petroleum macromolecules have many more repeating groups than two. However, remember that we are trying to arrive at the simplest approximation that describes the phenomenological data and not necessarily the best approximation to reality. Later, this approximation will be tested by comparing it with experimental data to determine its limitations.

Figure 3.1 shows how petroleum macromolecules of fractions of unconverted and converted heavy oil may be represented by a distribution of pendant and core building blocks. Later it will be shown how to estimate the molecular weight and hydrogen content of each building block. Nevertheless, although this model will not describe the millions of different molecules of a vacuum resid, it does introduce some distribution, and it enables aromatic fractions to undergo thermal cracking without converting to another fraction. In chapter 2, with the solvent-resid phase diagram, we saw that this is a property of resid conversion, and more details will be given in chapter 4 when thermal conversion kinetics will be discussed.

3.3 MODEL FOR CONRADSON CARBON RESIDUE (CCR)

Conradson carbon residue is a standard measure of coke formation tendency used by the petroleum industry. It is used extensively as a feed property to predict the yield of coke in petroleum processing by means of correlations. As a result, it is a very important property. However, it has been difficult to relate this measurement in a functional test to any molecular property of the resid or other heavy oil. There was no way to predict how the CCR would change during processing and when it should and should not be used to predict coke yield at conditions much different than it was measured. So far, the greatest success of the Pendent-Core Building Block Model is to provide this needed insight about CCR.

The CCR test[1] originally involved heating a crucible containing 10 g of the oil to cherry red with a Bunsen burner and measuring the residue as a weight percent of the starting oil. However, this has largely been replaced by the micro-carbon residue test[2] (see ASTM D4530). This test is designed to give the same values as the CCR test but with much smaller samples and a more defined thermal history. In this test, samples of the oil are weighed into glass vials and heated by a prescribed rate under a prescribed nitrogen flow to 500°C and held there for 15 min before cooling. The percentage of the sample not volatilizing out of the vial during the test is taken as the CCR. An instrument manufactured by Alcor (MCRT-120) enables measuring multiple samples simultaneously, of which one sample should be a standard to assure that the test was run reproducibly. Although the microcarbon test is similar to a miniature coker, the yield of residue, or coke, is typically less than that of commercial cokers, which recycle the heavier volatile liquid product back to the coker and operate under some pressure.

Despite its widespread use, little has been published on the chemical mechanism of CCR. Roberts[3] found the surprising result that, for resids, the CCR is a linear function of either hydrogen content or hydrogen-to-carbon (H/C) atomic ratio that is independent of resid source or process history. However, no mechanistic explanation was given for this correlation. Green et al.[4] showed that fractions of residua generated by liquid chromatography gave close to the same linear function of the CCR and hydrogen-to-carbon atomic ratio as the whole resid. They also measured the concentrations of sulfur, nitrogen, nickel, and vanadium in the residue from the microcarbon residue test and determined that the concentrations of these heteroatoms in the residue usually were much higher than in the various feeds and feed fractions. Wiehe[5] described the chemical mechanism of Conradson carbon, including the linear relationship to hydrogen content, based on the Pendant-Core Building Block Model as described here.

Figure 3.2 shows the reaction of a resid with a pendant-core distribution of macromolecules to form distillable liquids and CCR in the microcarbon residue test. Therefore, the Conradson carbon is pictured to give the weight fraction of the resid that is cores. A hydrogen balance around this reaction shows that the CCR should be a linear function of the hydrogen content of the resid.

The hydrogen content of the pendant is that at zero CCR on this line, and the hydrogen content of the core is that at 100% CCR. A test of this prediction is given in Figure 3.3, where the CCR is plotted against the hydrogen content for a number of different vacuum resids. These data include Conradson carbon data

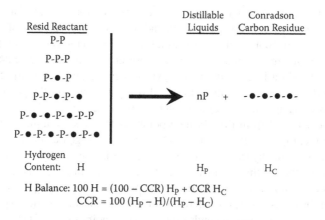

FIGURE 3.2 Model shows Conradson carbon residue (CCR) is a linear function of hydrogen content. (From IA Wiehe. *Energy and Fuels* 8: 536–544, 1994. Reprinted with permission. Copyright 1994. American Chemical Society.)

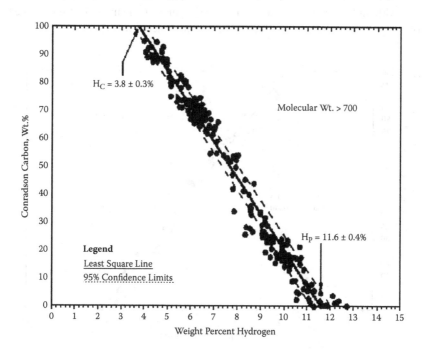

FIGURE 3.3 Conradson carbon is a linear function of the hydrogen content. (From IA Wiehe. *Energy and Fuels* 8: 536–544, 1994. Reprinted with permission. Copyright 1994. American Chemical Society.)

on fractions (saturates, small-ring aromatics, resins, asphaltenes) of unconverted resids; fractions of converted resids, including coke; and fractions of converted fractions. The only data excluded were on fractions with molecular weights below 700 amu (atomic mass unit), which will be presented separately. The 240 data in Figure 3.3 fall in a linear band and fit with a least-square line with 95% confidence limits as shown. This measures the hydrogen content of the pendants to be 11.6 ± 0.4 wt%, and the hydrogen content of the cores to be 3.8 ± 0.3 wt%. The averages calculated from the 114 data points of Roberts,[3] assuming an average carbon content of 84 wt%, are 12.1 wt% for the hydrogen content of pendants, and 4.0 wt% for the hydrogen content of cores. This pendant hydrogen content is only slightly outside the confidence limits of the present data, whereas the core hydrogen content is within the limits of the present data. Although the hydrogen content of the cores can be considered a "universal" constant within experimental error, that of the pendants depends on the boiling point range of the resid. Thus, the pendants of an atmospheric resid should have higher hydrogen content than the pendants of a vacuum resid.

To determine if there are feed variations in the pendant and core hydrogen contents, Figure 3.4 to Figure 3.6 are Conradson carbon versus hydrogen content plots for data on the vacuum resid feeds: Arabian Heavy, Cold Lake, and Hondo. Of course, these include fractions of the vacuum resid and its conversion products. Although there may be some slight variation of pendant and core hydrogen content with feed, they are within the 95% confidence limits of the total data.

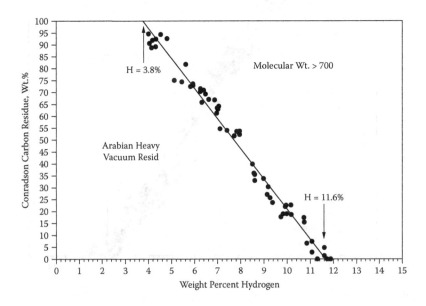

FIGURE 3.4 Conradson carbon is a linear function of the hydrogen content for Arabian Heavy vacuum resid. (From IA Wiehe. *Energy and Fuels* 8: 536–544, 1994. Reprinted with permission. Copyright 1994. American Chemical Society.)

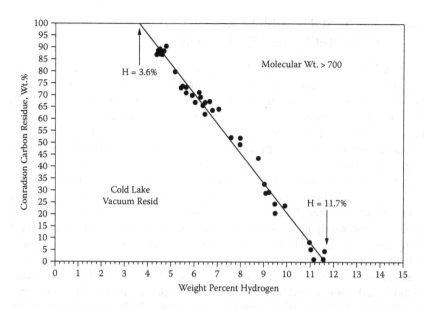

FIGURE 3.5 Conradson carbon is a linear function of the hydrogen content for Cold Lake vacuum resid. (From IA Wiehe. *Energy and Fuels* 8: 536–544, 1994. Reprinted with permission. Copyright 1994. American Chemical Society.)

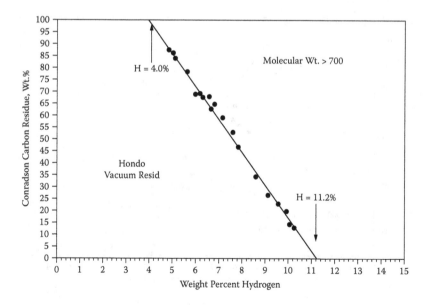

FIGURE 3.6 Conradson carbon is a linear function of the hydrogen content for Hondo vacuum resid. (From IA Wiehe. *Energy and Fuels* 8: 536–544, 1994. Reprinted with permission. Copyright 1994. American Chemical Society.)

3.4 CONRADSON CARBON RESIDUE OF LOW-MOLECULAR-WEIGHT FRACTIONS

The Conradson carbon versus hydrogen content is plotted in Figure 3.7 for fractions of resids and resid thermal conversion products with molecular weights between 350 and 700 amu. Included in these are coker scrubber bottoms, the fractions that volatilize out of cokers, but are condensed and recycled back to the cokers. Although some of the data are close to the least-square line on resid fractions with molecular weights above 700 amu, most of the data lie below this line. An explanation is that a substantial fraction of these moderate-molecular-weight fractions volatilizes out of the microcarbon residue test without reacting. By volatilizing part of the cores, the residue yield in the microcarbon residue test is lower than expected, based on the hydrogen content. However, if these moderate molecular-weight feeds are forced to completely react in a process, such as by recycling the coker bottoms, the hydrogen content and the least-square line should be a better predictor of expected coke yield than the measured Conradson carbon residue. Otherwise, the coke yield from these fractions cannot be predicted without adding a model for the vapor–liquid equilibrium to predict which fraction volatilizes without reacting and which fraction reacts before volatilizing.

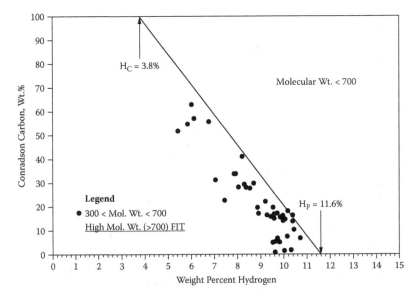

FIGURE 3.7 Conradson carbon is low for the hydrogen content for moderate-molecular-weight (350–700) fractions. (From IA Wiehe. *Energy and Fuels* 8: 536–544, 1994. Reprinted with permission. Copyright 1994. American Chemical Society.)

3.5 ELEMENTAL ANALYSIS OF CONRADSON CARBON RESIDUE

With the Pendant-Core Building Block Model successfully predicting the linear relationship between CCR and hydrogen content, it needs to be tested with a different type of data. Thus, we examine the elemental analysis of the residue from the microcarbon residue test. In particular, we are interested in determining if the hydrogen content of the residue falls within the 3.8 ± 0.3 wt% predicted by the linear relationship. Although Green et al.[4] measured the heteroatom content of the microcarbon residue, they did not report the hydrogen content. The vacuum resid feeds for this study have elemental analyses and molecular weights as given in Table 3.1. After running the microcarbon residue tests on these vacuum resids, elemental analyses of the residues were measured with the results shown in Table 3.2. Indeed, the hydrogen content of the microcarbon residues are within the 3.8 ± 0.3 wt% predicted by the linear relationship of CCR and hydrogen content. Therefore, the Pendant-Core Building Block Model successfully passes this test, and the hydrogen content of the cores can be considered as a resid-independent constant.

The carbon content of the CCRs does not vary significantly about the mean of 87 wt%. However, this is only slightly higher than the carbon content of the

TABLE 3.1
Properties of Vacuum Resids

Vacuum Resid	CCR (wt%)	C (wt%)	H (wt%)	S (wt%)	N (wt%)	V (ppm)	Ni (ppm)	VPO Mol. Wt (o-DCB)	Average Molecular Formula
Arabian Heavy	22.3	83.51	9.93	5.80	0.45	165	40	936	$C_{65} H_{93} S_{1.7} N_3 O_2$
Heptane solubles	15.0	84.07	10.73	4.51	0.33	140	33	778	$C_{55} H_{83} S_{1.1} N_2 O_2$
Cold Lake	25.8	82.45	9.70	5.75	0.62	289	102	1040	$C_{76} H_{107} S_2 N_5 O_5$
Hondo	24.6	82.02	9.85	7.00	1.23	475	216		
Heptane solubles	13.7	81.20	10.25	5.84	0.83	193	88		
European coker feed	23.7	85.24	9.95	4.19	0.41	209	82		
Venezuelan coker feed	23.4	85.03	10.02	3.19	0.69	683	84	1071	$C_{76} H_{106} S_{1.1} N_5 O_7$
Texas coker feed	25.9	84.37	10.11	4.63	0.66	375	112		

Source: IA Wiehe. *Energy and Fuels* 8: 536–544, 1994. Reprinted with permission. Copyright 1994. American Chemical Society.

TABLE 3.2
Elemental Analyses of Conradson Carbon Residues

Vacuum Resid	Resid CCR (wt%)	Conradson Carbon Residue					
		C (wt%)	H (wt%)	S (wt%)	N (wt%)	V (ppm)	Ni (ppm)
Arabian Heavy	22.3	85.76	3.73	8.56	1.30	808	165
Cold Lake	25.8	85.98	3.86	7.23	1.66	1290	542
Hondo	24.6	85.64	3.61	5.44	3.02	2040	848
European coker feed	23.7	89.28	3.73	5.58	1.31	846	309
Venezuelan coker feed	23.4	88.19	3.79	4.46	1.35	2880	390
Texas coker feed	25.9	89.14	3.73	4.85	2.03	1530	419
Average	24.3	87.33	3.74	6.02	1.78	1565	456

Source: IA Wiehe. *Energy and Fuels* 8: 536–544, 1994. Reprinted with permission. Copyright 1994. American Chemical Society.

feeds and, as discussed in chapter 2, the carbon content of all resid fractions and products do not vary much. As found by Green et al.,[4] nearly all of the vanadium and nickel in the resid end up in the CCR. Thus, the metal content of the residue is predicted to be the metal content of the feed divided by the fraction of the feed that is residue. As shown in Table 3.3, this provides a good prediction of vanadium and nickel in the CCR. If the measurements were accurate, the measured metals in the residue could not be higher than those predicted to be in the residue. Thus, when they are much higher, as for the nickel content of the Cold Lake residue, either the measurement was not accurate or the sample got contaminated.

TABLE 3.3
All of Feed Metals Are in the Conradson Carbon Residue

Vacuum Resid	Resid FCCR (Wt Fract.)	V_{CCR} (ppm)	V_{Feed}/FCCR (ppm)	Ni_{CCR} (ppm)	Ni_{Feed}/FCCR (ppm)
Arab Heavy	0.223	809	740	165	179
Cold Lake	0.258	1290	1120	542	395
Hondo	0.246	2040	1931	848	878
European coker feed	0.237	846	881	309	346
Venezuelan coker feed	0.234	2880	2919	390	359
Texas coker feed	0.259	1530	1448	419	432

The sulfur and nitrogen partition somewhat differently between volatiles and residue for each feed. Hondo resid even produces a coke with lower sulfur content than the resid, unlike the other five resids studied. Green et al.[4] also found this to be the case for Wilmington resid, which, like Hondo, is from a California crude. This results partly from a high fraction of the sulfur in the resid being as aliphatic sulfides, which are released as hydrogen sulfide during the microcarbon test. Indeed, George, Gorbaty, and Kelemen,[6,7,8] using XPS and XANES, have determined the aliphatic sulfide fraction of sulfur of Hondo resid (42%) to be much higher than that for Arabian Heavy (24 to 27%) and Cold Lake (32 to 33%). In addition, as will be shown later, the volatile liquid product of thermolysis of Hondo resid has higher sulfur content than the other resids. Nevertheless, it is clear that the cores from different resids have varying concentrations of sulfur, nitrogen, and metals.

One of the challenges in the processing of vacuum resids is predicting the properties of the products. Although measuring the Conradson carbon of resids is quite common, only Green et al.[4] and Wiehe[5] have reported measuring the elemental content of the residue from the microcarbon residue test. However, this would be a good starting point if one wanted to predict the elemental analysis of the coke in a coking process. Of course, one would need to make corrections for the different coking temperature and coke yield of the coking process compared with the microcarbon residue test. Actually, only the sulfur and nitrogen content need to be measured because the metals can be calculated, and the hydrogen and carbon content of the CCR are constant, independent of feed.

3.6   ELEMENTAL ANALYSIS OF DISTILLABLE LIQUID PRODUCTS

As a vacuum resid is thermally converted, and the gaseous and volatile liquid products are removed, the remaining vacuum resid becomes enriched in heteroatoms and depleted in hydrogen. Despite this, the Pendant-Core Building Block Model predicts that the volatile liquid products will not change properties with conversion and the hydrogen content will be 11.6 ± 0.4 wt%. Thus, to further test the Pendant-Core Building Block Model, we examine the distillable liquid product of the thermolysis of vacuum resids. Arabian Heavy vacuum resid and its heptane-soluble fraction were thermally converted at various times at temperatures from 400 to 470°C. in tubing bomb reactors. The elemental composition determined on the distillable (atmospheric boiling points below 524°C) liquid product are plotted in Figure 3.8 as a function of the yield of these distillable liquids. There is no systematic variation of the cumulative hydrogen, carbon, sulfur, or nitrogen content with yield of distillable liquids. Each cumulative elemental content can be described by a random variation about a mean. This was verified by regression analysis that gave a low slope (< I.005I) for each regression line with zero slope always well within the 95% confidence limits. In addition, there does not appear to be any difference between the elemental composition of distillable liquids from the heptane-soluble fraction and from the entire resid. Actually, the average elemental analysis of the distillable liquids from both reactants agree

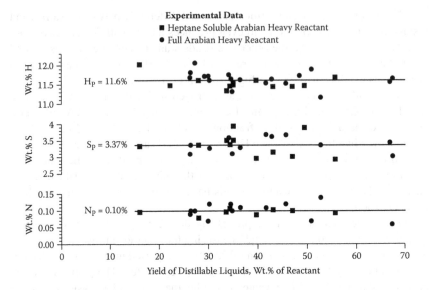

FIGURE 3.8 The properties of Arabian Heavy distillable liquid products are independent of conversion. (From IA Wiehe. *Energy and Fuels* 8: 536–544, 1994. Reprinted with permission. Copyright 1994. American Chemical Society.)

within 0.01 wt%. Thus, the Pendant-Core Building Block Model correctly predicts for Arabian Heavy vacuum residuum that the quality of the distillable liquid product is independent of conversion and is independent of which fraction of the resid is the reactant.

The elemental composition of the distillable liquids as a function of yield for Hondo vacuum resid and its heptane-soluble fraction is shown in Figure 3.9. The cumulative hydrogen and nitrogen content are independent of conversion within 95% confidence limits. However, the cumulative sulfur content of Hondo distillable liquids definitely decreases with increasing conversion. This is the same resid that has a higher sulfur content in the resid than in its CCR. The reason is a high fraction of the sulfur in Hondo pendants existing as aliphatic sulfides that have an increasing probability of forming hydrogen sulfide with increasing reaction time and, thus, not being part of the liquid or coke products. Although the heptane-soluble fraction has a higher hydrogen content and a lower sulfur content than the entire resid, there is a slight bias for the distillable liquids from the heptane-fraction to have lower hydrogen content and higher sulfur content than the distillable liquids from the whole resid. This is consistent with the heptane-insoluble fraction having a higher fraction of aliphatic sulfur (50%),[8] than the whole resid (42%)[7] because the evolution of hydrogen sulfide decreases sulfur content and increases hydrogen content of the remaining resid. However, the average nitrogen content of the distillable liquids from both reactants is the same. Therefore, the Pendant-Core Building Block Model is not quite as good

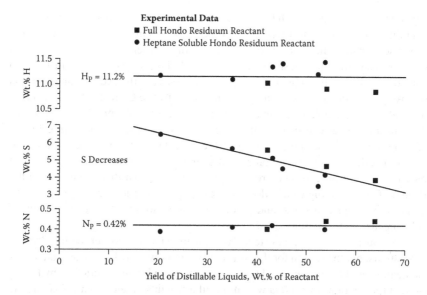

FIGURE 3.9 Hondo distillable liquid products decrease sulfur with conversion, but hydrogen and nitrogen remain constant. (From IA Wiehe. *Energy and Fuels* 8: 536–544, 1994. Reprinted with permission. Copyright 1994. American Chemical Society.)

a model for Hondo resid as for Arabian Heavy resid, with the largest deviation attributed to the high fraction of aliphatic sulfur. Nevertheless, for each resid, the hydrogen content of the pendants, determined by the average hydrogen content of the distillable liquids, agrees exactly with that determined by extrapolating the linear variation of Conradson carbon with hydrogen content to a Conradson carbon of zero as shown in Figure 3.4 and Figure 3.6. This indicates that the linear relationship between Conradson carbon and hydrogen content can be estimated without measuring the Conradson carbon of the resid. The hydrogen content at zero Conradson carbon can be estimated from the hydrogen content of the distillable liquid (C5 to 524°C atmospheric boiling points) product of thermolysis, and the hydrogen content at 100% Conradson carbon can be taken as the universal value of 3.8 wt%. On the other hand, the hydrogen content and Conradson carbon of a resid enables one to estimate the average hydrogen content of the distillable liquid products of the resid.

3.7 VARIATION OF CONRADSON CARBON RESIDUE WITH PROCESSING

One of the main objectives of resid processing is to convert or separate the coke precursors to make the lighter product suitable for more conventional downstream processing. Fuels deasphalting is an example process that is used to separate deasphalted oil with low concentrations of coke precursors from asphalt, containing high concentrations of coke precursors. On the other hand, coking processes separate the coke-forming functionality by thermal reaction into a solid coke product, simultaneously forming a volatile liquid product with little or no coke precursors. Only in hydroconversion processes are the overall concentration of coke precursors actually reduced as measured by the CCR. However, no commercial process is able to completely eliminate all the coke precursors in resids. As a result, all resid processes produce a heavy by-product containing higher concentrations of coke precursors than the resid feed. Because the understanding of the effect of processing on coke precursors is based on experience and the measure of CCR of feeds and products, it is important to utilize the insight of the Pendant-Core Building Block Model to explain past experience, recognize exceptions to the empirical rules, and predict new ways to alter the concentration of coke precursors through processing.

Previously it has been reported,[3,9] that the total CCR is conserved for physical separations, but the reason for this was elusive. However, hydrogen, as well as any element, should be conserved. Thus, as the CCR is a linear function of hydrogen content for fractions of residua with molecular weights greater than 700 amu, one should expect that the CCR is also conserved. This is shown by the following equations:

Hydrogen Balance:

$$\text{(wt of resid) } (H_R) = \Sigma \text{ (wt of fraction) } (H_{Fraction}) \tag{3.1}$$

Pendant-Core Building Block Model for resid or fraction of resid:

$$H = H_P - CCR \, (H_P - H_C)/100 \tag{3.2}$$

Substitution of these equations for the hydrogen content, H, of the resid and the fractions of the resid in the hydrogen balance gives

$$\text{(wt of resid) } (CCR_R) = \Sigma \text{ (wt of fraction) } (CCR_{Fraction}) \tag{3.3}$$

Hydroconversion is known to decrease the total CCR, and this is called Conradson carbon conversion.[10] Although the model compound results later in this chapter allow for the possibility for Conradson carbon conversion without hydrogen addition, verified by laboratory coking data in chapter 9, no previous case has been reported in the literature. Both of these results follow directly from the linear relationship between Conradson carbon and hydrogen content. This relationship even predicts the minimum hydrogen addition required for complete conversion of a resid to pendants:

$$(100) \, (H_{added} + H_R) = H_P \, (100 + H_{added}) \tag{3.4}$$

For Arabian Heavy vacuum resid:

$$H_R = 9.93 \text{ wt\%} \quad \text{and} \quad H_P = 11.6 \text{ wt\%}$$

Thus, $H_{added} = 1.92$ wt% of feed for 101.92 wt% conversion to pendants. However, hydrogen consumption can be much higher than this minimum even though the complete conversion of vacuum resids has not been practiced. The reason for such hydrogen consumption is the formation of significant yields of hydrocarbon gases (methane, ethane, etc.) containing high concentrations of hydrogen. In cracking a saturated ring, these hydrocarbon gases are the likely product. On the other hand, the formation of hydrogen sulfide from thiophenic and aliphatic sulfur during hydroconversion decreases the minimum hydrogen added because the hydrogen content in hydrogen sulfide (5.88 wt%) is much less than the hydrogen content in vacuum resids. Notwithstanding, the Pendant-Core Building Block Model provides no insight as to any barrier to complete hydroconversion of resids to volatile products. However, the addition of hydrogen to large-ring aromatics to crack off hydrocarbon gases and form a four-ring PNA by-product usually makes little economic sense. As a result, one of the challenges in resid hydroconversion is to determine the optimum conversion, as will be discussed later in chapter 10.

In thermal conversion, the CCR of the liquid product is typically found to be higher than that of the resid feed. An example is shown in Table 3.4 for eight separate reactions of Arabian Heavy vacuum resid in closed reactors at the same conditions of 60 min at 400°C. Even though the conversion is below the level at which toluene-insoluble coke is formed, there is a small but discernible increase in

TABLE 3.4
Thermal Conversion Increases the Total Conradson Carbon Residue of Arabian Heavy Vacuum Resid (400°C, 60 min)

	CCR (wt%)	H (wt%)
	22.6	9.83
	23.8	9.85
	23.5	9.80
	23.3	9.81
	23.1	9.84
	23.3	9.82
	23.7	9.83
	22.6	9.82
Average =	23.2	9.82
Feed =	22.3	9.93

Source: IA Wiehe. *Energy and Fuels* 8: 536–544, 1994. Reprinted with permission Copyright 1994. American Chemical Society.

Conradson carbon residue. This is paralleled by a small but discernible decrease in the hydrogen content of the liquid product because of the thermal cracking of a small amount of the distillable liquid product to form hydrocarbon gases of high hydrogen content. A hydrogen balance for this reaction of the resid to form thermally cracked resid (TCR) and gas is as follows:

$$100 \, H_R = (wt\% \, Gas) \, H_{Gas} + (100 - wt\% \, Gas) \, H_{TR} \tag{3.5}$$

The hydrogen balances for the microcarbon residue tests on the resid and the thermally cracked resid are as follows:

$$100 \, H_R = (100 - CCR) \, H_P + CCR \, H_C \tag{3.6}$$

$$(100 - wt\% \, Gas) \, H_{TR} = (100 - wt\% \, Gas - TCCR) \, H_P + TCCR \, H_C \tag{3.7}$$

Solving for the CCR after thermal cracking (TCCR):

$$TCCR = CCR + (wt.\% \, Gas) \frac{(H_{Gas} - H_P)}{(H_P - H_C)} \tag{3.8}$$

Thus, the Pendant-Core Building Block Model provides a way to estimate the increase in the CCR of the liquid and solid products of the thermal conversion of vacuum residua from the quantity and hydrogen content of the gas produced.

3.8 CONRADSON CARBON RESIDUE DISTRIBUTION

It is desirable not only to know the average CCR of resids, but also to understand how it is distributed in resids and how this distribution is altered by processing. As discussed in chapter 2, one way to measure the distribution of properties in resids is to divide them and their conversion products into pseudocomponents, such as saturates, aromatics, resins, asphaltenes, and coke, using solubility in various solvents and adsorption on solids, and then measuring the properties of the pseudocomponents. We saw in chapter 2 that each of these pseudocomponents occupies a unique area on a plot of molecular weight versus hydrogen content, the solvent-resid phase diagram (Figure 2.3). This diagram is useful in tracking how the pseudocomponents are altered and converted during processing. However, it is exactly these two properties, molecular weight and hydrogen content, that specify a species in the Pendant-Core Building Block Model. The hydrogen content specifies the ratio of the number of pendants to the number of cores, and the molecular weight determines the total number of building blocks. Thus, the solubility characteristics of the fractions of resids are consistent with the Pendant-Core Building Block Model. In addition, advantage can be taken of the relationship between CCR and hydrogen content for fractions of molecular weight over 700 amu to display the distribution of CCR directly on the solvent-resid phase diagram as shown in Figure 2.3. The CCR merely becomes an additional axis to the hydrogen content for the higher-molecular-weight fractions.

Figure 2.3 shows that the saturates have zero CCR, unlike the other pseudocomponents. The aromatics always have molecular weights less than 700 amu, showing that measured CCR values may not gauge their full coke-making potential. This figure also demonstrates that thermal conversion greatly increases the width of the distribution of CCR toward higher values and eventually includes values representative of toluene-insoluble coke. This, of course, is equivalent to cracking off volatile pendants to leave species with higher proportions of cores in the nonvolatile products until eventually only the cores remain. This analogy is carried further in Figure 3.1 where several pendant-core species are shown that fall in each of the solvent separation fractions. Thus, the path from aromatics to resins to asphaltenes to coke is produced by both molecular-weight growth at similar proportions of cores and by molecular-weight reduction to form species with increased proportions of cores.

3.9 PENDANT-CORE MODEL COMPOUNDS: DISCOTIC LIQUID CRYSTALS

One of the difficulties in investigating the thermal chemistry and mechanism of coke formation of petroleum resids is that they are composed of millions of different types of molecules, and the molecular structure of even one of these molecules cannot be specified with any reasonable degree of certainty. Progress in understanding could be made much more rapidly if we could identify pure compounds that mimic the chemistry and physical interactions of the molecules in resids. Because the Pendant-Core Building Block Model provides a good approximation to the thermal chemistry of resids, it is logical to begin the search for residuum model compounds with those that have aromatic cores and paraffinic pendants. The first compound selected is 2,3,6,7,10,11-hexaoctoxytriphenylene. As shown in Figure 3.10, it has a triphenylene core with six paraffinic side chains of eight carbons each. These side chains are connected to the aromatic core by ether bridges. Although ether bridges are not expected in resids, these were selected for their ease of synthesis to represent thermally unstable bonds, such as aliphatic sulfur. This compound not only was selected for its potential to model the thermal chemistry of resids, but also to mimic the association of asphaltenes in resids. The large-ring aromatics in asphaltenes are known to associate by stacking.[11] Likewise, hexaoctoxytriphenylene is a member of a family of discotic liquid crystals in which the molecules tend to stack in columns.[12] Another discotic liquid crystal and the second model compound selected is benzenehexa-n-hexanoate[13] with the structure also shown in Figure 3.10. This has only a single aromatic ring core with six paraffinic side chains of five carbons each. Ester bridges, which also are not expected in resids, are used as the thermally unstable bridges between pendants and core. Table 3.5

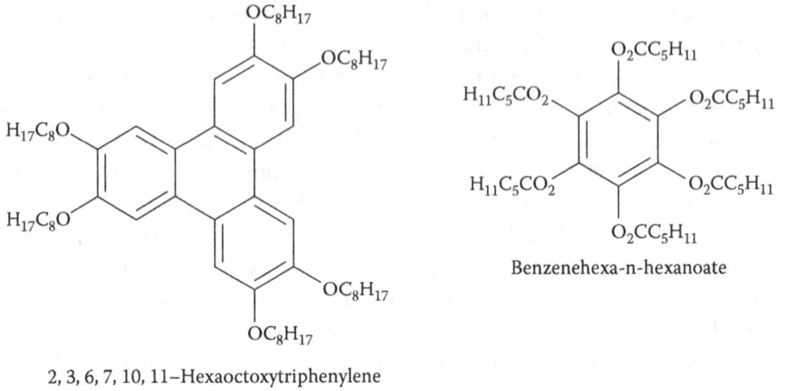

2, 3, 6, 7, 10, 11–Hexaoctoxytriphenylene

Benzenehexa-n-hexanoate

FIGURE 3.10 Discotic liquid crystals as Pendant-Core Building Block Model compounds. (From IA Wiehe. *Energy and Fuels* 8: 536–544, 1994. Reprinted with permission. Copyright 1994. American Chemical Society.)

shows the elemental analysis, molecular weight, and molecular formula of the two pendant-core model compounds that can be compared with the average properties of the resids in Table 3.2. Hexaoctoxytriphenylene has a molecular weight, carbon number, and hydrogen number in the range of a full resid, but much higher heteroatom content with the six oxygen atoms per molecule. Nevertheless, the measured CCR of 24.1 wt% is in the range of the vacuum resids. Although the bonds between the ether oxygens and the paraffinic side chains are expected to break thermally much easier than between the aromatic carbons and the oxygens, this CCR value suggests that only one of the six oxygens remain in the residue. Perhaps, some of the cores evaporated in the test before they combined chemically with other cores. However, the results with benzenehexa-*n*-hexanoate suggest that this evaporation of cores should not have been significant.

TABLE 3.5
Properties of Pendant-Core Model Compounds

Compound	CCR (wt%)	C (wt%)	H (wt%)	H/C (atomic)	O (wt%)	Mol. Wt (amu)	Mol. Formula
Hexaoctoxytriphenylene	24.1	79.46	10.91	1.64	9.62	998	$C_{66}H_{108}O_6$
Benzenehexa-*n*-hexanoate	14.0	65.66	9.35	1.57	24.99	768	$C_{42}H_{66}O_{12}$

Source: IA Wiehe. *Energy and Fuels* 8: 536–544, 1994. Reprinted with permission Copyright 1994. American Chemical Society.

Benzenehexa-*n*-hexanoate has a molecular weight similar to that of the heptane-soluble fraction of Arabian Heavy vacuum residuum, but lower carbon and hydrogen contents because of the 12 ester oxygens per molecule. The CCR of 14 wt% is close to the values (14 to 16 wt%) for the three heptane-soluble fractions of residua. In this case, the bond between the ester oxygen and the aromatic carbon is expected to break about as easily as between the oxygen and the ester carbon, but the CCR suggests that two out of six oxygens bonded to aromatic carbons remain in the residue. The surprising conclusion is that, with only a single ring core, few of the cores evaporated during the microcarbon residue test. Thus, the requirement that the core be nonvolatile should be relaxed. Instead, if small-aromatic-ring cores have multiple linkages to large structures, they can form CCR and, thus, coke, by remaining nonvolatile with the breakage of one or two linkages and giving time to combine with other core-containing compounds.

The HPLC results discussed in chapter 2 showed that a significant portion of resids contain one- to four-ring aromatics, and NMR data indicate that most of the aromatic carbons are substituted. Thus, the coke-formation mechanism demonstrated by benzenehexa-*n*-hexanoate is likely to occur in resids. This coke formed by the combination of one- to four-ring aromatic cores containing multiple substituted groups is called *extrinsic coke*. On the other hand, the coke formed by cores containing five or more aromatic rings is called *intrinsic coke* because there is no economical way to prevent its formation. This is in contrast to extrinsic coke that forms molecules containing five or more aromatic rings during the microcarbon residue test. As a result, most of the Conradson carbon conversion during hydroconversion is the consequence of terminating free radicals by donatable hydrogen and preventing the combination of smaller-ring aromatics to form molecules containing five-ring or greater aromatics. Thus, during hydroconversion, Conradson carbon is actually prevented from being formed, rather than being converted. Meanwhile, intrinsic coke can only be converted by hydrogenating rings and cracking until reducing the number of aromatic rings to less than five. This occurs by much slower kinetics than preventing the formation of five-ring or greater aromatics.

These results with model compounds indicate that the Pendant-Core Building Block Model is overly pessimistic as it assumes that all coke is intrinsic. Now we realize that a substantial fraction of the coke is extrinsic. This opens the possibility that the CCR of resids can be reduced by mechanisms other than hydrogenation of large-aromatic-ring cores. As a result, methods were sought to achieve this objective, and chapter 9 on coking will discuss two methods that were successful. On the other hand, because part of the coke is intrinsic, a carbonaceous by-product is still inevitable, preventing complete conversion of resids. A research challenge is to devise a method to determine what fraction of Conradson carbon is intrinsic and what part is extrinsic. One possible method will be discussed in chapter 4 when natural hydrogen donors are discussed.

3.10   LIMITATIONS AND FUTURE DEVELOPMENTS

3.10.1   Successes

The Pendant-Core Building Block Model has proved to be an excellent approximation, considering its simplicity. It passed one test by correctly predicting the linear relationship between CCR and hydrogen content. The hydrogen content of the pendants and cores, evaluated from this linear relationship, were independently verified by direct analysis of the CCR and the distillable liquids from thermal conversion. This model correctly predicted that the hydrogen content of distillable liquids remain constant with increasing conversion. As a result, the Pendant-Core Building Block Model provides insight into the cause of CCR, a petroleum characteristic that has been measured for over 90 years. Now we have a basis to estimate how and why the CCR varies with processing, including the large increases resulting from recycling volatile aromatic products. As a result, the aromatic core concept must capture the most important cause for the coke-forming and insolubility tendencies of petroleum macromolecules.

The success of the pendant-core model compounds at producing both coke-forming potential and aromatic stacking similar to macromolecules in residua encourages us to search further for resid model compounds. An obvious next step is to synthesize pendant-core compounds without heteroatoms, or with sulfur, in the links.

3.10.2   Limitations

The single-ring model compound results show that not all coke from residua comes from large-ring aromatic cores. Thus, a more detailed model should contain aromatic cores of varying sizes from one to at least five rings. In addition, the varying sulfur content of the distillable liquids from Hondo resid points to the need to include aliphatic sulfur as a third building block. Finally, the fact that distillable liquids contain liquids of varying boiling points and properties show that pendants should be modeled with a distribution of molecular weight and hydrogen content.

3.10.3   Innovations

The Pendant-Core Building Block Model provides a tool for predicting the coke-forming tendency of petroleum macromolecules and for predicting the elemental analysis of commercial coke. Because the Pendant-Core Building Block Model is capable of describing both the solubility and coke-forming characteristics of vacuum residua, a logical next step is to replace the pseudocomponents in the Phase-Separation Kinetic Model (chapter 4) for residua thermolysis with a distribution of species composed of various combinations of pendants and cores. A sample of such species has already been shown in Figure 3.1. Nevertheless, the greatest innovation actually results from the failure of the Pendant-Core Building Block Model to describe the coke formation of model compounds. This opens the door to the possibility of reducing the yield of coke or other carbonaceous by-product from resid conversion without consuming large amounts of expensive hydrogen.

REFERENCES

1. PH Conradson. Apparatus and method for carbon test and ash residue in petro-leum lubricating oils. *Ind Eng Chem* 4: 903–905, 1912.
2. F Noel. An alternative to the conradson carbon residue Test. *Fuel* 63: 931–934, 1984.
3. I Roberts. The chemical significance of carbon residue data. Prep Paper—Am Chem Soc, *Div Pet Chem* 34: 251–254, 1989.
4. JB Green, JY Shay, JW Reynolds, JA Green, LL Young, ME White. Microcarbon residue yield and heteroatom partitioning between volatiles and solids for whole vacuum resids and their liquid chromatographic fractions. *Energy and Fuels*, 6: 836–844, 1992.
5. IA Wiehe. The pendant-core building block model of petroleum residua. *Energy and Fuels* 8: 536–544, 1994.
6. GN George, ML Gorbaty. Sulfur k-edge x-ray adsorption spectroscopy of petro-leum asphaltenes and model compounds. *J Am Chem Soc*, 111: 3182–3186, 1989.
7. SR Kelemen, GN George, ML Gorbaty. Direct determination and quantification of sulfur forms in heavy petroleum and coals. *Fuel*, 69: 939–944, 1990.
8. GN George, ML Gorbaty, SR Kelemen. Sulfur k-edge x-ray adsorption spectros-copy of petroleum asphaltenes and model compounds. In: WL Orr, CM White, Eds. *Geochemistry of Sulfur in Fossil Fuels*, Washington, D.C.: American Chem-ical Society, 1990, pp. 220–230.
9. IA Wiehe. A solvent-resid phase diagram for tracking resid conversion. *Ind Eng Chem Res*, 31: 530–536, 1992.
10. E.C. Sanford. Conradson carbon residue conversion during hydrocracking of Athabasca bitumen: Catalyst mechanism and deactivation. *Energy and Fuels*, 9: 549–559, 1994.
11. LB Ebert, JC Scanlon, DR Mills. X-ray diffraction of n-paraffins and stacked aromatic molecules: Insights into the structure of petroleum asphaltenes. *Liq Fuel Technol*, 2: 257–286, 1984.
12. AM Levelut. Structure of a disk-like mesophase. *J Phys Lett*, 40: L81–L84, 1979.
13. S. Chandrasekhar, BK Sadashiva, KA Suresh. Liquid crystals of disc-like mol-ecules. *Pramana*, 9: 471–480, 1977.

4 Thermal Conversion Kinetics

4.1 INTRODUCTION

4.1.1 POTENTIAL APPLICATIONS

The thermal conversion kinetics of petroleum macromolecules is not only important for thermal conversion processes, such as visbreaking and coking, but it is also the primary cracking mode for hydroconversion and an important, undesired side reaction for fluid catalytic cracking. In addition, thermal conversion kinetics is required for distillation furnace design to ensure that little or no cracking occurs to prevent coking and gas formation. Except for coking, these operations do not allow for the accumulation of solid coke that insulates surfaces from heat transfer and eventually blocks flow. Before the research described here, the prevailing opinion was that coke is a direct reaction product. A certain degree of conversion was expected to result in a certain yield of coke. Instead, this research proved that coke is the result of a liquid–liquid phase separation. This opened the door for innovation—the ability to increase resid thermal conversion without forming coke.

This chapter will use resid thermal conversion kinetics to show the evolutionary process in the development of kinetic models. First, information in the literature and experimental data will challenge us until a kinetic model is devised to describe the experimental data. This model will be tested against additional experimental data. Although the kinetic model will meet some of these challenges, it will fail to describe some of the new data. This will encourage us to change the kinetic model to describe this new data, while reducing the number of adjustable parameters. As a result, new insight and understanding will be developed. Further challenging the kinetic model will bring to light inconsistencies. Removing these inconsistencies will enable the further reduction in the adjustable parameters and increase in power of the kinetic model. The path to still more powerful kinetic models will be outlined. Meanwhile, learnings from kinetic model development point the direction to new innovations in resid conversion. Some of these will be described in this and subsequent chapters.

4.1.2 WISH LIST FOR A KINETIC MODEL

The following list of desired features of a kinetic model for the thermal conversion of resids has been compiled:

1. Quantitatively describes kinetic data.
   a. Concentration versus time
   b. Effect of temperature

2. Convenient to use.
3. Applies to a wide variety of feeds.
4. Predicts the effect of changing initial concentrations.
5. Has a minimum number of adjustable parameters and/or evaluates parameters from characterization data.
6. Predicts the effect of changing reactor type.
7. Provides insight for new innovations.
8. Predicts properties of products.

These should be kept in mind as the progress of kinetic model development is described. At the end of this chapter, the current status of the kinetic model will be compared against these desirable features.

4.1.3 NEED FOR RESID THERMAL CONVERSION RESEARCH

The author joined Exxon Corporate Research early in 1977 during an energy crisis after holding teaching (University of Rochester) and industrial (Xerox) positions. It is important for the career of an experienced industrial researcher to select a research project that fits strategically and economically into his company's future. If successful, the research should have an excellent chance of being applied, adding much to the company's bottom line. Surprisingly, many industrial research projects have low probability of being applied, even if successful. In addition, the researcher should be uniquely qualified to conduct the research within the company, if not within the world. Although these requirements should be for inexperienced researchers, the manager of the inexperienced researcher should take the responsibility for ensuring that the job assignment is a good match.

The author was assigned to a reaction engineering group even though he had never done research in reaction kinetics or in catalysis. Thus, the uniqueness criteria appeared to be a challenge. Before joining Exxon, the author projected coal liquefaction to be a great technological area for research, a very active research area within Exxon at that time. However, after joining Exxon, his analysis of coal liquefaction revealed that the economics were not very favorable, even with optimistic assumptions. On the other hand, there was a plentiful supply of very heavy oils that were not being produced, and petroleum resids were under-converted at most refineries. The economics for conversion of resids and heavy oils were much more favorable. In addition, the author could take advantage of his background in organic polymers by doing research on the largest molecules in petroleum, the petroleum macromolecules. Therefore, the author decided to do research on petroleum macromolecules and resid conversion even though no one else was doing research in this area at Exxon Corporate Research at that time.

The author did an extensive literature search and talked with senior researchers who had done research on resids and heavy oils previously. Two aspects were being sought. First, what is the barrier that limits the conversion of heavy oils? Second, how can the author's background in solution thermodynamics and phase behavior make an impact? It soon became clear that resid conversion was limited by the formation of solid coke, either on reactor walls or in catalyst pores. Little was understood about how and why coke formed. The literature[1-3] included a proposal that coke formed by a phase separation step, but experimental proof was lacking. As a result, it was largely ignored by subsequent researchers. Bob Long, senior scientific advisor at Exxon Corporate Research, suggested that the author learn about the carbonaceous mesophase and talk with a group that was investigating making carbon fibers by spinning carbonaceous mesophase made from thermal treating of the distillation bottoms of fluid catalytic cracking. More reading and analysis of the literature led the author to propose the hypothesis that the liquid–liquid phase separation of converted asphaltenes to carbonaceous mesophase triggers the onset of coke formation during resid thermal conversion. Thus, the combination of the literature review, analysis, and hypotheses became the author's first report at Exxon, titled *Solution Thermodynamics of Petroleum Macromolecules at High Temperatures, A Research Proposal.* This proposed research met the criteria of high potential company impact and uniquely fitted the author's background. Although the author ended up spending 22 years working at Exxon on this proposal and extensions, the first critical step was testing the basic hypothesis by measuring the thermal conversion kinetics of resids.

4.2 PSEUDOCOMPONENT MODEL WITH STOICHIOMETRIC COEFFICIENTS

In order to approximate the thermal conversion kinetics of complex systems, one needs to model the reactants, model the reactions to products, and determine the reaction order and rate constant for each step. One approach is to represent the petroleum macromolecules with a set of actual compounds and evaluate the large number of rate constants from model compound data and estimation methods. Then, Monte Carlo simulations are done to track the distribution of species with time. Klein and Savage[4] have championed this approach. Although it is impressive that a formalism can be devised to handle huge numbers of species and even larger numbers of reactions, this does not fit our approach of applying the simplest model first, and adding complexity when required, using experimental data. In addition, as learned in chapter 2, we cannot specify the starting distribution of petroleum macromolecules with a reasonable degree of certainty. As a result, even this approach is quite approximate. However, as characterization methods improve for petroleum macromolecules and as computers continue to increase speed and storage, the Monte Carlo simulation approach is expected to become the method of choice for modeling the thermal conversion kinetics of petroleum macromolecules in the future.

The simplest model for the thermal conversion kinetics of petroleum macromolecules that allows for coke formation is a pseudocomponent model that treats solubility fractions as the reactants and products. However, the Solvent-Resid Phase Diagram in chapter 2 and the Pendant-Core Building Block Model in chapter 3 showed that solubility fractions can be partially converted to other fractions, especially to distillable liquids, but the major fraction can remain within the original solubility class. Thus, one needs a minimum of two different pseudocomponents to represent unreacted and converted molecules in each solubility class. Any species that is in between these two extremes can be represented by a combination of these two pseudocomponents. Because of the laborious laboratory work involved, it was important to keep the number of solubility classes to a minimum. Therefore, the resid and its thermal conversion products were separated into volatiles, coke (toluene insoluble), asphaltenes (toluene soluble, n-heptane insoluble), and heptane solubles (or maltenes). Of these, asphaltenes and heptane solubles require reactant and product pseudocomponents. Because volatiles and coke are not present in an unreacted vacuum resid, only a product pseudocomponent is required for each of these fractions.

As seen in chapter 3, a common reaction in resid thermal conversion is to crack volatile pendant groups off nonvolatile aromatic cores. Thus, one needs to represent the kinetics of such reactions. It was common practice of pseudocomponent kinetics to represent reactions as separate parallel reactions of a reactant to separate products. However, a closer approximation of the expected mechanism is to represent a cracking reaction by one reaction to both lighter and heavier products. Then, stoichiometric coefficients were used to determine the fractional split of converted reactant to each product. Because stoichiometric coefficients should be relatively independent of temperature, although rate constants are temperature-dependent, the reduction in temperature-dependent parameters is a clear advantage. Later in this chapter it will be shown that hydrogen balance can be used to evaluate stoichiometric coefficients, a third advantage of this approach.

Soon after the project on resid thermal conversion was initiated at Exxon, Bob Schucker joined the same group, and he decided to do research on the hydroconversion of resids. As a result, we learned from each other in developing experimental techniques and concepts about resid conversion. We had received permission to present our individual kinetic models at the same American Chemical Society session in 1980. However, at the last minute, the author was asked to withdraw his paper, and the Schucker paper was presented.[5] This was a result of the decision to keep the phase-separation step for coke formation as a trade secret. As hydroconversion greatly extends the coke induction period, Schucker could present kinetics that just covered the induction period. As a result, the use of stoichiometric coefficients and an asphaltene aromatic core reaction limit in a pseudocomponent kinetic model was first published by Schucker, even though first used by the author within Exxon. Meanwhile, the author was not given permission to present[6] and to publish[7] the phase-separation kinetic model until 1993, even though it was developed in 1978. Some Exxon managers opposed publication even after 15 years. In hindsight, the author could have been more clever in obtaining patent protection of the applications of the phase-separation step for coke formation. Once a concept is patented, it is usually an advantage for a company to publish the concept to inform others that they are excluded from practicing the patent. In addition, even if the patent is contested, publication can often give one the right to use without being excluded by others. In any event, this points out a disadvantage of doing research in an area that has a strong strategic and economic fit with one's company. If it is decided to keep the research results as a trade secret, the publication of one's research results can be delayed or prevented all together. However, the excitement and satisfaction of doing research that influences the direction and bottom line of a large corporation, such as Exxon, can be a compensation. This is certainly better than publishing research that is never applied. Nevertheless, it is still difficult to keep a concept a secret after it is widely publicized within a corporation and becomes part of the technical culture. The phase-separation mechanism for coke formation was even presented by an Exxon employee in a 1980 paper on resid hydroconversion.[8] Fortunately, without the complete research being presented, the phase-separation kinetic model for coke formation was not published by others during this delay period and was still well received when published.

4.3 PHASE-SEPARATION MECHANISM FOR COKE FORMATION

4.3.1 BACKGROUND

In 1977, the prevailing view, following the lead of Levinter,[9,10] was that coke was formed during resid thermolysis by a sequence of polymerization and condensation steps from the lightest to the heaviest fractions:

$$\text{Oils} \rightarrow \text{Resins} \rightarrow \text{Asphaltenes} \rightarrow \text{Carbenes} \rightarrow \text{Coke}$$

As we saw in chapter 2, this is also the approximate path taken by the aromatic fragments on the Solvent-Resid Phase Diagram. However, the aromatic fragments initially form lower molecular weight and more aromatic fragments, while remaining in the solubility class. Only when such fragments combine to form higher-molecular-weight species do they move to the lower solubility class. Nevertheless, most researchers in 1977 believed that this sequence followed direct chemical reactions with no phase-separation step for molecular weight growth. However, kinetic models based on direct reactions fail to predict the coke induction period before coke formation begins. This coke induction period had been observed by many investigators, including Levinter,[1,9–12] and has made visbreaking possible. One exception to this view was that of Magaril,[1–3] who was the first to postulate that coke formation is triggered by the phase separation of asphaltenes. Unfortunately, his scattered kinetic data led him to use linear variations of the concentration of each fraction with reaction time, or zeroth-order kinetics, rather than first-order kinetics expected for thermolysis. As a result, this work was largely ignored except by the author, who was looking for the application of phase behavior to resid kinetics. Later, Yan[13] in 1987 described coke in visbreaking as resulting from a phase-separation step, but did not include this step in his kinetic model for coke formation. Likewise, Takatsuka[14] in 1989 included a phase separation to form "dry sludge" in his hydroconversion kinetic model, but neglected this step to form coke in his thermal cracking model.[12]

4.3.2 EXPERIMENTAL PROCEDURE

Mike Lilga, who worked with the author for two years before going to graduate school and getting a doctorate in chemistry, developed this procedure and ran these experiments. Isothermal, batch reactors were used that may be open or closed to the atmosphere. The open reactor was a quartz test tube containing about 3 g of reactant that was heated to 95% of the 400°C reaction temperature within 4 min by inserting it in a preheated, vertical tube furnace. Cold Lake vacuum resid with normal boiling points above 566°C (1050°F) and its fractions were used as reactants for this study. Thus, the resid had to be thermally cracked to become volatile under the experimental conditions. Nitrogen flowing over the sample facilitated removal of volatile liquids when they formed, while preventing refluxing as well as oxidation of the hot resid. At the end of each reaction time, the quartz tube was removed from the tube furnace and inserted into ice water to cool the sample to below 200°C within 20 sec.

The asphaltene fraction could not be reacted in an open reactor because of its tendency to foam out of the reactor. For this reason, a closed tube reactor was used. The experimental methods for closed reactors were developed by Bob Schucker and applied by Jerry Machusak. The closed tube reactor was also later used for the full resid, as will be discussed later in this chapter. This reactor was a presulfided stainless steel tubing bomb, made with tube fittings and containing about 5 g of asphaltenes and 1.2 MPa of nitrogen. It was heated to 95% of the reaction temperature within 3 min by inserting it into a large preheated, fluidized sand bath. Again, a thermocouple inserted in the sample assured that rapid heat up, isothermal condition for the designated reaction time, and rapid cooling with cold water were obtained.

Figure 4.1 shows the separation scheme used to fractionate the reactants and the products. Gas and volatile liquids that were formed during the reaction were allowed to escape the open reactor, resulting in the first separation class. This was measured by the difference in weight of the reactor before and after the reaction. The remaining reaction mixture was separated into toluene-insoluble

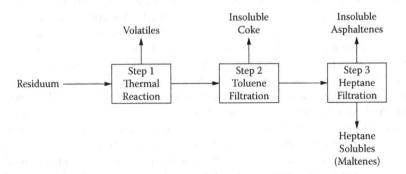

FIGURE 4.1 Separation scheme used to fractionate the reactants and products for kinetic studies. (From IA Wiehe. *Ind Eng Chem Res.* 32: 2447–2454, 1993. Reprinted with permission. Copyright 1993. American Chemical Society.)

coke; toluene-soluble, heptane-insoluble asphaltenes; and the nonvolatile heptane solubles or maltenes. In each of these solvent separations, the samples were allowed to set overnight in 15 parts of solvent per part heavy oil. This mixture was poured through a fine (4 to 5.5 μm pores), fritted glass filter. The solids on the filter were washed with at least 25 parts additional solvent, and this was continued until the solvent passed the filter without color. The insolubles were vacuum dried on the filter at 100°C for at least 8 h. The solvent was removed from the soluble oil by rotary evaporation at 50°C, followed by vacuum drying for 8 h at 50°C. The same procedure was followed for the closed reactor without removing the volatile oils. However, in the process of evaporating solvent, the more volatile oils also evaporated. Therefore, the closed reactor was used to measure the rate of change of asphaltenes and the rate of appearance of coke without any attempt to measure the rate of change of volatile and nonvolatile heptane solubles separately. Later, this problem was solved. A method was developed for distilling the volatile oils directly out of tubing bombs before exposing the oils to solvents.

4.3.3 Experimental Results

Cold Lake vacuum resid was separated into 25.0 wt% asphaltenes and 75.0 wt% heptane solubles using the procedure just described. The unreacted resid contains no toluene insolubles or volatiles. The heptane solubles and full resid were reacted at various times at 400°C in the open reactor, and the products were separated into fractions. For each of these reactants, the concentration of the solvent-volatility fractions were determined as a function of reaction time. These data have shown four common features of resid thermal conversion kinetics, which are described below.

4.3.3.1 Coke Induction Period

Figure 4.2 shows the formation of coke as a function of reaction time for the three reactants—heptane solubles (maltenes), asphaltenes, and the full resid. When the reactant is the asphaltene fraction, coke forms immediately at a high rate without an induction period. When the reactant is the maltene fraction, there is a 90 min induction period, after which it forms at a slow rate. We might expect the full resid, composed of 25 wt% asphaltenes, to form coke initially at about a quarter of the rate as the asphaltenes. Instead, there is a coke induction period of 45 min. This coke induction period is the first common feature of resid thermal conversion kinetics. It demonstrates that the maltenes inhibit the formation of coke by the asphaltenes. As pointed out earlier, many other investigators have observed this coke induction period in resid thermal conversion kinetics.

4.3.3.2 Asphaltene Maximum

Figure 4.3 shows the conversion of heptane-soluble reactant to the four classes as a function of reaction time. The asphaltenes increase from zero to a maximum and then decrease. The maximum occurs at the same reaction time as the end of

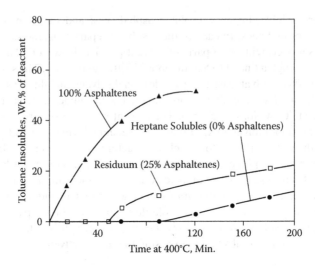

FIGURE 4.2 Coke formation from three reactants: asphaltenes, full resid, and heptane-soluble portion of resid for Cold Lake vacuum resid at 400°C showing different coke induction periods. (From IA Wiehe. *Ind Eng Chem Res.* 32: 2447–2454, 1993. Reprinted with permission. Copyright 1993. American Chemical Society.)

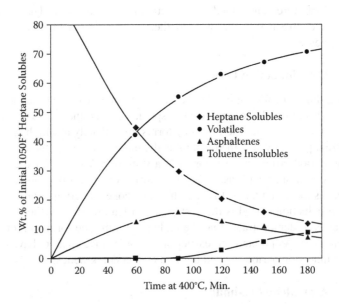

FIGURE 4.3 Plot of the weight fraction of four conversion products of heptane-soluble fraction of Cold Lake vacuum resid versus reaction time at 400°C in open reactors.

the coke induction period. This maximum is the second common feature of resid thermal conversion kinetics. It is a result of maltenes reacting to form asphaltenes, which in turn react to form coke. This maximum in the asphaltene concentration has also been observed by other investigators.[1,9,11,12]

4.3.3.3 Decrease of Asphaltenes Parallels Decrease of Heptane Solubles

As shown in Figure 4.3, during the period that coke is formed, the ratio of the asphaltenes to heptane solubles approaches a constant. This approach to a constant ratio is the third common feature of resid thermal conversion kinetics, but this was not observed by previous investigators. It will be suggested that this ratio is the solubility limit of converted asphaltenes in the heptane solubles.

4.3.3.4 High Reactivity of Unconverted Asphaltenes

A first-order reaction rate constant of 0.013 min^{-1} was obtained by fitting the decrease in heptane solubles with reaction time, as is shown in Figure 4.4. While

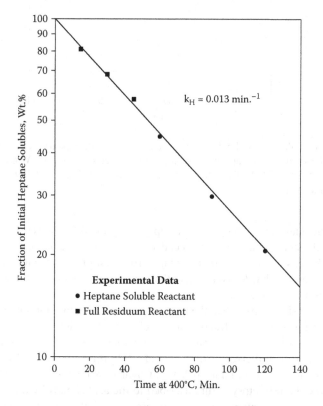

FIGURE 4.4 Evaluation of the first-order rate constant for disappearance of heptane solubles of Cold Lake vacuum resid at 400°C in open-tube reactors, using the data for which the coke concentration was less than 3%. (From IA Wiehe. *Ind Eng Chem Res.* 32: 2447–2454, 1993. Reprinted with permission. Copyright 1993. American Chemical Society.)

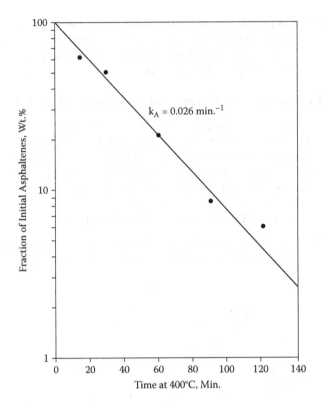

FIGURE 4.5 Evaluation of the first-order rate constant for disappearance of asphaltenes of Cold Lake vacuum resid at 400°C in closed-tube reactors. (From IA Wiehe. *Ind Eng Chem Res.* 32: 2447–2454, 1993. Reprinted with permission. Copyright 1993. American Chemical Society.)

this data includes the full resid and the heptane-soluble fraction as reactants, data for which the coke concentration exceeded 3 wt% were excluded. Once significant coke begins to form, the data deviates from first-order behavior. As will be discussed later, this has led to the hypothesis of a heptane-soluble by-product for the coke-forming reaction.

Figure 4.5 shows that when the reactant is the asphaltene fraction, the disappearance of the asphaltene fraction in the closed tube reactor can be described with a first-order kinetic model using a reaction rate constant of 0.026 min^{-1}. This is within experimental error of the value of 0.025 min^{-1} measured by Schucker and Keweshan[5] at 400°C under 6 MPa of hydrogen for the same asphaltenes. Thus, unlike the refractory nature ascribed to the asphaltenes by most previous investigators, the unconverted asphaltenes are actually the most thermal reactive fraction of the vacuum resid. The asphaltenes just have a limited extent of reaction. Nevertheless, the high reactivity of unconverted asphaltenes provides the fourth common feature of resid thermal conversion kinetics.

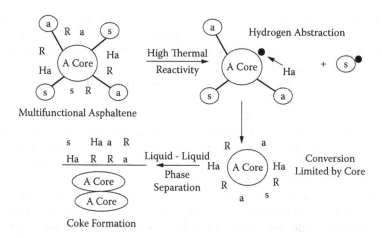

Multifunctional Asphaltene

Coke Formation

FIGURE 4.6 Phase-separation mechanism of the formation of coke from the thermolysis of resids. (From KA Gould and IA Wiehe. *Energy and Fuels* 21: 1199–1204, 2007. Reprinted with permission. Copyright 2007. American Chemical Society.)

4.3.4 PHASE-SEPARATION KINETIC MODEL

A kinetic model was developed that describes these four common features of resid thermal conversion kinetics. It is based on the mechanism shown in Figure 4.6. Based on the Pendant-Core Building Block Model, the asphaltenes have a thermally stable, polynuclear aromatic core with saturate and aromatic pendants. These pendants are thermally cracked with high reactivity to form free radicals. It is hypothesized that the resid contains natural hydrogen donors (proof given in section 4.7), saturated rings adjacent to aromatic rings, that can donate hydrogen and terminate free radicals. This explains why the asphaltene molecular weight decreases during thermal conversion, as observed in chapter 2. However, the solubility of the asphaltenes decreases as they lose pendants and approach the reaction limit of the asphaltene aromatic core. At some point, the asphaltenes become insoluble in the reacting medium and undergo a liquid–liquid phase separation to form a phase lean in hydrogen donors. In this heavy phase, asphaltene free radicals combine by addition and recombination reactions to form high-molecular-weight coke. This mechanism explains why the phase separation of asphaltenes triggers the onset of coke formation and why, when asphaltenes are reacted alone, they form coke without an induction period. In chapter 2, we saw direct evidence that the coke that is formed from this reaction is higher in molecular weight. However, there was no evidence at that time to show that resids contained natural hydrogen donors, although an explanation was needed as to why asphaltenes in solution decreased in molecular weight, but in a separate phase they increased molecular weight. Previously, Langer[15] of Exxon used partially hydrogenated refinery process streams to provide donor hydrogen and, as a result, inhibit coke formation during resid thermal conversion. In 1978, Exxon was developing a coal liquefaction process using similar hydrogen donor solvents. As a result, the natural hydrogen donor explanation was a convenient one. Nevertheless, the testing of this and other parts of the mechanism and kinetic model were done as part of the process of challenging and looking for inconsistencies. The results of these challenges, including the natural hydrogen donor hypothesis, will be discussed later in this chapter (section 4.7).

One should recognize that the development of a kinetic model is a long trial and error process. Approximately 50 different kinetic models were tried before arriving at the one presented here. This one was obtained by focusing on the simplest model that described the common features, contained a liquid–liquid phase separation step, and contained the aromatic core reaction limit for both asphaltenes and heptane solubles. Finally, minimum complexity is added to describe the experimental data. The resulting kinetic model, shown here, contains two parallel first-order reactions for the thermolysis of unreacted heptane solubles and unreacted asphaltenes. These are the only reactions that occur during the coke induction period when only lower molecular weight products are formed.

Phase-separation kinetic model	**Nomenclature**
$M^+ \xrightarrow{k_M} m\,A^* + (1-m)\,V$	M^+ = unreacted heptane solubles, wt%
	M^* = heptane soluble cores, wt%
$A^+ \xrightarrow{k_A} a\,A^* + b\,M^* + (1-a-b)\,V$	A^+ = unreacted asphaltenes, wt%
	A^* = asphaltene cores, wt%
Solubility Limit: $A^*_{max} = S_L\,(M^+ + M^*)$	V = volatiles, wt%
	TI = toluene-insoluble coke, wt%
$A^*_{Ex} = A^* - A^*_{max}$	S_L = solubility limit
	A^*_{max} = maximum asphaltene cores that can be held in solution
$A^*_{Ex} \xrightarrow{\infty} y\,H^* + (1-y)\,TI$	A^*_{Ex} = excess asphaltene cores beyond the solubility limit
	k_M = first-order rate constant for the thermolysis of heptane solubles, min^{-1}
	k_A = first-order rate constant for the thermolysis of asphaltenes, min^{-1}
	$m, a, b,$ and y = stoichiometric coefficients

As long as the asphaltene cores remain dissolved, the heptane solubles can provide sufficient donor hydrogen to terminate asphaltene-free radicals. As the conversion proceeds, the concentration of asphaltene cores continues to increase, and the heptane solubles continue to decrease until the solubility limit, S_L, is reached. Beyond the solubility limit, the excess asphaltene cores, A^*_{Ex}, phase separate to form a second liquid phase that is lean in donor hydrogen. As a result, in this new phase, asphaltene radical recombination is quite frequent, causing a very rapid reaction to form solid coke and a by-product of heptane-soluble cores.

4.3.5 DERIVATION OF EQUATIONS FOR THE PHASE-SEPARATION KINETIC MODEL

The nomenclature is the same as previously given, with the addition of A_0 = initial asphaltene concentration. Thus, on the basis of 100 parts of resid initially, the initial heptane-soluble concentration is $100 - A_0$. The concentration of unreacted asphaltenes, A^+, unreacted heptane solubles, M^+, and volatiles, V, are unaffected by whether coke is forming or not. Because the thermolysis of asphaltenes and heptane solubles are by first-order reactions:

$$\frac{dA^+}{dt} = k_A A^+ \qquad \frac{dM^+}{dt} = k_M M^+$$

Integrating these equations from time zero to time t yields

$$A^+ = A_0 e^{-k_A t} \qquad M^+ = (100 - A_0) e^{-k_M t}$$

$$V = (1 - m)(100 - A_0)(1 - e^{-k_M t}) + (1 - a - b)A_0(1 - e^{-k_A t})$$

During the coke induction period,

$$A^* = m(100 - A_0)(1 - e^{-k_M t}) + aA_0(1 - e^{-k_A t})$$

$$A = A^+ + A^*$$

$$M^* = bA_0(1 - e^{-k_A t})$$

$$M = M^+ + M^*$$

$$TI = 0$$

The coke induction period ends when

$$A^* = S_L(M^+ + M^*)$$

Thereafter, A^* is given by the preceding solubility equation, and

$$M^* = \frac{y\,TI}{(1 - y)} + bA_0(1 - e^{-k_A t})$$

$$TI = [1 - y]\left[m(100 - A_0)(1 - e^{-k_M t}) + aA_0(1 - e^{-k_A t}) - S_L(M^+ + M^*) \right]$$

Substituting for M* and solving for TI:

$$TI = \left[\frac{1-y}{1+yS_L}\right]\left[m(100 - A_0) - (m + S_L)(100 - A_0)e^{-k_M t}\right.$$

$$\left. + (a - bS_L)A_0(1 - e^{-k_A t})\right]$$

For Cold Lake vacuum resid at 400°C, the following constants are used:

$$k_H = 0.013 \text{ min}^{-1} \qquad k_A = 0.026 \text{ min}^{-1}$$

$$m = 0.221 \qquad\qquad y = 0.30$$

$$a = 0.825 \qquad\qquad b = 0.02$$

For the full resid, $A_0 = 25.0$, $M_0 = 75.0$, and $S_L = 0.49$.
For the heptane-soluble fraction as the reactant, $A_0 = 0$, $M_0 = 100$ and $S_L = 0.61$.

4.3.6 COMPARISON OF KINETIC MODEL WITH QUANTITY DATA

To describe the temporal variation in the four classes when the reactant is the heptane-soluble fraction, only four constants are required. The rate constant k_M at 400°C was determined as described previously (Figure 4.4). The other three constants, S_L, m, and y, were evaluated to fit the experimental data on the thermolysis of Cold Lake vacuum resid and its heptane-soluble fraction using nonlinear least square software. The resulting values of the four constants are as shown previously. Figure 4.7 shows that the kinetic model describes quite well the experimental data on the variation in quantity of each of the four compound classes with reaction time. Because the four curves are all nonlinear, the agreement is more than a curve fit.

As discussed previously, the asphaltene thermolysis data was used to evaluate the asphaltene rate constant, k_A, at 400°C. As the closed reactor conversion could not provide the split between volatile and nonvolatile heptane solubles, it could not be used to evaluate stoichiometric coefficients. Instead, the remaining stoichiometric coefficients and a different solubility limit were evaluated, using the full resid thermolysis data. Again, the agreement between the experimental data of the variation of the four compound classes with reaction time and the nonlinear kinetic model calculation is quite good, as is shown in Figure 4.8. However, the dependence of the solubility limit on the initial asphaltene concentration indicates that a portion of the mechanism is not well-modeled. Nevertheless, this kinetic model is a large advance beyond that of Magaril[3] and shows that a realistic kinetic model with a phase-separation step for coke formation is consistent with the experimental data. This is a condition necessary, but not sufficient to justify the hypothesized mechanism for resid thermolysis. Considering that the model

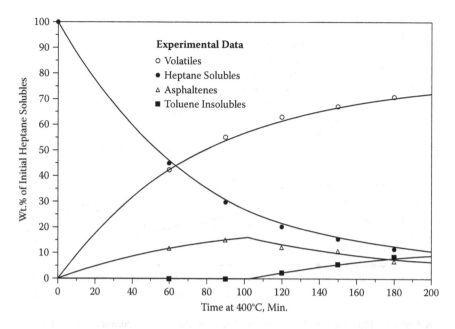

FIGURE 4.7 Replot of the weight fraction of four products of conversion of the hep-tane-soluble fraction of Cold Lake vacuum resid versus reaction time at 400°C in open-tube reactors, but with curves calculated from the phase-separation kinetic model. (From IA Wiehe. *Ind Eng Chem Res.* 32: 2447–2454, 1993. Reprinted with permission. Copyright 1993. American Chemical Society.)

was still fitted to the data with eight parameters, independent data is required to justify the mechanism.

4.3.7 COMPARISON OF THE KINETIC MODEL WITH QUALITY DATA

4.3.7.1 Asphaltene Hydrogen Content and Molecular Weight

Figure 4.8 shows that asphaltene concentration varies little in the coke induction period for the thermolysis of the full resid and then decreases once coke begins to form. As a result, if thermolysis data were collected only on the full resid, one might incorrectly conclude that asphaltenes are unreactive. On the contrary, it is the high reactivity of the asphaltenes down to the asphaltene core that offsets the generation of asphaltene cores from the maltenes to keep the overall asphaltene concentration nearly constant. In addition, only by postulating two asphaltene species, unreacted and cores, can the kinetic model simultaneously describe the thermolysis data of the two resid fractions and the whole resid. Therefore, further evidence was sought in the asphaltene quality data for the two asphaltene species and high asphaltene reactivity.

Chapters 2 and 3 pointed out the importance of the combination of molecular weight and hydrogen content as characterization tools for the Solvent-Resid Phase

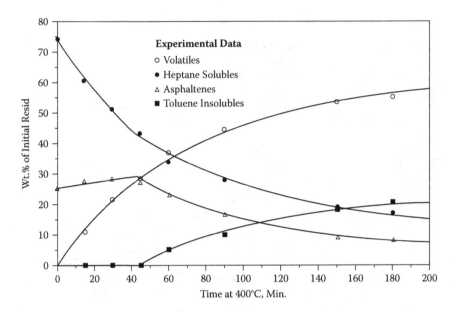

FIGURE 4.8 Plot of the weight fraction of four products of the conversion of full Cold Lake vacuum resid versus reaction time at 400°C in open-tube reactors. Curves were calculated from the phase-separation kinetic model. (From IA Wiehe. *Ind Eng Chem Res.* 32: 2447–2454, 1993. Reprinted with permission. Copyright 1993. American Chemical Society.)

Diagram and the Pendant-Core Building Block Model. Because the kinetic model provides the relative amount of the different type of asphaltenes, the overall hydrogen content and molecular weight of the asphaltenes can be calculated as a function of reaction time:

$$H_A = \frac{A^+ H_A^+ + A^* H_A^*}{A^+ + A^*} \qquad M_A = \frac{A^+ M_A^+ + A^* M_A^*}{A^+ + A^*}$$

where

H_A = overall asphaltene hydrogen content, wt%

H_A^+ = hydrogen content of reacted asphaltenes, wt%

H_A^* = hydrogen content of asphaltene cores, wt%

M_A = overall asphaltene molecular weight, wt%

M_A^+ = molecular weight of reacted asphaltenes, wt%

M_A^* = molecular weight of asphaltene cores, wt%

A^+, A^* = asphaltene-type concentrations as predicted by the kinetic model, wt%

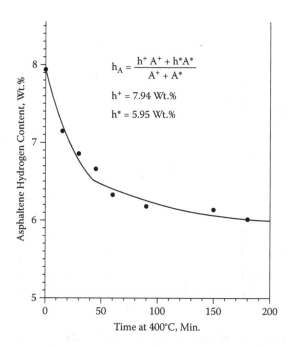

FIGURE 4.9 Temporal variation in the hydrogen content of asphaltenes from the thermolysis of the full Cold Lake vacuum resid at 400°C in open-tube reactors. Curve was calculated from the phase-separation kinetic model using $H_A^+ = 7.94$ wt% and $H_A^* = 5.95$ wt%. (From IA Wiehe. *Ind Eng Chem Res*. 32: 2447–2454, 1993. Reprinted with permission. Copyright 1993. American Chemical Society.)

In Figure 4.9 and Figure 4.10, the hydrogen content and molecular weight (VPO in o-dichlorobenzene at 130°C) are compared with the kinetic model for asphaltenes from the thermolysis of the full resid. Although the properties of the unreacted asphaltenes can be directly measured, the properties of the asphaltene core have to be fitted to the data at long reaction times. The asphaltene core is expected to have much higher hydrogen content at 400°C than the core determined at 500°C as shown in chapter 3. Nevertheless, having fixed the initial and long time property of the asphaltenes, Figure 4.9 shows that the kinetic model provides a quantitative description of the hydrogen content between these limits. Although the molecular weight data is more scattered, Figure 4.10 shows a reasonable description of molecular weight versus reaction time by the kinetic model. Thus, the ability of the kinetic model to describe asphaltene hydrogen content data with a sharp initial decrease and long time approach to a constant provides independent verification of several parts of the underlying mechanism. This includes a high rate constant for asphaltene thermolysis, the asphaltene core reaction limit, and the use of two species of asphaltenes to represent the distribution that is actually present.

The kinetic model predicts that the asphaltenes from thermolysis of heptane solubles will have a constant hydrogen content and molecular weight—those of

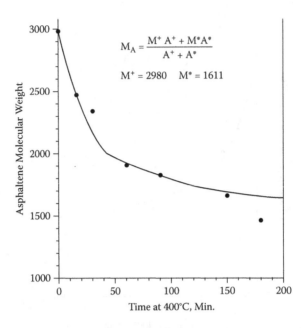

FIGURE 4.10 Temporal variation in the number average molecular weight of asphaltenes from the thermolysis of the full Cold Lake vacuum resid at 400°C in open-tube reactors. Curve was calculated from the phase-separation kinetic model using $M_A^+ = 2980$ and $H_A^* = 1611$. (From IA Wiehe. *Ind Eng Chem Res*. 32: 2447–2454, 1993. Reprinted with permission. Copyright 1993. American Chemical Society.)

asphaltene cores. Table 4.1, with example data at a reaction time of 90 min, shows that this prediction is not verified. Instead, the hydrogen content and molecular weight of asphaltenes from thermolysis of heptane solubles are in between those of unreacted asphaltenes and asphaltene cores—more similar to those from thermolysis of the full resid when compared at the same reaction time. This leads to the conclusion that the asphaltenes formed from thermolysis of heptane solubles are better approximated by unreacted asphaltenes than asphaltene cores. These asphaltenes

TABLE 4.1

Comparison of Properties of Asphaltenes Formed from Cold Lake Vacuum Resid Maltenes and from the Full Resid with Unreacted Asphaltenes and with Asphaltene Aromatic Cores after 90 min at 400°C in Open Reactors

Property	Maltene Product	Full Resid Product	Unreacted Asphaltenes	Asphaltene Aromatic Core
H Content, wt%	6.44	6.17	7.94	5.95
Mol. Wt.	1902	1820	2980	1611

then would continue to react similar to unreacted asphaltenes. This insight led to the series reaction kinetic model discussed in section 4.4.

4.3.7.2 Asphaltene Association Factor

In chapter 2, we learned that the measurement of molecular weight of petroleum asphaltenes gives different values depending on the technique, the solvent, and the temperature. Small-angle x-ray and neutron scattering showed that this phenomenon is a result of asphaltenes tending to self-associate and form aggregates. Eventually, it is this association tendency that causes reacted asphaltenes to phase separate and form coke. Thus, it is of interest to investigate the relative tendency for asphaltene association with thermolysis reaction time. In order to measure the relative tendency for asphaltene association, advantage is taken of the effect of the environment on the average molecular weight measured by vapor pressure osmometry. Thus, the asphaltene association factor, α, is defined by

$$\alpha = [\text{Mol. Wt in toluene at } 50°C] / [\text{Mol. Wt in } o\text{-dichlorobenzene at } 130°C]$$

This gives the associated molecular size of the asphaltene measured in toluene, the poorest solvent required to dissolve asphaltenes, at a relatively low temperature divided by the best measure of the unassociated molecular weight (chapter 2).

Figure 4.11 shows how this asphaltene association factor varies with reaction time for the thermolysis of the full resid. Also, in Figure 4.11, the variation of

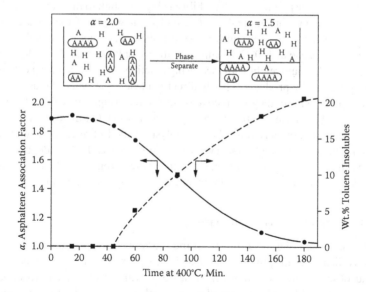

FIGURE 4.11 Temporal variation of the asphaltene association factor for asphaltenes from the thermolysis of the full Cold Lake vacuum resid at 400°C in open-tube reactors. It shows that associated asphaltenes preferentially phase separate to form coke. (From IA Wiehe. *Ind Eng Chem Res.* 32: 2447–2454, 1993. Reprinted with permission. Copyright 1993. American Chemical Society.)

toluene-insoluble coke with reaction time is replotted. During the 45 min coke induction period, the association factor is nearly constant at 1.9. However, once coke begins to form, the remaining asphaltenes have less tendency to associate. As more and more coke is formed, the association factor continually decreases until, at 180 min, the remaining asphaltenes show almost no tendency to associate ($\alpha = 1$). The explanation is that some asphaltenes are associated (denoted by circled As in Figure 4.11) and some asphaltenes are not associated (denoted by individual As). Both are distributed in a matrix of heptane solubles (denoted by Hs). The associated asphaltenes tend to preferentially phase separate to form coke as is indicated by the product below the line. Thus, the asphaltenes that remain (those above the line) in solution have, on average, a lower association factor. This process continues with increasing reaction time until all of the associated asphaltenes are phase separated to form coke, causing the association factor to reduce to unity. Thus, this data is consistent with asphaltene insolubility initiating the formation of coke.

4.3.7.3 Coke Molecular Weight

Once sufficient quantity of toluene-insoluble coke is formed from the thermolysis of either the heptane-soluble fraction or the full resid, the coke is not completely soluble in the o-dichlorobenzene at 130°C used to measure the molecular weight. However, because coke forms at high rates at short reaction times during the thermolysis of asphaltenes, its solubility is high enough to measure its molecular weight, as shown in Figure 4.12. The molecular weight of the coke continuously increases up to 60 min, until at 90 min the coke is not completely soluble in o-dichlorobenzene at 130°C. Meanwhile, the asphaltenes that do not form coke decrease in molecular weight with increasing reaction times by cracking off lower-molecular-weight by-products. This is direct evidence that when the asphaltenes are in a separate phase, they form coke by molecular weight growth that is better characterized by oligimerization than as polymerization. On the other hand, during the coke induction period for the thermolysis of the full resid, we saw that the asphaltene molecular weight underwent a rapid decrease. This evidence is consistent with a phase separation triggering the switch from asphaltene molecular weight decrease to molecular weight increase, resulting in coke formation.

4.3.7.4 Carbonaceous Mesophase

Toluene-insoluble coke, resulting from the thermolysis of the full resid for 180 min, was dispersed in quinoline (an excellent solvent for carbonaceous materials) and observed in transmitted light at 600× with an optical microscope, as shown in Figure 4.13. All the quinoline-insoluble coke is in the form of spheres or agglomerates of spheres. This is the most direct proof that at least part of the coke was formed by a liquid–liquid phase separation and that interfacial tension forced them into the spherical shape, much like oil dispersed in water, before they formed a solid by oligimerization. Because only ordered structures appear bright

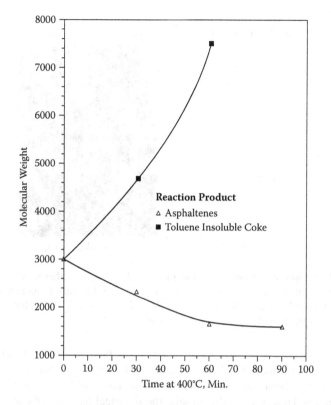

FIGURE 4.12 Temporal variation of the molecular weight of the coke and of the asphaltenes from the thermolysis of the asphaltene fraction of Cold Lake vacuum resid at 400°C in closed-tube reactors. (From IA Wiehe. *Ind Eng Chem Res.* 32: 2447–2454, 1993. Reprinted with permission. Copyright 1993. American Chemical Society.)

under cross polarized light, it is significant that these quinoline-insoluble particles are anisotropic or ordered. Books and Taylor[16] were the first to observe the carbonaceous mesophase, or liquid crystalline coke, which has had an enormous impact on the technology of producing carbon-based products. However, the feeds are usually smaller-ring aromatics with fewer side chains and contain less heteroatoms than vacuum resids. These provide the carbonaceous mesophase precursors with much lower reactivity than for resids, enabling the carbonaceous mesophase to remain liquid at elevated temperatures for a long enough time to form carbon products in the desired shape. Nevertheless, we can learn from carbon research that the carbonaceous mesophase is caused by the preferential parallel alignment of large-ring aromatics. The drops of carbonaceous mesophase grow by the coalescence of smaller drops. Oberlin,[17] using transmission electron microscopy, found that the carbonaceous mesophase only grows to sizes large enough (0.5 μm and larger) to resolve by optical microscopy if its viscosity and the viscosity of the

| Cold Lake
Vacuum Residuum | $\xrightarrow[\text{180 Min.}]{\text{400°C}}$ | Extract Toluene
Insolubles | $\xrightarrow{\hspace{2cm}}$ | Dispersed
in Quinoline |

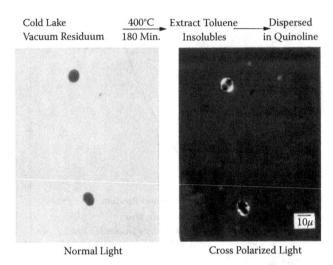

Normal Light   Cross Polarized Light

FIGURE 4.13 Optical micrographs of toluene-insoluble coke from thermolysis of the full Cold Lake vacuum resid for 180 min at 400°C in open-tube reactors and dispersed in quinoline. (From IA Wiehe. *Ind Eng Chem Res.* 32: 2447–2454, 1993. Reprinted with permission. Copyright 1993. American Chemical Society.)

medium remain low. Thus, the coke that appears to be quinoline-soluble under an optical microscope may be submicron spheres of ordered structures.

Later in this book, the carbonaceous mesophase will help us to understand fouling causes. However, for the present, the spherical particles of carbonaceous mesophase are significant for providing direct proof of the liquid–liquid phase separation mechanism for coke formation. This verified the basic hypothesis that the author made in his research proposal, not only about the phase separation but also the role of the carbonaceous mesophase. By making this hypothesis in advance, the author had justified the purchase of an expensive polarizing microscope and the receiving of specialized training in optical microscopy. Otherwise, optical microscopy would not normally be part of a study of resid thermal kinetics. This turned out to be an extremely valuable tool. Although not quantitative, optical microscopy has provided qualitative insight to resid conversion and solubility mechanisms. For instance, by observing only a drop of resid conversion product, one can tell whether or not that coke is present by looking for the carbonaceous mesophase. Thus, coke induction periods can be measured without mixing the thermal product with toluene and filtering it. Likewise, during pilot plant runs of resid conversion, one can determine if coke is forming in the reactor by the presence of carbonaceous mesophase in the heavy product. This enables the change of the conditions before filling the reactor with coke. As a result, the development of resid conversion processes was made much more rapid. Finally, by looking at a foulant from a refinery under an optical microscope, one can determine if the cause was thermal coking by the presence of the carbonaceous mesophase. These applications and more will be covered later in this book.

4.3.8 SUMMARY OF PHASE-SEPARATION KINETIC MODEL AND MECHANISM

It has been shown that most of the quantity and quality experimental data are consistent with the Phase-Separation Kinetic Model and mechanism. The kinetic model describes how the maltenes, asphaltenes, volatiles, and coke vary with reaction time during the thermolysis of Cold Lake vacuum resid and its heptane soluble fraction. The high thermal reactivity of asphaltenes to an aromatic core reaction limit was verified by the kinetic model based on these hypotheses describing the variation of the hydrogen content and molecular weight of asphaltenes. The liquid–liquid phase separation step triggering the onset of coke formation was directly proved by the presence of spherical carbonaceous mesophase in the coke. This was also consistent with the coke molecular weight increasing when asphaltenes are thermally reacted alone, but decreasing during the coke induction period when thermally reacted with the heptane solubles in the full resid. In addition, the more associated asphaltenes preferentially forming coke is consistent with the phase-separation mechanism. The test of natural hydrogen donor hypothesis will be discussed in section 4.7.

There were two inconsistencies in the comparison of the Phase-Separation Kinetic Model with experimental data. First, different solubility limits had to be used when applying this model to the thermolysis of the heptane soluble fraction and full resid. Second, the hydrogen content of the asphaltenes from the thermolysis of the heptane solubles showed that these asphaltenes initially are more similar to the unreacted asphaltenes than the asphaltene cores. This indicates that the two reactions in the kinetic model should be in series, rather than in parallel. In addition, the concept of a heptane soluble by-product of the reaction of asphaltene cores to coke seemed contrary to our view that any fragments that can crack off asphaltene cores would be small enough to be volatile.

Before changing the kinetic model, the present one is a close enough approximation to be tested against a much wider range of data: longer reaction times, different reactor type, more reaction temperatures, different initial asphaltene concentrations, and different resids. This will test how general the present model is and, if not general, it will provide both the motivation and direction of change. This follows the research strategy of the author. Once a model describes the experimental data, push that model to describe a greater range of data until it fails. In order to develop a model to describe the greater range of data, one must learn more about the mechanism. At each of these steps of learning, one should investigate the practical significance toward new innovation. Although these innovations are covered later in this book, in reality, they were pursued immediately. The phase-separation mechanism for coke formation meant that resids could be converted much more than previously thought without forming coke by keeping the converted asphaltenes soluble. This had tremendous potential impact on resid conversion and was being exploited almost immediately once the research results in this section were publicized within Exxon. This also explains why the permission to publish this research was withdrawn until 15 years after it was completed.

4.4 SERIES REACTION KINETIC MODEL

4.4.1 SOME NEW RESID THERMAL CONVERSION DATA

This section describes new resid thermal conversion data on Cold Lake vacuum resid at 400°C using the open reactor and the comparison with that predicted by the Phase-Separation Kinetic Model of section 4.3. In order not to be confused with a new kinetic model that will be developed in this section, the previous kinetic model will be called the "Parallel Reaction Kinetic Model" in this section. Later, additional data at 370°C and 420°C will be described along with the data using closed reactors and data on different resids. However, it makes sense to focus first on obtaining a kinetic model that describes the thermal conversion data on Cold Lake vacuum resid at 400°C using the open reactor and expanding it to different temperatures, reactors, and resids after this is accomplished.

4.4.1.1 Cold Lake Vacuum Resid with 50% Initial Asphaltenes

A master batch was prepared by blending two parts by weight Cold Lake vacuum resid (25 wt% asphaltenes) with one part Cold Lake asphaltenes at 250°C under nitrogen to yield a homogeneous mixture containing 50 wt% asphaltenes. All of the open tube reactions were done on samples from this master batch. No problems were experienced with foaming as obtained from open reactor thermal reactions on 100% asphaltenes. Figure 4.14 shows a comparison of the resulting thermal conversion of this mixture at 400°C with the Parallel Reaction Kinetic Model.

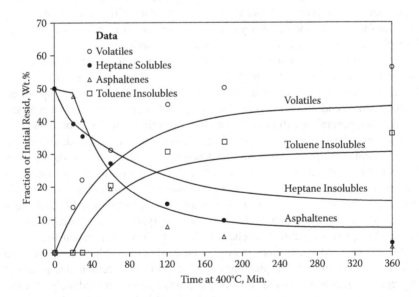

FIGURE 4.14 Deviations from the experimental data (points) for the thermolysis of Cold Lake vacuum resid with 50 wt% asphaltenes shown by the Parallel Reaction Kinetic Model calculations (curves).

TABLE 4.2

Comparison of the Parallel Reaction Kinetic Model and 400°C Data at 360 Min Reaction Time in Open-Tube Reactors

Initial Asphaltene (wt%)	Calculated Kinetic Model (wt%)				Experimental Data (wt%)			
	Volatil.	Heptane Sol.	Asph.	Toluene Insol.	Volatil.	Heptane Sol.	Asph.	Toluene Insol.
0	77.3	6.2	3.8	12.7	77.4	3.8	2.6	15.9
25	61.1	11.0	6.7	21.3	66.3	5.2	3.6	24.9
50	44.9	16.6	6.5	32.0	56.7	3.7	2.3	36.8

Although previously a different solubility limit was used for 0 and 25 wt% initial asphaltenes, the same solubility limit as for 25 wt% initial asphaltenes ($S_L = 0.50$) was used. The coke induction period predicted by the model is less than that shown by the data, indicating that the solubility limit should be closer to the value of 0.61 used for 0 wt% initial asphaltenes. However, even greater deviations between the model and data exist for all fractions at times beyond the coke induction period.

4.4.1.2   Longer Reaction Times at 400°C

The longest reaction time was doubled from 180 min to 360 min at each of the initial asphaltene concentrations. A comparison between experimental concentrations of each of the fractions and that calculated with the Parallel Reaction Kinetic Model at 360 min for each initial asphaltene concentration is given in Table 4.2. The volatiles and coke are predicted by the model to be lower than the data, whereas the maltenes and asphaltenes are predicted to be higher than the data.

4.4.2   DERIVATION OF THE SERIES REACTION MODEL

The new data provided further motivation to change the kinetic model, but not the direction. Therefore, advantage of the insight is gained from the hydrogen content and molecular weight of the asphaltenes formed in the maltene thermolysis by making unreacted asphaltenes as a product of this process rather than asphaltene cores. In addition, the coke was made as a direct reaction product of asphaltene cores without the maltene core by-product. Thus, the result is the Series Reaction Model.[18]

Series reaction kinetic model

$$M^+ \xrightarrow{\;k_M\;} m\,A^+ + (1-m)\,V$$

$$A^+ \xrightarrow{\;k_A\;} a\,A^* + b\,M^* + (1-a-b)\,V$$

Solubility Limit: $A^*_{max} = S_L\,(M^+ + M^*)$

Nomenclature

M^+ = unreacted maltenes, wt%

M^* = maltene cores, wt%

A^+ = unreacted asphaltenes, wt%

A^* = asphaltene cores, wt%

V = volatiles, wt%

TI = toluene insoluble coke, wt%

S_L = solubility limit

$A^*_{Ex} = A^* - A^*_{max}$

$A^*_{Ex} \xrightarrow{\infty} TI$

A^*_{max} = maximum asphaltene cores that can be held in solution

A^*_{Ex} = excess asphaltene cores beyond the solubility limit

k_M = first-order rate constant for the thermolysis of maltenes, min^{-1}

k_A = first-order rate constant for the thermolysis of asphaltenes, min^{-1}

m, a, and b = stoichiometric coefficients

This kinetic model contains one fewer stoichiometric coefficient than the Parallel Reaction Kinetic Model, but the series reactions makes the Series Reaction Kinetic Model more mathematically complex.

Based on the reactions being first order:

$$\frac{dM^+}{dt} = -k_M M^+$$

Integrating from time equal 0 to time equal t:

$$M^+ = (100 - A_0)e^{-k_M t}$$

Likewise, for the rate of change of unreacted asphaltenes:

$$\frac{dA^+}{dt} = -m\frac{dM^+}{dt} - k_A A^+$$

$$\frac{dA^+}{dt} = mk_M M^+ - k_A A^+$$

$$\frac{dA^+}{dt} + k_A A^+ = mk_M(100 - A_0)e^{-k_M t}$$

$$e^{k_A t}dA^+ + k_A A^+ e^{k_A t}dt = mk_M(100 - A_0)e^{(k_A - k_M)t}dt$$

$$\int_{A_0} d(e^{k_A t}A^+) = mk_M(100 - A_0)\int_0 e^{(k_A - k_M)t}dt$$

$$e^{k_A t}A^+ - A_0 = \frac{mk_M(100 - A_0)}{k_A - k_M}[e^{(k_A - k_M)t} - 1]$$

$$A^+ = A_0 e^{-k_A t} + \frac{mk_M(100 - A_0)}{k_A - k_M}[e^{-k_M t} - e^{-k_A t}]$$

$$A^+ = \frac{mk_M(100 - A_0)}{k_A - k_M}e^{-k_M t} - \left[\frac{mk_M(100 - A_0)}{k_A - k_M} - A_0\right]e^{-k_A t}$$

$$A^+ = A_0 + \left[\frac{mk_M(100 - A_0)}{k_A - k_M} - A_0 \right][1 - e^{-k_A t}] - \frac{mk_M(100 - A_0)}{k_A - k_M}[1 - e^{-k_M t}]$$

The volatiles, V, are by-products of both reactions:

$$V = (1 - m)(100 - A_0 - M^+) + [1 - a - b][m(100 - A_0 - M^+) + A_0 - A^+]$$

$$V = (100 - A_0 - M^+)(1 - m\{a + b\}) + [1 - a - b][A_0 - A^+]$$

$$V = (1 - m\{a + b\})(100 - A_0)(1 - e^{-k_M t}) - [1 - a - b]$$

$$\times \left[\begin{array}{l} \left[\dfrac{mk_M(100 - A_0)}{k_A - k_M} - A_0 \right][1 - e^{-k_A t}] \\[3mm] - \dfrac{mk_M(100 - A_0)}{k_A - k_M}[1 - e^{-k_M t}] \end{array} \right]$$

$$V = \left[1 - m + m(1 - a - b)\frac{k_A}{k_A - k_M} \right][100 - A_0][1 - e^{-k_M t}]$$

$$+ [1 - a - b]\left[A_0 - \frac{mk_M(100 - A_0)}{k_A - k_M} \right][1 - e^{-k_A t}]$$

The heptane soluble cores, M*, are by-products of the asphaltene reaction:

$$M^* = b[m(100 - A_0 - M^+) + A_0 - A^+]$$

$$M^* = b\left[\begin{array}{l} m(100 - A_0)(1 - e^{-k_M t}) - \left[\dfrac{mk_M(100 - A_0)}{k_A - k_M} - A_0 \right][1 - e^{-k_A t}] \\[3mm] - \dfrac{mk_M(100 - A_0)}{k_A - k_M}[1 - e^{-k_M t}] \end{array} \right]$$

$$M^* = b\left[\left\{ A_0 - \frac{mk_M(100 - A_0)}{k_A - k_M} \right\}\{1 - e^{-k_A t}\} + \left\{ \frac{mk_A(100 - A_0)}{k_A - k_M} \right\}\{1 - e^{-k_M t}\} \right]$$

$$M = M^+ + M^*$$

$$M = (100 - A_0)e^{-k_M t} + b\left[\left\{ A_0 - \frac{mk_M(100 - A_0)}{k_A - k_M} \right\}\{1 - e^{-k_A t}\} \right.$$

$$\left. + \left\{ \frac{mk_A(100 - A_0)}{k_A - k_M} \right\}\{1 - e^{-k_M t}\} \right]$$

$$M = 100 - A_0 + b\left[A_0 - \frac{mk_M(100 - A_0)}{k_A - k_M}\right][1 - e^{-k_A t}] + (100 - A_0)$$

$$\times \left[\frac{mbk_A}{k_A - k_M} - 1\right][1 - e^{-k_M t}]$$

During the coke induction period when all of the asphaltenes remain in solution $(A^* < S_L H)$,

$$A^* = a[m(100 - A_0 - M^+) + A_0 - A^+]$$

$$A^* = a\left[\left\{A_0 - \frac{mk_M(100 - A_0)}{k_A - k_M}\right\}\{1 - e^{-k_A t}\} + \left\{\frac{mk_A(100 - A_0)}{k_A - k_M}\right\}\{1 - e^{-k_M t}\}\right]$$

$$A = A^+ + A^*$$

$$A = A_0 - [1 - a]\left[A_0 - \frac{mk_M(100 - A_0)}{k_A - k_M}\right][1 - e^{-k_A t}] + \frac{m(100 - A_0)(ak_A - k_M)}{k_A - k_M}$$

$$\times [1 - e^{-k_M t}]$$

After the coke induction period,

$$A^* = S_L(H^+ + H^*)$$

$$A = A_0 + S_L(100 - A_0) - [1 - bS_L]\left[A_0 - \frac{mk_M(100 - A_0)}{k_A - k_M}\right][1 - e^{-k_A t}]$$

$$-[100 - A_0]\left[\frac{S_L k_A(1 - bm) + (m - S_L)k_M}{k_A - k_M}\right][1 - e^{-k_M t}]$$

$$TI = A^*_{\text{Produced}} - A^*_{\text{Soln}}$$

$$TI = a\left[\left\{A_0 - \frac{mk_M(100 - A_0)}{k_A - k_M}\right\}\{1 - e^{-k_A t}\} + \left\{\frac{mk_A(100 - A_0)}{k_A - k_M}\right\}\{1 - e^{-k_M t}\}\right]$$

$$-S_L\left[100 - A_0 + b\left[A_0 - \frac{mk_M(100 - A_0)}{k_A - k_M}\right][1 - e^{-k_A t}] + (100 - A_0)\right.$$

$$\times \left[\frac{mbk_A}{k_A - k_M} - 1\right][1 - e^{-k_M t}]\right]$$

$$TI = [a - bS_L]\left[A_0 - \frac{mk_M(100 - A_0)}{k_A - k_M}\right][1 - e^{-k_A t}] - S_L(100 - A_0)$$

$$+ \left[\frac{mak_A + S_L(k_A - k_M - mbk_A)}{k_A - k_M}\right][100 - A_0][1 - e^{-k_M t}]$$

If the reaction is allowed to go to completion, the Series Reaction Kinetic Model predicts the following results:

$$V_\infty = 100 - (a + b)[A_0 + m(100 - A_0)]$$

$$M_\infty = b[A_0 + m(100 - A_0)]$$

$$A_\infty = S_L b[A_0 + m(100 - A_0)]$$

$$TI_\infty = [a - bS_L][A_0 + m(100 - A_0)]$$

4.4.3 COMPARISON OF THE SERIES REACTION MODEL WITH 400°C DATA

The first step in utilizing the kinetic model is evaluating the parameters in the model. However, the rate constants, k_A and k_M, were already determined, independent of the model specifics. The three stoichiometric coefficients were determined by fitting all the open reactor data at 400°C for the thermolysis of Cold Lake 1050°F+ fractions, containing 0, 25, and 50 wt% initial asphaltenes, using a nonlinear least squares fitting program. These stoichiometric coefficients were fitted to the experimental volatiles, heptane solubles, and the sum of asphaltenes and coke. Then, the solubility limit was determined that best described the split of asphaltene cores into asphaltenes and coke. It was found that only one solubility limit, S_L, independent of initial asphaltene concentration, was sufficient. The resulting set of constants is given in the following text.

Constants in Series Reaction Model at 400°C for Cold Lake Vacuum Resid

$k_A = 0.026$ min^{-1}	$k_M = 0.013$ min^{-1}	$S_L = 0.362$
$m = 0.302$	$a = 0.602$	$b = 0.108$

It should be noted that the Series Reaction Kinetic Model has the advantage over the Parallel Reaction Model of at least two less constants. One stoichiometric coefficient, y, was eliminated by removing the maltene core by-product for the conversion of insoluble asphaltene cores to coke. At least one solubility limit was eliminated because previously a different solubility limit was used for the thermolysis of the maltenes and the full resid. Because it was not clear what the solubility limit should be in the Parallel Reaction Kinetic Model at other initial

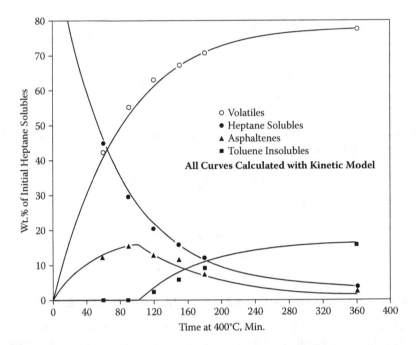

FIGURE 4.15 Description of the weight fraction of four products of the thermolysis of the heptane-soluble fraction of Cold Lake vacuum resid (points) versus reaction time (up to 360 min) at 400°C in open-tube reactors by the Series Reaction Kinetic Model (curves).

asphaltene concentrations, additional solubility limit parameters may have been eliminated by the use of the new kinetic model.

Despite having fewer constants, Figure 4.15 to Figure 4.17 show that the Series Reaction Kinetic Model quantitatively describes the experimental data quite well for the thermolysis of Cold Lake 1050°F⁺ at three initial asphaltene concentrations at 400°C in an open reactor. This data includes the 50% initial asphaltene data not described well by the Parallel Reaction Kinetic Model. In addition, Table 4.3 shows that the calculated results for the newer model describe the experimental data at the longest reaction time of 360 min much better than the previous model (Table 4.2). Thus, the changes from reactions in parallel to reactions in series and the elimination of a heptane soluble core by-product in the asphaltene to coke reaction have been substantiated by comparison with the experimental data.

4.4.4 Linear Relationships between the Fractions

In this section, we follow the recommended strategy to look for independent evidence of the reactions being in series, rather than in parallel. Previous to the development of the Series Reaction Kinetic Model, it was observed that when the fractions, maltenes, asphaltenes, and coke were plotted against volatiles, straight

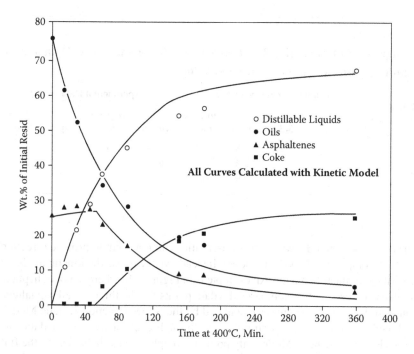

FIGURE 4.16 Description of the weight fraction of four products of the thermolysis of the full Cold Lake vacuum resid (points) versus reaction time (up to 360 min) at 400°C in open-tube reactors by the Series Reaction Kinetic Model (curves).

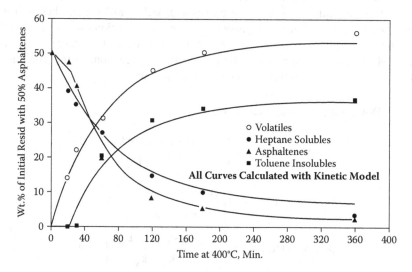

FIGURE 4.17 Description of the weight fraction of four products of the thermolysis of the Cold Lake vacuum resid with 50 wt% asphaltenes (points) versus reaction time (up to 360 min) at 400°C in open-tube reactors by the Series Reaction Kinetic Model (curves).

TABLE 4.3

Comparison of the Series Reaction Kinetic Model and 400°C Data at 360 Min Reaction Time in Open-Tube Reactors

Initial Asphaltene (wt%)	Calculated Kinetic Model (wt%)				Experimental Data (wt%)			
	Volatil.	Heptane Sol.	Asph.	Toluene Insol.	Volatil.	Heptane Sol.	Asph.	Toluene Insol.
0	77.9	4.0	1.8	16.3	77.4	3.8	2.6	15.9
25	65.7	5.7	2.4	26.3	66.3	5.2	3.6	24.9
50	53.5	7.4	2.9	36.2	56.7	3.7	2.3	36.8

lines were obtained. However, this observation did not seem to provide direction to the development of a kinetic model. Instead, we try to use the kinetic model to describe the linear behavior. An example is in Figure 4.18 where a plot of heptane solubles versus volatiles compares the data points for the three initial asphaltene concentrations with the curves calculated by the Series Reaction Kinetic Model. Although we might draw a straight line through each of the sets of data, the Series Reaction Kinetic Model only predicts complete linear behavior for the full resid (25% initial asphaltenes). This model also predicts curved behavior at short reaction times (low volatiles) for 0% and 50% initial asphaltenes that approaches linear behavior at longer times (higher volatiles), which is parallel to the line that is predicted for 25% initial asphaltenes.

Let us determine why all three curves approach parallel linear behavior at large reaction times. Because k_A is twice k_M, at large reaction times $[\exp(-k_A t)]$ approaches zero much before $[\exp(-k_M t)]$. In other words, the generation of A^+ from M^+ decomposition becomes the rate-controlling reaction. At this condition the weight percent of each fraction becomes as follows:

When exp(-k_At) approaches zero

$$V = V_\infty - [100 - A_0]\left[\frac{(1 - ma - mb)k_A - k_M(1 - m)}{k_A - k_M}\right]e^{-k_M t}$$

$$M = M_\infty + [100 - A_0]\left[\frac{(1 - mb)k_A - k_M}{k_A - k_M}\right]e^{-k_M t}$$

$$A = A_\infty + [100 - A_0]\left[S_L - \frac{m(bS_L k_A - k_M)}{k_A - k_M}\right]e^{-k_M t}$$

$$TI = TI_\infty - [100 - A_0]\left[S_L + \frac{m(a - bS_L)k_A}{k_A - k_M}\right]e^{-k_M t}$$

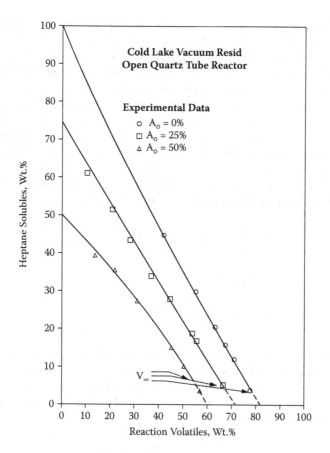

FIGURE 4.18 Description of the variation in heptane solubles with reaction volatiles (points) for the thermolysis of the Cold Lake vacuum resid with three different initial asphaltene concentrations at 400°C in open-tube reactors by the Series Reaction Kinetic Model (curves).

Because each of these fractions are linear functions of [exp(-k_Mt)], any one can be expressed as a linear function of any other fraction. For instance, the heptane solubles can be expressed as a linear function of volatiles:

When exp(-k_At) approaches zero

$$M = M_\infty + (V - V_\infty)\left[\frac{(mb-1)k_A + k_M}{(1-ma-mb)k_A - (1-m)k_M}\right]$$

$$\text{Slope} = \frac{(mb-1)k_A + k_M}{(1-ma-mb)k_A - (1-m)k_M}$$

$$V_{\text{Intercept@M=0}} = (1-m)(100 - A_0)\left(a + \frac{mb(ak_A + k_M)}{(1-mb)k_A + k_M}\right)$$

This predicts, as observed, that at long times the heptane solubles will approach a linear function of the volatiles with a slope that is independent of initial asphaltene concentration. Substitution of the values of the constants shows that this slope is −1.06. Although the concentration of heptane solubles will never be predicted to reach zero because of thermally stable heptane soluble cores, if the lines are extrapolated to zero heptane solubles, the intercepts depend on the initial asphaltene concentration. These intercept volatile values are calculated to be 81.6, 71.0, and 60.4 for initial asphaltene concentrations of 0, 25, and 50%, respectively.

The Series Reaction Kinetic Model explains the parallel linear behavior at long reaction times as a result of the second reaction in the series being much faster than the first. After sufficient time, the second reaction is forced to match the rate of the first reaction. When this happens, the ratio of A^+ to M^+ becomes a constant:

$$A^+ = \frac{mk_M(100 - A_0)}{k_A - k_M} e^{-k_M t} - \left[\frac{mk_M(100 - A_0)}{k_A - k_M} - A_0 \right] e^{-k_A t}$$

$$M^+ = (100 - A_0)e^{-k_M t}$$

$$\text{Lim } e^{-k_A t} \longrightarrow 0 \quad \frac{A^+}{M^+} = \frac{mk_M}{k_A - k_M}$$

However, what happens if the initial ratio of asphaltenes to maltenes is fixed at this constant? A quick look at the model equations for each fraction shows that at this condition all the terms involving $[1 - \exp(-k_A t)]$ vanish and each fraction becomes a linear function of any other fraction. For Cold Lake vacuum resid at 400°C, this condition occurs for

$$\frac{A_0}{100 - A_0} = \frac{mk_M}{k_A - k_M} = 0.321$$

For the full Cold Lake vacuum resid with A_0 equal to 25, this ratio is 0.333 and close enough to 0.321 to result in linear behavior over the entire set of reaction times. Thus, it was only by chance that the full resid had the weight fraction of asphaltenes to give complete linear behavior. In the general case, the compositions change during the reaction so that A^+/M^+ is directed toward this constant value. Only when the ratio becomes sufficiently close is the linear behavior obtained.

The Parallel Reaction Kinetic Model also predicts a linear relationship among the products at long reaction times. Because k_A is much greater than k_M, the asphaltene decomposition reaches its conversion limit, whereas the heptane solubles are still reacting. Thus, during this period of only one active reaction, a linear relationship among the products is obtained. However, for nonzero values of the stoichiometric coefficients, the Parallel Reaction Kinetic Model cannot predict a linear relationship among the products at all reaction times as the Series Reaction Kinetic Model can. Thus, this is the independent evidence for a series of reactions that we have been seeking.

4.4.5 COMPARISON OF THE SERIES REACTION KINETIC MODEL WITH DATA AT 370°C AND 420°C

Data were obtained at 370°C and 420°C using the open quartz tube reactor on Cold lake 1050°F+ and its maltene fraction to determine if the Series Reaction Kinetic Model could describe the data over a range of temperatures. The stoichiometric coefficients, m, a, and b, were assumed to be temperature independent and, thus, were fixed at the values determined using the 400°C data described in the previous section. Thus, the rate constants, k_A and k_M, were fitted to the data on volatiles, maltenes, and the sum of asphaltenes and coke, using a nonlinear least square fitting program. The resulting rate constants, k_A and k_H, at the three temperatures fell on straight lines when plotted versus the reciprocal of absolute temperature on semilogarithmic graph paper (Figure 4.19). The activation energies

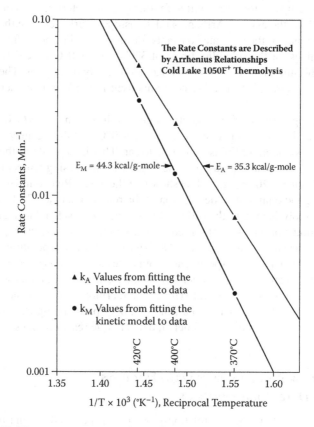

FIGURE 4.19 The two rate constants in the Series Reaction Kinetic Model for the thermolysis of Cold Lake vacuum resid, described by Arrhenius relationships.

TABLE 4.4

Values of the Six Parameters Used in the Series Reaction Kinetic Model for Cold Lake 1050°F⁺ Thermolysis in Open Reactors

T (°C)	k_A (min⁻¹)	k_M (min⁻¹)	S_L	m	a	b	A_0 Range of Data (wt%)
370	0.00754	0.00286	0.349	0.302	0.602	0.108	0, 25
400	0.260	0.0134	0.362	0.302	0.602	0.108	0, 25, 50
420	0.0553	0.0350	0.284	0.302	0.602	0.108	0, 25

for asphaltene decomposition, 35.3 kcal/g-mol, and for maltene decomposition, 44.3 kcal/g-mol, seem a bit low. However, a wider temperature range with data at many more temperatures are required for the accurate determination of activation energies. Nevertheless, the Arrhenius relationships clearly describe the temperature dependency of the two rate constants for the data collected so far.

The values for the solubility limits at 370°C and 420°C were determined to best split the sum of asphaltenes and coke into these two fractions. The values of the solubility limits and the other constants used in the kinetic model are given in Table 4.4.

Surprisingly, the solubility limit is higher at 400°C than at 370°C or 420°C. Because the difference in solubility limit at 370°C and 400°C may not be significant, the value at 420°C is definitely lower. This is opposite to the direction expected because solubility usually increases with increasing temperature. Nevertheless, Figure 4.20 to Figure 4.23 show that the Series Reaction Kinetic Model provides a good quantitative description of the reaction data at 370°C and 420°C for the thermolysis of both the full Cold Lake vacuum resid and its maltene fraction. The model provides an observed decrease in the coke induction period with increasing thermolysis temperature. Otherwise, at each temperature once coke formation is initiated, the experimental data is described by asphaltenes and heptane solubles decreasing in parallel. The one case where deviations appear to be larger than random experimental error is the thermolysis of heptane solubles at 420°C. Figure 4.20 shows that the kinetic model predicts higher volatiles and lower asphaltenes than the experimental data at large reaction times.

4.4.6 LINEAR RELATIONSHIPS BETWEEN FRACTIONS AT DIFFERENT TEMPERATURES

It is informative to revisit the relationships among the fractions and include the temperature dependency. Surprisingly, Figure 4.24 shows that the data for the three temperatures all fall on the same line when the heptane soluble fraction is

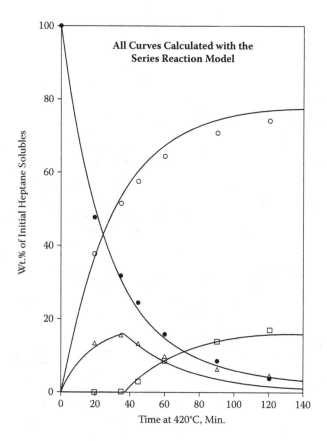

FIGURE 4.20 Description of the weight fraction of four products of the thermolysis of the heptane-soluble fraction of Cold Lake vacuum resid (points) versus reaction time at 420°C in open-tube reactors by the Series Reaction Kinetic Model (curves) with the following data: o = volatiles, • = heptane solubles, Δ = asphaltenes, and □ = toluene-insoluble coke.

plotted versus volatiles for the thermolysis of the full resid and a different line when the reactant is the heptane soluble fraction of Cold Lake resid. The coke (Figure 4.25) also shows temperature independent linear behavior when the full resid is the reactant. However, when the heptane soluble fraction is the reactant (Figure 4.26), the data at 400 and 420°C fall on separate lines. This is what may have required different solubility limits for these two temperatures. Nevertheless, Figure 4.27 compares the data on the thermolysis of the full resid at the three temperatures with that calculated by the Series Reaction Kinetic Model in terms of toluene insoluble coke versus reaction volatiles. The model gives three different coke induction values of volatiles for the three different temperatures, whereas the data only justifies one.

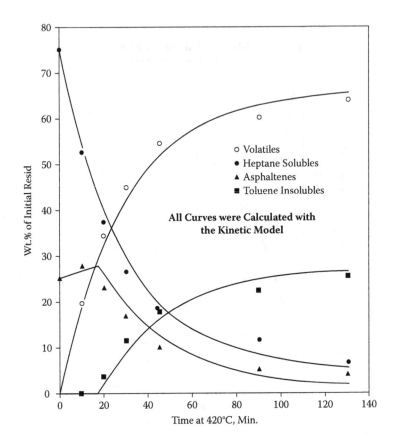

FIGURE 4.21 Description of the weight fraction of four products of the thermolysis of full Cold Lake vacuum resid (points) versus reaction time at 420°C in open-tube reactors by the Series Reaction Kinetic Model (curves).

With increasing volatiles beyond the induction values, the three slightly non-linear curves converge. Thus, the data show a simpler relationship than the kinetic model exists, but the kinetic model only approximates this simpler relationship. These are clues that solubility limits should be temperature independent as shown in section 4.4.9 after hydrogen balance constraints are applied, and the kinetic model can be further simplified as shown in section 4.4.10.

4.4.7 Open Reactor Data on Arabian Heavy Vacuum Resid

To further test the Series Reaction Kinetic Model, it is applied to the data collected on a different resid, Arabian Heavy 1050°F+ ($A_0 = 23.8$ wt%) and its heptane soluble fraction, at 400°C in an open quartz tube reactor. The Parallel Reaction

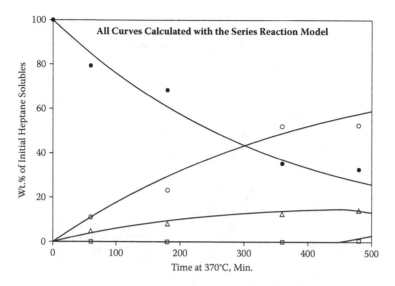

FIGURE 4.22 Description of the weight fraction of four products of the thermolysis of the heptane-soluble fraction of Cold Lake vacuum resid (points) versus reaction time at 370°C in open-tube reactors by the Series Reaction Kinetic Model (curves) with the following data: o = volatiles, • = heptane solubles, Δ = asphaltenes, and □ = toluene insolubles.

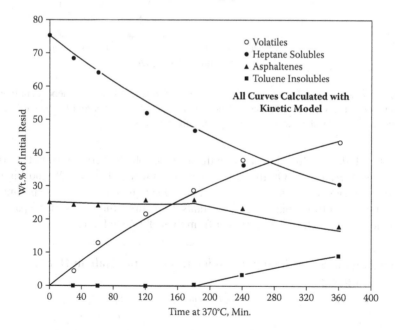

FIGURE 4.23 Description of the weight fraction of four products of the thermolysis of full Cold Lake vacuum resid (points) versus reaction time at 370°C in open-tube reactors by the Series Reaction Kinetic Model (curves).

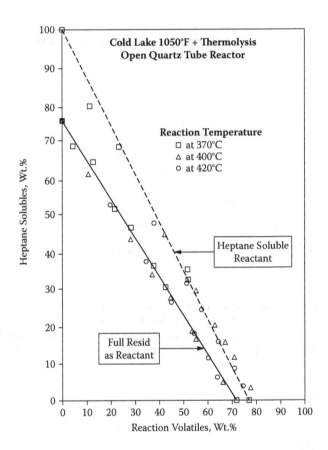

FIGURE 4.24 The yield of heptane solubles as different temperature-independent linear functions of the yield of volatiles for the heptane-soluble reactant and for the full resid reactant for Cold Lake vacuum resid in open-tube reactors.

Kinetic Model was found to describe the heptane soluble thermolysis, but not the full resid thermolysis. On the other hand, as shown in Figure 4.28 and Figure 4.29, the Series Reaction Kinetic Model described quantitatively the experimental data with both Arabian Heavy reactants. Of course, the values of the parameters for Arabian Heavy are different from those for Cold Lake:

Constants in Series Reaction Model at 400°C for Arabian Heavy Vacuum Resid

$k_A = 0.0204$ min^{-1}	$k_M = 0.00773$ min^{-1}	$S_L = 0.291$
$m = 0.383$	$a = 0.611$	$b = 0.0194$

Arabian Heavy vacuum resid is slower reacting than Cold Lake vacuum resid, but it has a larger asphaltene core and a lower solubility limit. Arabian Heavy vacuum

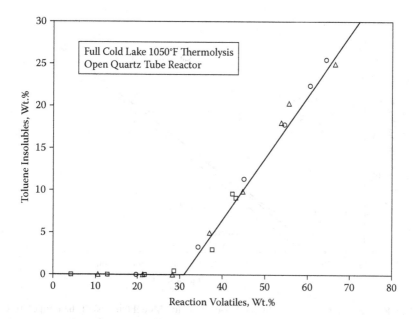

FIGURE 4.25 The yield of coke as a temperature-independent linear function of the yield of volatiles for the thermolysis of full Cold Lake vacuum resid in open-tube reactors with the following data: □ at 370°C, Δ at 400°C, and o at 420°C.

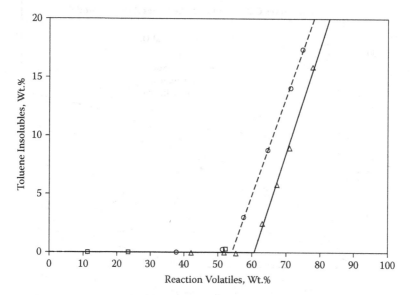

FIGURE 4.26 The yield of coke (toluene insolubles) as different linear functions of the yield of volatiles at 400°C and at 420°C for the thermolysis of the heptane-soluble fraction of Cold Lake vacuum resid in open-tube reactors with the following data: □ at 370°C, Δ at 400°C, and o at 420°C.

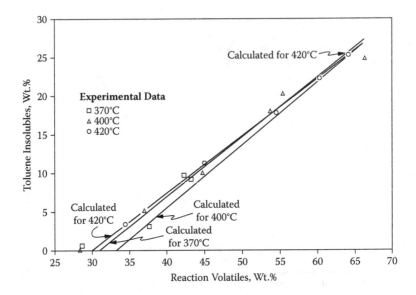

FIGURE 4.27 Approximation by the Series Reaction Model (curves) of the temperature-independent linear relationship between coke and volatile yields (points) for the thermolysis of full Cold Lake vacuum resid in open-tube reactors.

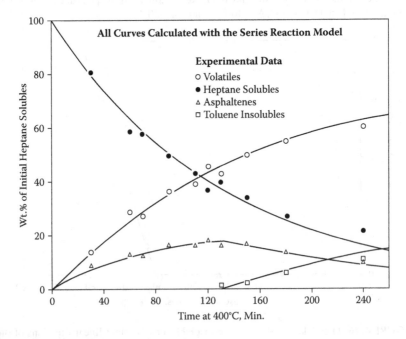

FIGURE 4.28 Description of the weight fraction of four products of the thermolysis of the heptane-soluble fraction of Arabian Heavy vacuum resid (points) versus reaction time at 400°C in open-tube reactors by the Series Reaction Kinetic Model (curves).

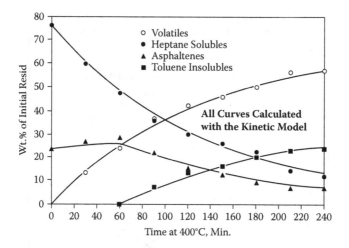

FIGURE 4.29 Description of the weight fraction of four products of the thermolysis of the full Arabian Heavy vacuum resid (points) versus reaction time at 400°C in open-tube reactors by the Series Reaction Kinetic Model (curves).

resid also forms very little heptane soluble cores ($b = 0.0194$) that probably could be approximated to be zero.

In Figure 4.30 and Figure 4.31, the heptane solubles, asphaltene, and the toluene insolubles are plotted against the reaction volatiles for the thermolysis

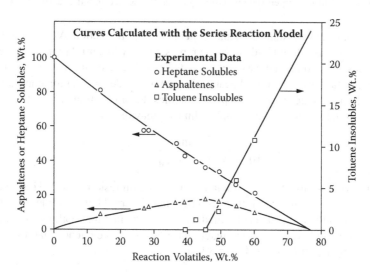

FIGURE 4.30 The Series Reaction Kinetic Model (curves): Prediction of the linear variation in the yield of heptane solubles, asphaltenes, and coke (toluene insolubles) as a function of the yield of volatiles (points) beyond an initial amount of volatiles for the thermolysis of the heptane-soluble fraction of Arabian Heavy vacuum resid at 400°C in open-tube reactors.

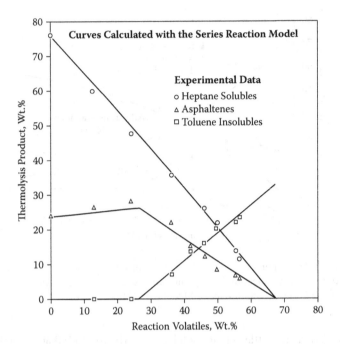

FIGURE 4.31 Prediction of the linear variation in the yield of heptane solubles, asphaltenes, and coke (toluene insolubles) as a function of the yield of volatiles (points) beyond an initial amount of volatiles for the thermolysis of the full Arabian Heavy vacuum resid at 400°C in open-tube reactors by the Series Reaction Kinetic Model (curves).

of Arabian Heavy vacuum resid and its heptane soluble fraction. As for Cold Lake vacuum resid, much of the data can be represented by linear relationships. Again, the Series Reaction Kinetic Model predicts linear behavior at long reaction times (or high volatiles production) because the asphaltene decomposition constant, k_A, is again much larger than the heptane solubles decomposition constant, k_M. The kinetic model would predict complete linear behavior when

$$\frac{A_0}{100 - A_0} = \frac{mk_M}{k_A - k_M} = 0.234$$

but this constant for the full Arabian Heavy vacuum resid is 0.310. Thus, unlike for Cold Lake vacuum resid, there is a significant period before a linear relationship between the products is obtained. However, the fact that both vacuum resids approach linear behavior at long reaction times provides evidence that this might be a common feature of resid thermolysis.

4.4.8 CLOSED REACTOR DATA

This section describes experimental data applied to closed reactors. These reactors are enclosed tubing bombs with a small amount of gas space that force volatile liquid products to remain in the liquid phase during thermolysis. This is the opposite vapor–liquid condition of the open reactor that allows volatile liquid products to escape the reactor when they form. Although coke is triggered by a liquid–liquid phase separation, it will be interesting to see if the change in vapor–liquid equilibrium will affect the coke induction period.

4.4.8.1 Thermolysis of Cold Lake Vacuum Resid at 400°C

The same experimental procedure was used, as described previously, for reacting Cold Lake asphaltenes in a closed tubing bomb reactor. In both cases, part of the volatile liquids was lost during the evaporation of separating solvents. For this reason, mass balances were normally only 76 to 99 wt%, and no attempt was made to separate the volatile and nonvolatile heptane solubles. As a result, only the asphaltenes and toluene insoluble coke were measured accurately as a function of reaction time. Later, an improved experimental procedure will be used where the volatile liquids will be distilled out of the tubing bombs before the products are exposed to solvent separation. This procedure yields both good mass balances and a way to directly measure volatile and nonvolatile heptane solubles.

Because the volatile liquid products remain in the liquid phase for closed reactors, the Series Reaction Kinetic Model needs to include volatile liquids, V, as part of the solvent for asphaltene cores:

4.4.8.2 Closed Reactor Solubility Relationship

$$A^*_{sol} = S'_L (H^+ + H^* + V)$$

The solubility limit, S'_L , is used rather than S_L for the open reactor model to allow for the possibility that making the volatile liquids part of the reactor solvent could alter the quality of the solvent. Otherwise, the kinetic model for the closed reactor should have the same reaction rate constants and stoichiometric coefficients as for the kinetic model for the open reactor. The best value of S'_L was determined that fit closed reactor kinetic data on Cold Lake vacuum resid thermolysis at 400°C:

$$S'_L = 0.137$$

This is less than half the value of the solubility limit of 0.379 determined previously for the same conditions, but an open reactor. Thus, the volatile liquid products are not nearly as good as a solvent for asphaltene cores as the nonvolatile heptane solubles. This is because the volatile liquid products are more saturated, and they can even be nonsolvents for asphaltene cores. This will be covered in detail later in this book. At present, the important point is that the change in the solubility relationship for closed reactors provides a good description of the asphaltene and coke data as a function of reaction time (Figure 4.32).

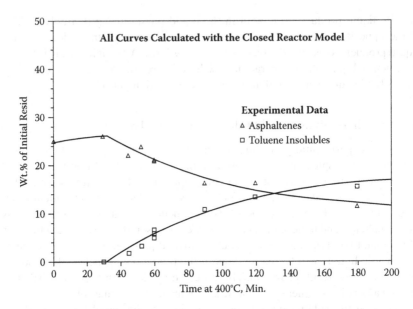

FIGURE 4.32 Description of the weight fraction of the asphaltenes and of coke (toluene insolubles) from the thermolysis of full Cold Lake vacuum resid (points) versus reaction time at 400°C in closed-tube reactors by the Series Reaction Kinetic Model (curves).

In Figure 4.33, the coke formation curves are compared for closed reactor (data points and kinetic model with solid curve) and the open reactor (kinetic model with dashed curve) models. The closed reactor has a shorter induction period because holding the volatile liquids in the liquid phase produces a solvent with much lower ability to dissolve asphaltene cores. However, at longer reaction times the curves cross and the closed reactor forms less coke than the open reactor. This is because the open reactor continuously decreases the quantity of solvent (nonvolatile heptane solubles) with increasing reaction time, whereas the quantity of solvent for the closed reactor remains almost constant. Eventually, the high quantity of a poor solvent (closed reactor) dissolves more asphaltene cores than the small amount of a good solvent (open reactor). Nevertheless, the earlier reaction time data highlights the most practical way to delay the onset of coke formation, preferentially evaporating the volatile liquid products out of the reactor. As will be discussed, this has applications to visbreakers, delayed coker heaters, and hydroconversion reactors. Although adding a solvent to the reactor feed is the most obvious way to delay the precipitation of asphaltene cores, removing the nonsolvent in the form of volatile liquid products is usually the most practical way.

No advantage has been devised for the lower coke yield at large reaction times for closed reactors as compared to open reactors. Although commercial cokers are open reactors that operate at long reaction times, they measure coke yield as

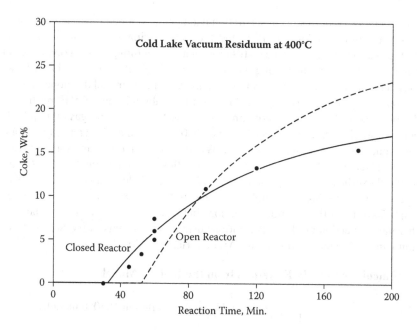

FIGURE 4.33 Coke induction period and coke yield in open and closed reactors. The coke induction period is shorter in a closed reactor than in an open reactor, but the yield of coke is less for a closed reactor at long reaction times for the full Cold Lake vacuum resid at 400°C. The points are only for the closed reactor data, but the two curves are calculated for the open and closed reactors by the Series Reaction Kinetic Model.

total nonvolatiles, not as toluene insolubles. Thus, a closed reactor coker would be a huge disadvantage because it would produce more coke as nonvolatiles, even though the yield of toluene insolubles would decrease. In addition, at the high temperatures of cokers, as discussed later in this chapter, the valuable volatile liquid products tend to react to form low-valued coke and gas. This provides more incentive to remove volatile liquid products from the reactor as quickly as possible.

4.4.8.3 Thermolysis of Arabian Heavy Vacuum Resid at 400°C

Two separate Arabian Heavy vacuum resids have been thermally converted in closed tubing bomb reactors. An Arabian Heavy 1050°F+ resid ($A_0 = 23.8$ wt%) was reacted at 400°C as a continuation of the open reactor thermolysis study discussed in section 4.4.6. As for the closed reactor thermolysis of Cold Lake vacuum resid, some volatile liquids escaped during solvent separation with the exception of the one data point at 60 min. Later, an Arab Heavy 975°F resid ($A_0 = 20.2$ wt%) from a different source was also thermally reacted in closed tubing bomb reactors at 400°C. However, when the thermal conversion study on this resid was initiated, a routine procedure of vacuum distilling volatile liquids directly out of tubing bombs prior to solvent separation had been developed. For vacuum resids, this distillation was done at 315°C and 1.4 mmHg using a small laboratory sand bath and a vacuum pump. The distillation enabled obtaining both a good material balance and directly measuring the conversion of volatile liquid products. However, a method needed to be developed to apply a kinetic model developed for the conversion of Arabian Heavy 1050°F+ resid to Arabian Heavy 975°F+ resid. The Arabian Heavy 975°F+ resid contains a fraction, V_0, of vacuum gas oil that boils at atmospheric pressure between 975 and 1050°F and remain as a part of the volatile liquids during thermal conversion. When unconverted heptane solubles from Arabian Heavy 975°F+ resid was vacuum distilled out of a tubing bomb, 5.6 wt% distilled overhead. This represents 4.5 wt% of the whole Arabian Heavy 975°F+ resid because the heptane solubles are the lightest 79.8 wt% of the unreacted resid. Thus, the following transformations in the model were used to calculate the thermal conversion product distributions using the constants in the Series Reaction Kinetic Model determined for Arabian Heavy 1050°F+ resid.

Calculation of 975°F+ Results from the 1050°F+ Model

$$A_0 = \frac{20.2}{1-V_0} = \frac{20.2}{1-0.045} = 21.15 \text{ (value inserted in } 1050^0\text{F model)}$$

$$D_{975} = 0.955 \, V_{1050} + 4.5$$

$$H_{975} = 0.955 \, H_{1050}$$

$$A_{975} = 0.955 \, A_{1050}$$

$$TI_{975} = 0.955 \, TI_{1050}$$

The thermolysis data on both Arab Heavy 1050°F+ and 975°F+ were used to determine the closed reactor solubility limit in fitting asphaltene and coke data. The value of S'_L so determined was

$$S'_L = 0.126$$

This value is less than half of the value of the solubility limit for the open reactor. Again, this indicates that the volatile liquids are much poorer solvents for asphaltene cores than the nonvolatile heptane solubles.

A comparison between the experimental data and values of the reaction fractions calculated by the kinetic model at each of the experimental reaction times is given in Table 4.5. The closed reactor model again gives a good description of the experimental data with only one different parameter than that used in the open reactor kinetic model.

4.4.8.4 Thermolysis of Hondo Vacuum Resid at 400°C

The thermolysis of Hondo 975°F+ ($A_0 = 29.5$ wt%) and its heptane soluble and asphaltene fractions were done in closed tubing bomb reactors at 400°C. As

TABLE 4.5

Comparison of the Kinetic Model with Closed Reactor Data for Arabian Heavy Resids

Arabian Heavy Reactant	Reaction Time (min)	Experimental Data (wt%)				Calculation with Kinetic Model (wt%)			
		Dist.	Heptane Sol.	Asph.	Toluene Insol.	Dist.	Heptane Sol.	Asph.	Toluene Insol.
1050°F+	60	26.9	47.6	21.4	4.2	25.5	48.3	22.3	3.9
	90	—	—	21.2	7.0	34.2	38.6	19.0	8.3
	90	—	—	18.8	10.3	34.2	38.6	19.0	8.3
	120	—	—	17.0	12.1	41.0	30.8	16.6	11.7
975°F+	60	28.1	47.7	22.2	1.7	28.8	47.7	20.9	2.6
	70	34.9	40.2	20.1	3.9	31.8	44.2	19.9	4.0
	120	44.1	31.2	15.7	9.2	43.8	30.4	16.0	9.8
Hept. sol.	60	28.0	60.2	11.3	0.0	29.4	59.5	11.1	0.0
of 975°F+	120	46.4	37.2	14.9	0.5	46.4	37.6	16.0	0.0

discussed in chapter 3, section 3.6, Hondo is an offshore California oil that is high in aliphatic sulfur. In this case, no thermal reactions were done in open reactors so that the six constants in the closed reactor form of the Series Reaction Kinetic Model were determined to fit the closed reactor data as shown in the following table:

Constants in Series Reaction Model at 400°C for Hondo Vacuum Resid

$k_A = 0.0315$ min.$^{-1}$	$k_M = 0.0353$ min.$^{-1}$	$S_L' = 0.144$
$m = 0.263$	$a = 0.640$	$b = 0.315$

Hondo is an extremely thermally reactive resid with the heptane solubles being even more reactive than the asphaltenes. Thus, Hondo reactivity is different than the reactivity of Cold Lake and Arabian Heavy that have the rate constant of asphaltenes and about twice that for heptane solubles. Still, the Series Reaction Kinetic Model gives a good description of the Hondo thermolysis data (Figure 4.34), including describing the very short coke induction period (17.8 min). The aliphatic carbon–sulfur–carbon bond is typically the weakest in resids and, thus, should provide resids their thermal reactivity by initiating free radicals. This conclusion is consistent with the data shown in Table 4.6.

4.4.8.5 Thermolysis of Refinery Blend Atmospheric Resid in a Closed Reactor

The thermolysis of a 650°F+ fraction ($A_0 = 9.2$ wt%) of a refinery blend and its heptane soluble and asphaltene fractions were done in closed tubing bomb reactors at 400°C. The thermal kinetics of atmospheric resids can only be done

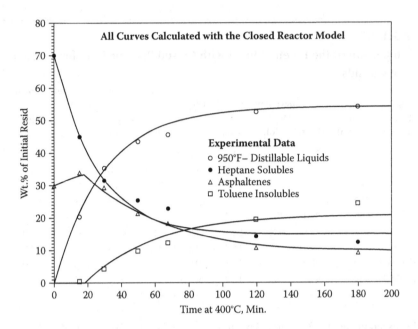

FIGURE 4.34 Description of the weight fraction of four products of the thermolysis of the full Hondo vacuum resid (points) versus reaction time at 400°C in closed-tube reactors by the Series Reaction Kinetic Model (curves).

in closed reactors. Otherwise, in open reactors part of the resid can evaporate without thermal reactions, and one cannot separate vaporization by reaction from evaporation without reaction. Instead, the 650°F⁻ fraction was distilled out each of the tubing bombs after the thermal reaction and before solvent separation. The distillation was done at 215°C and 1.4 mmHg using a laboratory sand bath and a vacuum pump to simulate a 650°F cut-point. The six constants in the closed reactor form of the Series Reaction Kinetic Model were determined to fit the closed reactor data on the refinery blend atmospheric resid, as shown below.

TABLE 4.6
Relation of Rate Constants to Aliphatic Sulfur

Vacuum Resid	XANES and XPS[19,20,21] Aliphatic S (wt%)	400°C Heptane Soluble k (min⁻¹)	400°C Asphaltene k (min⁻¹)
Arab Heavy	1.51	0.00773	0.0204
Cold Lake	1.84	0.0134	0.0260
Hondo	2.94	0.0353	0.0315

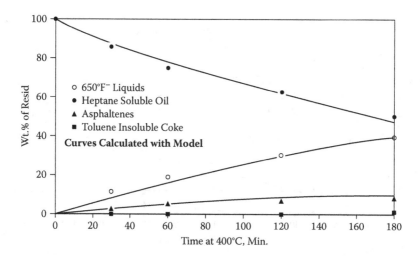

FIGURE 4.35 Description of the weight fraction of four products of the thermolysis of the heptane-soluble fraction of refinery blend atmospheric resid (points) versus reaction time at 400°C in closed-tube reactors by the Series Reaction Kinetic Model (curves).

Constants in Series Reaction Kinetic Model at 400°C for Refinery Blend Atmospheric Resid

$k_A = 0.0321$ min^{-1}	$k_M = 0.00432$ min^{-1}	$S_L' = 0.0966$
$m = 0.262$	$a = 0.716$	$b = 0.209$

As shown in Figure 4.35 to Figure 4.37, the Series Reaction Kinetic Model provides a good description of the thermolysis data of the refinery blend atmospheric resid and its asphaltene and heptane soluble fractions.

4.4.9 HYDROGEN BALANCE CONSTRAINTS ON STOICHIOMETRIC COEFFICIENTS

In this section, as part of a continuing effort to simplify and to improve the kinetic model for resid thermolysis, hydrogen balance constraints on stoichiometric coefficients are added to reduce the number of adjustable parameters. Although pseudocomponent or lumped models are commonly used to describe the reaction kinetics of heavy hydrocarbon streams, the application of elemental balance constraints appears to be the first.[22] This is surprising because for reactions of known compounds it is atomic balance of the elements that determines the stoichiometry. With no limits on the proposed products from a given reactant, lumped models often are merely curve fits of conversion data with little or no predictive power. For example, one or more light hydrocarbon streams are often proposed to be the only products from a heavy aromatic stream. However, by only requiring

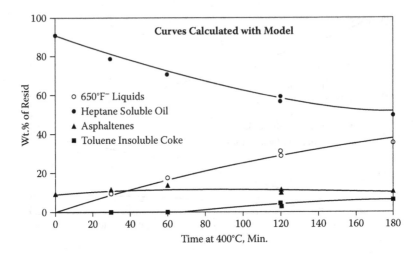

FIGURE 4.36 Description of the weight fraction of four products of the thermolysis of the full refinery blend atmospheric resid (points) versus reaction time at 400°C in closed-tube reactors by the Series Reaction Kinetic Model (curves).

elemental balance, particularly hydrogen balance, it becomes clear that every lighter, higher hydrogen content product requires an even heavier, more aromatic by-product. Unfortunately, in resid conversion there is a limit in the number of elements that can be balanced. Only carbon, hydrogen, and sulfur are of high

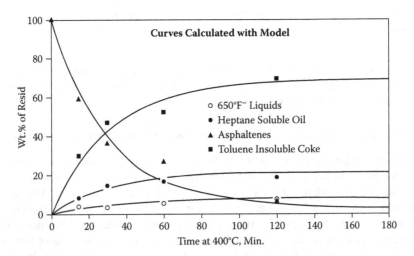

FIGURE 4.37 Description of the weight fraction of four products of the thermolysis of the asphaltene fraction of refinery blend atmospheric resid (points) versus reaction time at 400°C in closed-tube reactors by the Series Reaction Kinetic Model (curves).

enough concentration to be used. However, the carbon content does not vary significantly between reactant and product fractions, and sulfur is very difficult to balance experimentally. The hydrogen sulfide by-product reacts with steel and readily dissolves in both hydrocarbons and water. This leaves only hydrogen balance to reduce the number of constants and to provide a reality check.

4.4.9.1 Evaluation of Parameters

The three reactions in the Series Reaction Kinetic Model are now required to balance hydrogen:

$$M^+ \xrightarrow{\;k_M\;} mA^+ + (1-m)V$$

$$A^+ \xrightarrow{\;k_A\;} aA^* + bM^* + (1-a-b)V$$

$$A^*_{Ex} \xrightarrow{\;\infty\;} TI$$

Hydrogen Balance Equations

$$H_{M^+} = mH_{A^+} + (1-m)H_{VL}$$

$$H_{A^+} = aH_{A^*} + bH_{M^*} + (1-a-b)H_V$$

$$H_{A^*} = H_{TI}$$

Nomenclature

H_{M^+} = hydrogen content of an unreacted heptane soluble.

H_{A^+} = hydrogen content of an unreacted asphaltene.

H_{VL} = hydrogen content of a volatile liquid product.

H_{A^*} = hydrogen content of an asphaltene core.

H_{M^*} = hydrogen content of a heptane-soluble core.

H_V = hydrogen content of a mixture of gas and volatile liquid products.

H_{TI} = hydrogen content of toluene-insoluble coke.

The hydrogen content of the unconverted heptane solubles and the unconverted asphaltenes can be directly measured to give

$$H_{M+} = 10.34 \text{ wt\%} \qquad H_{A+} = 7.94 \text{ wt\%}$$

The hydrogen content of the asphaltene aromatic cores, by the earlier equation, is equal to the hydrogen content of the toluene insoluble coke. However, the hydrogen content of the coke slowly decreases with increasing thermal reaction time. This is because small fragments continue to crack off coke to form hydrocarbon gases. As the hydrocarbon gas mixture has a hydrogen content of about 20 wt%, the reduction of the hydrogen content does not represent a significant reduction in the weight of coke. Therefore, this reaction was ignored and the hydrogen content of the coke was obtained by averaging the data over the coke formed during the first 180 min of reaction time:

$$H_{A*} = H_{TI} = 5.31 \ wt\%$$

The volatile products were not collected, but were measured by weighing the quartz tube reactor before and after the reaction. However, the hydrogen content of the volatiles was calculated by hydrogen balance, using the measured hydrogen contents of the heptane solubles, asphaltenes, and coke fractions. These showed that the volatiles had a hydrogen content that was nearly independent of reaction time, but increased with increasing initial asphaltene concentration. The volatiles are mostly distillable liquids with the remaining being gases of much higher hydrogen content. Therefore, it was reasoned that the reaction of unconverted heptane solubles could be best approximated by forming only volatiles that were liquids, whereas the reaction of unconverted asphaltenes produced volatiles that were a constant ratio of gases to distillable liquids. Previous thermal reactions of Cold Lake vacuum resid in closed reactors (chapter 3) showed that the volatile liquids have a hydrogen content that is independent of reaction time:

$$H_{VL} = 11.6 \ wt\%$$

When all the hydrogen contents in the equation for unconverted heptane solubles are evaluated, the stoichiometric coefficient, m, can be calculated:

$$m = (H_{VL} - H_{M+})/(H_{VL} - H_{A+})$$

$$m = (11.6 - 10.34)/(11.6 - 7.94)$$

$$m = 0.344$$

The hydrogen content of the volatiles formed from unconverted asphaltenes, H_V, and the hydrogen content of maltenes cores, H_{M*}, could not be determined directly. Fortunately, their values are not very sensitive to the details of the kinetic model. Therefore, they could be determined by a few trial-and-error cycles by fitting volatile and hydrogen content data, and substituting in the hydrogen balance for unconverted asphaltene decomposition:

$$H_V = 14.1 \ wt\%$$

$$H_{M*} = 6.40 \ wt\%$$

$$H_{A+} = 7.94 = 5.31a + 6.40b + 14.1(1 - a - b)$$

$$a = 0.703 - 0.876b$$

With the hydrogen balance constraints only two parameters, the stoichiometric coefficient, b, and the solubility limit, S_L, remain to be determined. The value of b was determined to minimize the deviation of the model from the experimental volatiles and maltenes data, and the solubility limit was determined to best split the remaining portion into asphaltenes and coke. The result is as follows:

$$b = 0.110 \qquad a = 0.607$$
$$S_L = 0.400$$

Unlike previously, a temperature independent solubility limit, S_L, was sufficient to describe the asphaltene and coke data.

4.4.9.2 Comparison of Model with Experimental Data

Figure 4.38, Figure 4.40, and Figure 4.42 present the comparison between the experimental data and the model for data at 400°C at initial asphaltene concentrations of 0, 25, and 50 wt%. Figure 4.39, Figure 4.41, and Figure 4.43 show how well the model describes hydrogen content data at the same conditions. Figure 4.44 to Figure 4.47 demonstrate that the model with the same stoichiometric coefficients and solubility limit, determined at 400°C, can describe experimental thermal conversion kinetic data at 370°C and 420°C. Previously, the Series Reaction Kinetic Model described a greater range of experimental data than the original Phase-Separation (Parallel Reaction) Kinetic Model with one less stoichiometric coefficient and a solubility limit that was independent of initial asphaltene concentration, but dependent on temperature. Now, by using hydrogen balance and hydrogen content data, only one of the remaining three

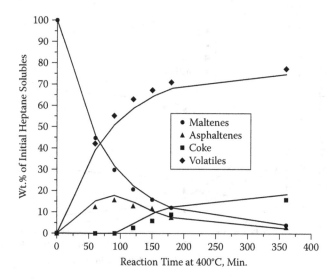

FIGURE 4.38 Comparison of curves calculated by the Series Reaction Kinetic Model constrained by hydrogen balance with experimental open-tube reactor data points at 400°C for the thermal conversion of the heptane-soluble fraction of Cold Lake vacuum resid.

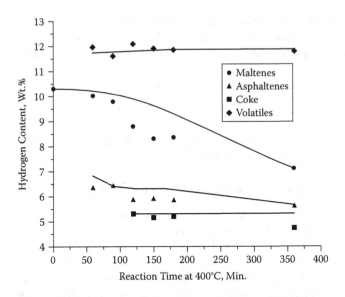

FIGURE 4.39 Comparison of curves calculated by the Series Reaction Kinetic Model constrained by hydrogen balance with experimental hydrogen content data points for the thermal conversion of the heptane-soluble fraction of Cold Lake vacuum resid at 400°C in open-tube reactors.

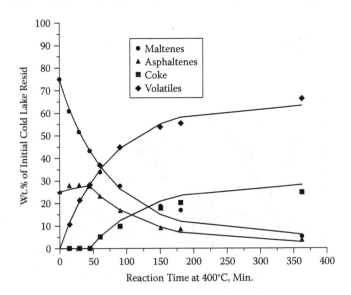

FIGURE 4.40 Comparison of curves calculated by the Series Reaction Kinetic Model constrained by hydrogen balance with experimental open-tube reactor data points at 400°C for the thermal conversion of Cold Lake vacuum resid with 25 wt% asphaltenes.

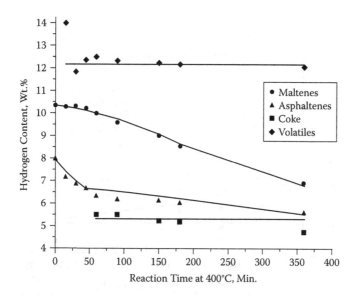

FIGURE 4.41 Comparison of curves calculated by the Series Reaction Kinetic Model constrained by hydrogen balance with experimental hydrogen content data points for the thermal conversion of Cold Lake vacuum resid with 25 wt% asphaltenes at 400°C in open-tube reactors.

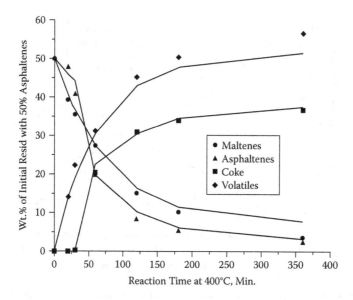

FIGURE 4.42 Comparison of curves calculated by the Series Reaction Kinetic Model constrained by hydrogen balance with experimental open-tube reactor data points at 400°C for the thermal conversion of Cold Lake vacuum resid with 50 wt% asphaltenes.

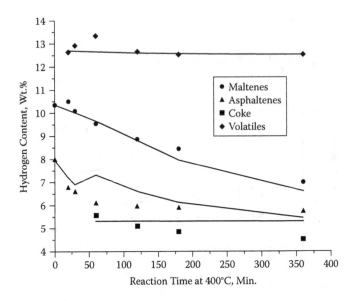

FIGURE 4.43 Comparison of curves calculated by the Series Reaction Kinetic Model constrained by hydrogen balance with experimental hydrogen content data points for the thermal conversion of Cold Lake vacuum resid with 50 wt% asphaltenes at 400°C in open-tube reactors.

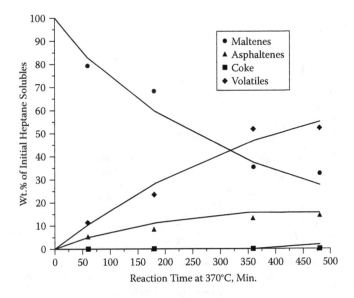

FIGURE 4.44 Comparison of curves calculated by the Series Reaction Kinetic Model constrained by hydrogen balance with experimental open-tube reactor data points at 370°C for the thermal conversion of the heptane-soluble fraction of Cold Lake vacuum resid.

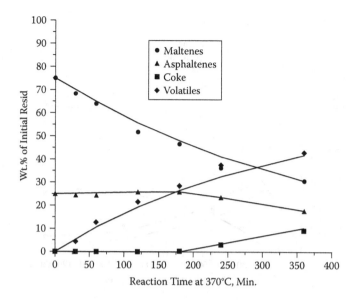

FIGURE 4.45 Comparison of curves calculated by the Series Reaction Kinetic Model constrained by hydrogen balance with experimental open-tube reactor data points at 370°C for the thermal conversion of Cold Lake vacuum resid with 25 wt% asphaltenes.

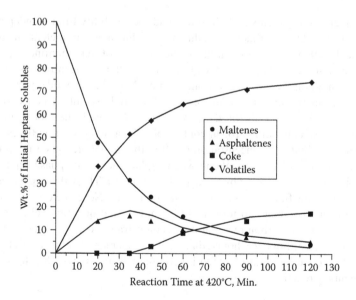

FIGURE 4.46 Comparison of curves calculated by the Series Reaction Kinetic Model constrained by hydrogen balance with experimental open-tube reactor data points at 420°C for the thermal conversion of the heptane-soluble fraction of Cold Lake vacuum resid.

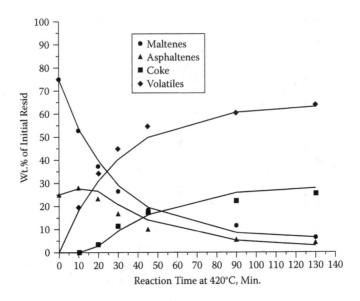

FIGURE 4.47 Comparison of curves calculated by the Series Reaction Kinetic Model constrained by hydrogen balance with experimental open-tube reactor data points at 420°C for the thermal conversion of Cold Lake vacuum resid with 25 wt% asphaltenes.

stoichiometric coefficients are adjustable and the solubility limit is independent of temperature. Despite this large reduction in adjustable parameters, the resulting model still provides a quantitative description of very nonlinear thermal kinetic data over the complete range of initial asphaltene concentrations and over a 50°C temperature range. In addition, the model provides a good description of hydrogen content data for the full resid (25% initial asphaltene concentration) and for the volatiles of all the data. As previously noted, the model does not describe the decrease in hydrogen content of coke at long reaction times. Also, at an initial asphaltene concentration of 50%, the model predicts a jump in the hydrogen content of the asphaltene when asphaltenes cores phase-separate to form coke that is not verified by the experimental data. These discrepancies provide direction for future improvements in the model. Nevertheless, the present significant improvements to form a simple model that quantitatively describes very complex hydrocarbon reaction data demonstrate the value of using hydrogen balance constraints. It is hoped that the use of elemental balance will become expected practice for future lumped kinetic models as it now is for the kinetics of elementary reactions.

4.4.10 SIMPLIFIED KINETIC MODEL

Although the Series Reaction Kinetic Model is not very complex, there are applications that would benefit from a simpler set of equations. These include cases where reaction kinetics is only a small part of a larger, more complex mathematical simulation of a refinery used for optimization. Consider the case of an atmospheric resid that usually has a low concentration of asphaltenes and a much higher rate constant for the asphaltene decomposition than the heptane solubles decomposition. These suggest two approximations:

$$k_A \cong \infty \quad \text{and} \quad b \cong 0$$

This was applied to the data in section 4.4.8.4 on the thermolysis of a refinery blend atmospheric resid ($A_0 = 9.2$ wt%) at 400°C in closed reactors plus additional data on the same resid at 420 and 440°C. The resulting model after applying the approximation and fitting the constants is as follows:

Simplified Kinetic Model

$$M \xrightarrow[\text{1}^{\text{st}}\text{ order}]{k_M} mA^* + (1-m)D$$

$$A^+ \xrightarrow[\text{1}^{\text{st}}\text{ order}]{\infty} aA^* + (1-a)D$$

$$A^*_{\text{soln}} = S'_L (D + M)$$

$$A^*_{\text{insol}} = A^* - A^*_{\text{soln}}$$

$$A^*_{\text{insol}} \xrightarrow{\infty} TI$$

$$M = (100 - A_0) e^{-k_M t}$$

$$D = (1-m)(100 - A_0 - M) + (1-a)A_0 = (1-m)(100-M) + (m-a)A_0$$

The end of the coke induction period occurs when

$$A^* = S'_L (M + D)$$

$$A = A^* = m(100 - A_0 - M) + a A_0 = m(100 - M) + (a - m)A_0$$

$$S'_L (M + D) = S'_L [M + (1 - m)(100 - M) + (m - a)A_0]$$

$$M_{\text{Induction}} = \frac{100m + (a - m)A_0 - S'_L[100 - (1 - m) + A_0(m - a)]}{m (S'_L + 1)}$$

When $M > M_{\text{Induction}}$

$$A = m(100 - M) + (a - m)A_0$$

$$TI = 0$$

When $M < M_{Induction}$

$$A = S'_L (M + D) = S'_L [M + (1 - m)(100 - M) + (m - a)A_0]$$

$$TI = m(100 - M) + (a - m) A_0 - S'_L [M + (1 - m)(100 - M) + (m - a)A_0]$$

$$TI = [m - S'_L (1 - m)][100 - M] - S'_L M + (a - m)(1 + S'_L) A_0$$

Constants in Simplified Kinetic Model for Refinery Blend Atmospheric Resid

$$m = 0.241 \quad a = 0.716 \quad S'_L = 0.135$$

$$k_M = 0.00402 \ \exp\left[\frac{54.8}{0.001987}\left(\frac{1}{673} - \frac{1}{T^0K} \right)\right]$$

at 400°C: $k_M = 0.00402$ min^{-1}

at 420°C: $k_M = 0.0401$ min^{-1}

at 440°C: $k_M = 0.131$ min^{-1}

These equations are certainly simplified. They predict a temperature independent linear relationship between the fractions as was implied by even Cold Lake
vacuum resid thermolysis data in section 4.4.5. Figure 4.48 shows that indeed this
temperature independent linear behavior in terms of heptane solubles versus distillable liquids is a good approximation for the thermolysis of the refinery blend
atmospheric resid. The exception is at short reaction times where the instanta-
neous reaction of initial asphaltenes into asphaltene cores and distillable liquids
is not a good approximation.

Table 4.7 shows that the simplified kinetic model describes the thermolysis
both of the full refinery blend atmospheric resid and of its heptane soluble fraction as well as the Series Reaction Kinetic Model beyond early reaction times.
However, the model also describes the thermolysis of resids with initial asphaltene concentrations as high as 100%. Nevertheless, in applications where it is
desired to predict thermolysis product distributions of the full atmospheric resid
or lower initial asphaltene concentrations, the simplified kinetic model should be
a good approximation.

4.4.11 THERMOLYSIS OF COLD LAKE VACUUM RESID AT 475°C

The kinetic models for resid thermolysis in closed reactors at temperatures of
370 to 440°C did not allow for distillable liquids recracking after being formed
from cracking the resid. The good descriptions of the experimental data by the
kinetic models justify this assumption. In addition, the kinetic models do not
allow for significant hydrocarbon gas formation. Section 4.4.8 showed that in
order to describe hydrogen content data hydrocarbon gases must be a product of
asphaltene and coke thermolysis. However, this amount of gas is never more than
a percent or two. Thermodynamics tells us that if hydrocarbons are kept at high

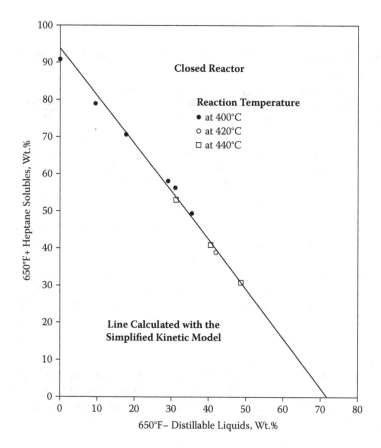

FIGURE 4.48 Comparison of the line calculated with Simplified Kinetic Model with experimental data points of the yield of heptane solubles versus 650°F⁻ distillable liquids for refinery mix atmospheric resid at 400°C in closed-tube reactors.

temperatures long enough all will form only coke and gas. Thus, in this section we determine if 475°C is a high enough temperature to cause secondary cracking of the distillable liquids and to increase the production of coke and gas.

Just two closed tubing bomb runs, at 5 and 15 min, were done at 475°C on Cold Lake vacuum resid. Gas was vented from the tubing bombs prior to distillation, and the amount of gas was determined by the weight lost by the tubing bomb. As is shown in Figure 4.49, the amount of gas formed was quite significant. Because the amount of distillable liquids formed at 15 min was less than that formed at 5 min, we conclude that the secondary cracking of distillable liquids is quite significant at 475°C. By 15 min at 475°C, the asphaltenes disappear and only a few percent heptane solubles remain. Meanwhile, the amount of coke is above 40 wt% and increasing, when at 400 to 420°C the coke never reached 30 wt% at long reaction times. Therefore, we conclude that at elevated temperatures, such as 475°C, secondary

TABLE 4.7

Description of Refinery Blend 650°F⁺ Thermolysis Data by the Simplified Kinetic Model

Initial Asphalt. (wt%)	Reaction Temperature (°C)	Reaction Time (min)	wt% as Calculated by the Model				wt% as Measured Data			
			D	M	A	TI	D	M	A	TI
0	400	30	8.6	88.6	2.7	0.0	11.4	86.1	2.5	0
0	400	60	16.3	78.6	5.2	0.0	19.1	75.6	5.3	0
0	400	120	29.0	61.7	9.2	0.0	30.8	62.3	6.9	0
0	400	180	39.1	48.5	11.8	0.6	39.7	50.4	8.5	1.4
9.2	400	30	10.4	80.5	9.1	0.0	9.4	78.8	11.8	0
9.2	400	60	17.4	71.3	11.3	0.0	17.4	70.5	13.4	1.2
9.2	400	120	29.0	56.0	11.5	3.5	30.6	56.1	10.6	2.7
9.2	400	120	29.0	56.0	11.5	3.5	28.7	58.0	9.7	3.6
9.2	400	180	38.1	44.0	11.1	6.8	35.0	49.0	10.1	5.6
9.2	420	60	40.2	41.3	11.0	7.5	41.4	38.8	9.6	6.5
9.2	440	10	25.4	60.8	11.6	2.1	30.8	52.9	12.8	2.5
9.2	440	20	40.6	40.7	11.0	7.6	40.0	40.8	10.7	6.2
9.2	440	27	48.2	30.8	10.7	10.4	48.1	30.8	10.2	9.7

cracking of distillable liquids to produce coke and hydrocarbon gases is quite significant. This secondary cracking of distillable liquids is especially important for coking processes. One direction that will be discussed in chapter 9 for improving the yield from coking is to minimize this secondary cracking.

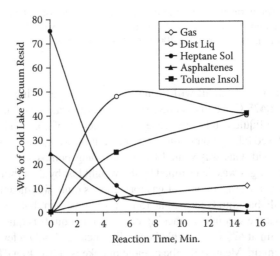

FIGURE 4.49 Thermal cracking of distillable liquid product to gas and coke: At 475°C, significant in closed-tube reactors for Cold Lake vacuum resid.

4.5 TGA KINETICS OF RESIDS

Thermal gravimetric analysis (TGA) is an instrumental method to heat a sample in a small pan by a programmed temperature change when flushing with an inert gas and at the same time continuously weighing the pan. Thus, one may suspect that thermal gravimetric analysis should provide a convenient method to measure coking kinetics. However, attempts[23] to apply this to resids have required suspect kinetic models, such as rate constants that change value with conversion or are second order. On the other hand, isothermal TGA of asphaltenes gives normal first-order reactions.[24] Therefore, the author proposed the hypothesis that under the high flow of inert gas in commercial TGAs at coking temperatures, vacuum resids can partially evaporate without reacting, but the nonvolatile asphaltenes require chemical reactions before volatilizing fragments. Thus, TGA data on vacuum resids cannot be simulated by a purely reaction model. As a result, two colleagues of the author, Bill Olmstead and Howard Freund,[25,26] investigated the TGA of Arabian Heavy 1300°F+, short-path distillation (Distact) bottoms, the less volatile 50% of Arabian Heavy 975°F+. By only following the generation of volatiles, at most one rate constant can be measured. Therefore, the Simplified Kinetic Model is used, despite the short-path distillation bottoms containing a very high fraction of asphaltenes.

$$V = (1-m)(100-M) + (m-a)A_0$$

$$V_\infty = (1-m)(100-A_0) + (1-a)A_0$$

$$\frac{dV}{dt} = -(1-m)\frac{dM}{dt} = (1-m)(100-A_0)\,k_M\,e^{-k_Mt}$$

$$\int_V^{V_\infty} dV = (1-m)(100-A_0)\,k_M \int_t^\infty e^{-k_Mt}dt$$

$$V_\infty - V = (1-m)(100-A_0)e^{-k_Mt}$$

$$\ln\left[1-\frac{V}{V_\infty}\right] = \ln\left[\frac{(1-m)(100-A_0)}{(1-m)(100-A_0)+(1-a)A_0}\right] - k_Mt$$

Thus, if the log of $[100 - 100\,V/V_\infty]$ is plotted against time as in Figure 4.50, the expected linear behavior is obtained after an initial heat-up time to the isothermal temperature of 404°C and the rate constant, k_M, is the negative of the slope. The volatiles at infinite time, V_∞, are easily measured with a TGA as the weight loss obtained after a long time. Therefore, the Simplified Kinetic Model is a good approximation to the TGA data on resid short-path distillation bottoms. However, the rate constant so measured is not necessarily the same as the rate constant for the thermolysis of the full resid. This is not only because the full resid contains more and different molecules, but the TGA is for conversion at a much higher

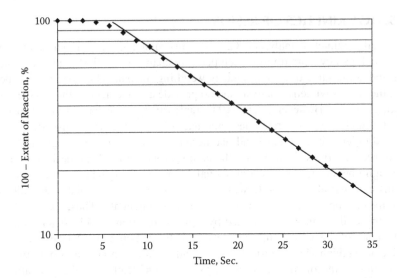

FIGURE 4.50 A first-order kinetic plot for TGA data of Olmstead and Freund[25] collected on Arabian Heavy 696°C⁺ at 513°C.

cut-point. In addition, the TGA data cannot separately determine the two stoichiometric constants, a and m. Nevertheless, the TGA rate constant can be a better approximation for coking processes where the lighter portion of the vacuum resid is evaporated and only the higher boiling fraction is completely converted.

Olmstead and Freund found that rate constants determined by TGA on Arabian Heavy Distact bottoms, using both isothermal and constant heating rate conditions, fit an Arrhenius relationship (Figure 4.51) with a single activation energy of 51.5 kcal/g-mol over more than a 100°C temperature range and three orders of magnitude of rate constants. Similar data on Cold Lake short-path distillation bottoms determined an activation energy of 50.9 kcal/g-mol. These are much more accurate activation energies than those determined from open and closed reactor data because there is much more TGA data and it spans a much greater range of temperatures. Freund and Kelemen[27] have combined TGA data with low temperature data on Green River kerogen to find that first order kinetics with a constant activation energy of 51.3 kcal/g-mole applies from 209 to 500°C and over nine orders of magnitude of rate constants. This has significance in developing models for kerogen maturation[28] in petroleum formation by being able to extrapolate to geological times and temperatures (50 to 160°C). It also provides evidence that a thermal activation energy of about 51 kcal/g-mol may be common for petroleum macromolecules.

Previous studies[29–31] have hypothesized that complex hydrocarbons, such as petroleum resids, must have a distribution of activation energies and rates and, as such, their thermal kinetics should exhibit second order behavior. This chapter has displayed considerable evidence that thermal kinetics of petroleum resids is

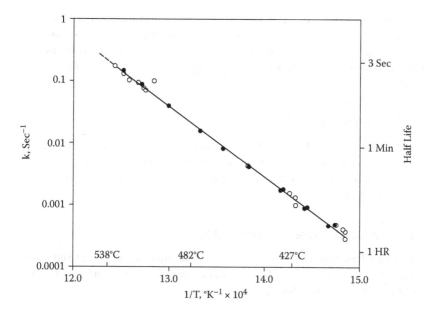

FIGURE 4.51 An Arrhenius plot for the thermolysis of Arabian Heavy 696°C+ using TGA data of Olmstead and Freund[25] giving an activation energy of 51.8 kcal/g-mol. Open points were constant rate TGA data and solid points were isothermal TGA data.

actually first order, and this section shows that a single activation energy applies for a given feed with little variation among feeds. According to the theoretical analysis of Golikeri and Luss,[31] a single activation energy over a wide range of conversions for a lump of species can only happen if the activation energies of the individual species are about equal. Thus, the thermal kinetics of resids is a lot simpler than most have imagined. As one proceeds through the book, this will be a common theme when applied to other properties of petroleum macromolecules.

4.6 COMPARISON WITH WISH LIST

In section 4.1.2, a wish list was given of the desirable attributes of kinetic models for complex hydrocarbons. In this section, the kinetics models in the chapter will be compared with the wish list.

4.6.1 QUANTITATIVELY DESCRIBES KINETIC DATA

There is no doubt that the Series Reaction Model with a phase-separation step quantitatively describes the kinetic data in terms of the weight percent of volatiles, heptane solubles, asphaltenes, and coke versus reaction time over a modest range of temperatures. Evaluation of all but one stoichiometric coefficient by hydrogen balance and having a temperature independent solubility limit reduces the quantitative fit, but that is expected for the large reduction in adjustable parameters.

The simplified kinetic model is even a good approximation in applications where it is desired to predict thermolysis product distributions of an atmospheric resid, or lower initial asphaltene concentrations beyond very early reaction times. In addition, the Simplified Kinetic Model is quantitative for TGA data of Distact distillation bottoms.

4.6.2 Convenient to Use

The Series Reaction Model with a phase-separation step is certainly more convenient to use as compared with Monte Carlo simulations of Klein and Savage.[4] If these equations seem too complex, for many applications the Simplified Kinetic Model may be used.

4.6.3 Applies to Wide Variety of Feeds

The Series Reaction Model with a phase-separation step has been successfully applied to three vacuum resids, Cold Lake, Arabian Heavy, and Hondo, and to one atmospheric resid, refinery mix.

4.6.4 Predicts the Effect of Changing Initial Concentrations

The Series Reaction Model with a phase-separation step has been successfully applied to initial asphaltene concentrations of 0, 25, 50, and 100 for Cold Lake vacuum resid, three initial asphaltene concentrations for Arabian Heavy vacuum resid, and for refinery mix atmospheric resid.

4.6.5 Model Has a Minimum Number of Adjustable Parameters and/or Evaluates Parameters from Characterization Data

It was shown that the stoichiometric coefficients for the Series Reaction Model with a phase-separation step are temperature-independent over 370 to 420°C for Cold Lake vacuum resid and all but one could be evaluated by hydrogen balance. The solubility limit was shown also to be temperature independent over 370 to 420°C. This is close to the minimum number of adjustable parameters. Chapter 5 will provide a method to independently predict the coke induction period and, thus, the solubility limit. Therefore, one independent stoichiometric coefficient remains to be evaluated by some, as yet unknown, characterization method. As rate constants cannot typically be predicted with high accuracy for single compounds, one should not expect to determine rate constants for petroleum macromolecules from characterization data. However, one should never underestimate the simplicity of petroleum resids without trying. At least, all data indicate that the temperature dependency of the rate constants of petroleum macromolecules fit an Arrhenius relationship. The TGA data in Figure 4.51 shows a fit of the Arrhenius relationship over a greater range of temperatures than the reactions of most single compounds.

4.6.6 Predicts the Effect of Changing Reactor Type

It was shown that the Series Reaction Model with a phase-separation step could quantitatively describe the two extremes of reactors: open and closed. It could not predict the closed reactor data from open reactor data because some closed reactor data are needed to evaluate the different value of the solubility limit. However, all the other parameters are the same. As yet there is no way to calculate the yields if the reactor is in between being open or closed. A vapor–liquid equilibrium model would be required to determine the split between heptane solubles in the liquid phase and volatiles in the vapor phase.

4.6.7 Provides Insight for New Innovations

Of course, the Series Reaction Kinetic Model with a phase-separation step can be used as a part of a process simulator to predict conditions when coke is expected to form so that it can be avoided. However, now that it has been proved that coke formation is triggered by a phase-separation step, it is clear that the onset of coke formation can be delayed by improving the solvency of converted asphaltenes. The comparison of open-tube and closed-tube reactors shows that the easiest method to delay the onset of coke is by removing the volatile liquids from the reacting medium because the volatile liquids are often nonsolvents for converted asphaltenes. Later, we will see that this is even more important for hydroconversion than thermal conversion. One also can add a refinery oil to the feed, which acts as a solvent for reacted asphaltenes. However, in the next section, we will find that there are hydrogen donor requirements on such refinery oils. Another innovation that resulted from this kinetic model is the presence of carbonaceous mesophase when coke forms from the phase-separation mechanism. This can be used to detect coke formation in the heavy product during thermal conversion and hydroconversion so that process conditions can be modified before reactors are plugged up with coke. Also, the presence of the carbonaceous mesophase in a foulant can be used to determine the cause of fouling and lead to mitigation methods, as will be discussed in chapter 6.

4.6.8 Predicts Properties of Products

The Series Reaction Kinetic Model with a phase-separation step has not yet been used to predict the properties of resid thermal conversion products. However, it has been shown that this model can describe hydrogen content of the four reaction products as a function of reaction time. In addition, this model was shown to describe the average molecular weight of the asphaltene fraction with reaction time. If the description of the molecular weight could be extended to the other fractions, the combination of hydrogen content and molecular weight could be used to predict many properties of the products. However, prediction of the boiling point distribution of the volatile liquid product would require greatly increasing the number of pseudocomponents.

4.7 NATURAL HYDROGEN DONORS

As was described in section 4.3.4, in order to rationalize why the phase separation of converted asphaltenes triggered the formation of coke, the hypothesis was made that the heptane soluble fraction of resids contained high concentrations of natural hydrogen donors[32] and the asphaltene fraction contained little or no hydrogen donors. Oddly, after it was published in 1993[6] and had been generally accepted, no one but the author demanded proof of this hypothesis. Because many naturally occurring substances, such as cholesterol, are hydrogen donors,[33] the hypothesis of natural hydrogen donors in petroleum is not out of the blue. Nevertheless, the testing of the natural hydrogen donor hypothesis in this section continues the author's practice of testing every element of the kinetic model to determine how the model and the basic concepts need to be improved and revised.

4.7.1 MEASUREMENT OF DONOR HYDROGEN CONCENTRATION

The methods for measuring donor hydrogen concentration and relative reactivity were devised by the author's former colleague, Ken Gould. Because organic chemistry is a field that relies heavily on observations, his accomplishments as a legally blind, organic chemist are remarkable. (His hobby of making his own furniture is another example of his practice of overcoming his challenged sight.) Gould used a chemical dehydrogenation agent, 2-3-dichloro-5,6-dicyano-p-benzoquinone (DDQ), commonly used in organic synthesis,[34] to abstract hydrogen from model compounds and petroleum oils to form the corresponding hydroquinone.

Figure 4.52 shows this reaction of DDQ with tetralin, a common model compound hydrogen donor. However, the procedure was made more complex because DDQ formed strong charge transfer complexes with polynuclear aromatics in the petroleum oils. Therefore, 1 to 2 g of sample was reacted with 3 to 4 g (an excess) of DDQ in 200 mL of dry toluene at reflux temperature for 2 h. To this mixture, 2.78 g (an excess) of 9,10-dihydroanthracene was added and reacted at reflux temperature for an additional 2 h. Thus, the amount of unreacted DDQ could be measured by how much 9,10-dihydroanthracene was dehydrogenated. The product mixture was centrifuged at room temperature to remove solids that might be formed because of the dehydrogenation of petroleum oils. The solid-free solution was analyzed by gas chromatography for 9,10-dihydroanthracene and the

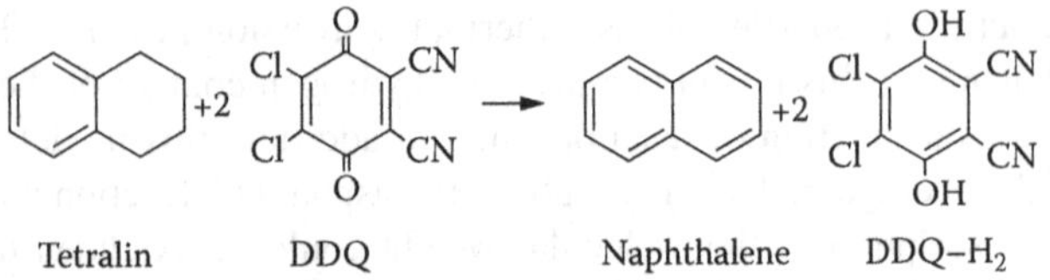

Tetralin   DDQ   Naphthalene   DDQ–H$_2$

FIGURE 4.52 DDQ abstraction of two donor hydrogens per molecule. (From K A Gould and I A Wiehe. *Energy and Fuels* 21: 1199–1204, 2007. Reprinted with permission. Copyright 2007. American Chemical Society.)

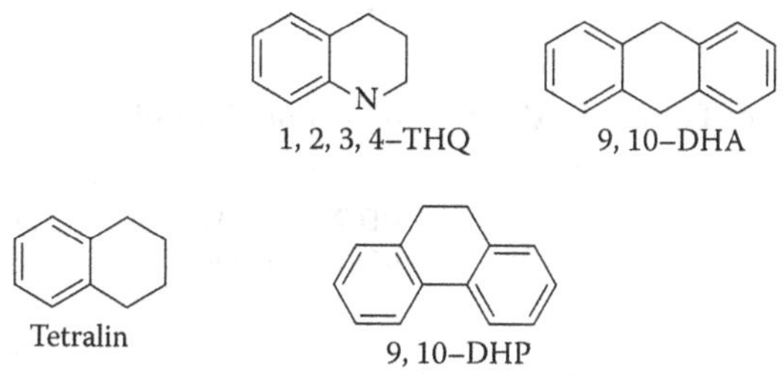

FIGURE 4.53 Chemical structures of model compound hydrogen donors: tetralin; 1,2,3,4-tetrahydroquinoline; 9,10-dihydroanthracene; and 9,10-dihydrophenanthrene. (From KA Gould and IA Wiehe. *Energy and Fuels* 21: 1199–1204, 2007. Reprinted with permission. Copyright 2007. American Chemical Society.)

dehydrogenation product, anthracene. Because the moles of unreacted DDQ equals the moles of anthracene formed and the moles of donatable hydrogen in the feed equals two times the moles of DDQ reacted, the grams of donatable hydrogen per 100 g of sample and the percent hydrogen that is donatable could be calculated. This procedure was successfully checked with the model compound hydrogen donors: tetralin; 1,2,3,4-tetrahydroquinoline; 9,10-dihydroanthracene (using tetralin in the procedure); and 9,10-dihydrophenanthrene (chemical structures in Figure 4.53). Then the procedure was applied to other model compounds, various resids, and the saturate, aromatic, resin, and asphaltene fractions of Arabian Heavy vacuum resid, using the separation procedure described in chapter 2. The results are shown in Table 4.8.

Clearly, the natural hydrogen donor hypothesis in the phase-separation mechanism of coke formation is validated because the six resids in Table 4.8 have higher concentrations of donor hydrogen than model compound hydrogen donors: dihydroanthracene and dihydrophenanthrene. Hondo vacuum resid even has almost as high a concentration of donor hydrogen as tetralin and tetrahydroquinoline, much higher than expected. However, Arabian Heavy asphaltenes have as high a concentration of donor hydrogen as the total Arabian Heavy resid, and Hondo has a concentration of donor hydrogen in its asphaltene fraction only slightly less than its full resid fraction. This high donor hydrogen in the asphaltenes contradicts the phase-separation mechanism of coke formation that assumed that the asphaltene fraction has little or no donor hydrogen. With 25% of the hydrogens in Arabian Heavy asphaltenes and 32% of the hydrogens in Hondo asphaltenes being on saturated rings fused to aromatic rings, those who worry about the average asphaltene chemical structure will probably have to add these to their average asphaltene chemical structure. In comparison, through proton NMR (nuclear magnetic resonance) only 10% of the hydrogens in Arabian Heavy asphaltenes are bonded to the 54% aromatic carbons (^{13}C NMR).

Completely saturated compounds cannot donate hydrogen to free radicals in agreement with DDQ dehydrogenation values of zero for decalin and cholestane

TABLE 4.8

Concentrations of Donor Hydrogen in Various Model Compounds and Resids

Sample	DDQ Donor H (g/100 g Sample)	DDQ Donor H (% of Total H)
Tetralin	3.0	33
1,2,3,4-Tetrahydroquinoline	3.0	33
9,10-Dihydroanthracene	1.1	17
9,10-Dihydrophenanthrene	1.1	17
Cholesterol	1.3	11
Cholestane	0.0	0.0
Deacalin	0.0	0.0
Brent atmospheric resid	1.2	11
Dan-Gorm atmospheric resid	1.6	14
Hondo vacuum resid	2.9	30
Hondo asphaltenes	2.7	32
Arabian Heavy vacuum resid	1.9	19
Arabian Heavy vacuum resid saturates	1.1	9.6
Arabian Heavy vacuum resid aromatics	2.1	19
Arabian Heavy vacuum resid resins	2.0	21
Arabian Heavy vacuum resid asphaltenes	1.9	25
Arabian Light vacuum resid	1.6	16
Brent vacuum resid	1.4	13

Source: KA Gould and IA Wiehe. *Energy and Fuels* 21: 1199–1204, 2007. Reprinted with permission. Copyright 2007. American Chemical Society.

(fully hydrogenation product of cholesterol). However, the saturate fraction of Arabian Heavy contains donor hydrogen, albeit a lower concentration than the other fractions. This is a result of the separation procedure producing a fraction more saturated than the other fractions, but still containing small ring aromatics, as discussed in chapter 2.

4.7.2 MEASUREMENT OF HYDROGEN DONOR RELATIVE REACTIVITY

Ken Gould measured the reactivity of hydrogen donors relative to the hydrogen donor reactivity of tetralin by setting up a direct competitive reaction with DDQ. Thus, the sample (1 to 2 g), an equivalent amount of donor hydrogen of tetralin, and a large excess of dry toluene for complete solubility (50 to 200 mL) were mixed. Sufficient DDQ to abstract 25% of the total donor hydrogen was added and this mixture was reacted at reflux temperature for 2 h. The product mixture was centrifuged at room temperature and the solid-free solution was analyzed by

TABLE 4.9
Hydrogen Donor Relative Reactivities of Various Model Compounds and Resids

Sample	Hydrogen Donor Relative Reactivity
Tetralin	1.0
1,2,3,4-Tetrahydroquinoline	50
9,10-Dihydroanthracene	28
9,10-Dihydrophenanthrene	0.63
Dan-Gorm atmospheric resid	6.0
Hondo vacuum resid	7.9
Arabian Heavy vacuum resid	2.7
Arabian Heavy vacuum resid saturates	1.2
Arabian Heavy vacuum resid aromatics	2.0
Arabian Heavy vacuum resid resins	2.9
Arabian Heavy vacuum resid asphaltenes	4.2
Arabian Light vacuum resid	5.4
Brent vacuum resid	5.7

Source: KA Gould and IA Wiehe. *Energy and Fuels* 21: 1199–1204, 2007.
Reprinted with permission. Copyright 2007. American Chemical Society.

gas chromatography for tetralin and naphthalene. The hydrogen donor reactivity relative to tetralin, R_{HD}, was calculated as follows:

$$R_{HD} = [50 - \% \text{ Conversion of Tetralin}]/[\% \text{ Conversion of Tetralin}]$$

The resulting hydrogen donor relative reactivities are shown in Table 4.9 for many of the model compounds, resids, and resid fractions in Table 4.8. Surprisingly, tetralin is not an ideal hydrogen donor in terms of reactivity. All the resids and resid fractions, and two of the other three model hydrogen donors, are more reactive hydrogen donors than tetralin. Only dihydrophenanthrene is a less reactive hydrogen donor than tetralin of the samples measured. On the other hand, tetrahydroquinoline (THQ) is a superreactive hydrogen donor being 50 times as reactive as tetralin. Hondo vacuum resid with the highest concentration of donor hydrogen of the resids tested also has the highest hydrogen donor reactivity.

The hydrogen donor reactivity of the fractions of Arabian Heavy increases from saturates to aromatics to resins to asphaltenes. This is in the same direction as increasing aromaticity and probably, as increasing ring size of the aromatic fused to the saturated ring with the donor hydrogen. Because the hydrogen donors in asphaltenes are more reactive than those in the other resid fractions, the fast depletion of donor hydrogen in asphaltenes is likely the cause of asphaltenes forming higher molecular weight coke when in a separate phase than the other resid fractions. In addition, when an asphaltene donates hydrogen, the ring size of

TABLE 4.10
Rapid Decrease in H-Donor Concentration of Asphaltenes Due to Thermolysis of Arabian Heavy Vacuum Resid

Time at 400°(min)	DDQ Donor H (g/100 g Sample)	Percent Aromatic Carbon
Unreacted	1.89	54
0[a]	1.78	63
55	1.64	67
90	1.51	71

[a] Taken to temperature and quenched.

Source: KA Gould and IA Wiehe. *Energy and Fuels* 21: 1199–1204, 2007. Reprinted with permission. Copyright 2007. American Chemical Society.

the aromatic core in the asphaltene increases and the possible extent of thermal conversion decreases. Of all the fractions of the resid, we would have preferred asphaltenes to have the least donor hydrogen.

4.7.3 EFFECT OF THERMOLYSIS ON DONOR HYDROGEN OF RESIDS

Arabian Heavy vacuum resid was reacted at various times at 400°C in closed tubing bomb reactors and the asphaltenes were separated. The donor hydrogen concentration was measured on the asphaltene fraction as well as the percent aromatic carbon by ^{13}C NMR. The result is in Table 4.10. Only heating the resid to 400°C and immediately quenching caused the asphaltenes to decrease in hydrogen donor concentration and increase in aromaticity. Further reaction time at 400°C caused the asphaltenes to decrease hydrogen donor concentration rapidly while increasing aromaticity. In contrast, the full Arabian Heavy vacuum resid as in Table 4.11 only slightly decreased the concentration of hydrogen donors after reacting for 55 min at 400°C even though the aromaticity increased. Reacting for

TABLE 4.11
Decrease in Donor Hydrogen Concentration Due to Prolonged Thermolysis of Arabian Heavy Vacuum Resid

Time at 400°(min)	DDQ Donor H (g/100 g Sample)	Percent Aromatic Carbon
Unreacted	1.94	31
55	1.90	38
120	1.70	—
180	1.55	—

Source: KA Gould and IA Wiehe. *Energy and Fuels* 21: 1199–1204, 2007. Reprinted with permission. Copyright 2007. American Chemical Society.

TABLE 4.12

In Situ **Generation of H-Donors during Thermolysis of Arabian Heavy Vacuum Resid Saturates**

Time at 400°(min)	DDQ Donor H (g/100 g Sample)	Percent Aromatic Carbon
Unreacted	1.10	13
100	1.29	18

Source: KA Gould and IA Wiehe. *Energy and Fuels* 21: 1199–1204, 2007. Reprinted with permission. Copyright 2007. American Chemical Society.

longer times caused the full resid to decrease hydrogen donor concentration more rapidly, but not as fast as the asphaltene fraction. The decrease in hydrogen donor concentration of the full resid after 180 min and the asphaltene fraction after 90 min was about the same. Because one fraction of the resid decreased hydrogen donor concentration rapidly, but the full resid barely changed hydrogen donor concentration for the first 55 min, donor hydrogen must have been generated by at least one fraction.

Indeed, the saturates do generate hydrogen donors as the hydrogen donor concentration increases when Arabian Heavy saturates are reacted for 100 min at 400°C while the aromaticity increases (Table 4.12). An example of how this might happen is shown in Figure 4.54. Fully saturated rings are not hydrogen donors. A two-ring saturated structure (2-decalyl phenyl sulfide) is attached to a pendant group by a sulfur bridge that is expected to break thermally and form a free radical at the saturated ring. Donating a hydrogen could form an olefin and donating four more hydrogens would form tetralin that can donate four more hydrogens to form naphthalene. Therefore, the conversion of a substituted saturated ring structure into an aromatic can generate substantial donor hydrogen. This example mechanism was confirmed by thermally reacting 2-decalyl phenyl sulfide in a tubing bomb reactor for 60 min at 400°C and forming 36 wt% tetralin and 5 wt% naphthalene, as analyzed by gas chromatography.

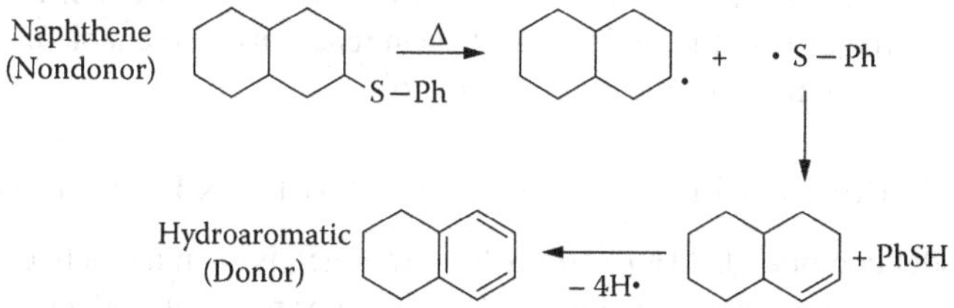

FIGURE 4.54 Mechanism for generating hydrogen donors during thermolysis of 2-decalyl phenyl sulfide. (From KA Gould and IA Wiehe. *Energy and Fuels* 21: 1199–1204, 2007. Reprinted with permission. Copyright 2007. American Chemical Society.)

TABLE 4.13

Generation of H-Donors during Thermolysis of Naphthenic Vacuum Resid

Time at 400°(min)	DDQ Donor H (g/100 g Sample)
Unreacted	1.50
45	1.96
65	1.68
80	1.60

Source: KA Gould and IA Wiehe. *Energy and Fuels* 21: 1199–1204, 2007. Reprinted with permission. Copyright 2007. American Chemical Society.

It is also demonstrated by the thermolysis at 400°C of a vacuum resid of a crude oil used for making naphthenic lube base stocks (Table 4.13). The concentration of donor hydrogen increased at short reaction times and then slowly decreased with further reaction time. However, even after a reaction time of 80 min, the hydrogen donor concentration was still higher than in the unreacted vacuum resid. Therefore, saturated ring structures (naphthenes) in resids can generate donor hydrogen during thermolysis. As a result, the combination of the generation of donor hydrogen by saturates and the high hydrogen donor reactivity of asphaltenes causes the difference between the hydrogen donor concentration in the reacting resid and in the asphaltene fraction to increase rapidly with reaction time. It is this difference in hydrogen donor concentration that is responsible for asphaltenes to increase molecular weight to form coke when in a separate phase, but to decrease molecular weight when soluble in the rest of the reacting resid.

Although the improved understanding of the role of natural hydrogen donors does not require changing the Series Reaction Kinetic Model, it improves our insight in how to delay the onset of coke formation. It is now clear that adding an aromatic oil with better solvency for asphaltenes than the resid is not sufficient. This coprocessing oil also must have at least the concentration of donor hydrogen as the resid. As the resid already contains a high concentration of donor hydrogen and the capacity to generate more, this is not an easy condition to meet. Determining the effect of high donor hydrogen reactivity will enable us to define further the requirements of the coprocessing oil.

4.7.4 EFFECT OF VERY REACTIVE HYDROGEN DONORS ON RESID THERMOLYSIS

As the model compound, THQ (tetrahydroquinoline), is such a reactive hydrogen donor, it was used to determine how hydrogen donor reactivity affects resid thermolysis. Given that the reactivity measurements showed that very little other donor hydrogen reacts in the presence of THQ, little molecular growth reactions or dehydrogenation reactions of the resid is expected. Therefore, equal weight mixtures of the asphaltenes and heptane solubles of Arabian Heavy vacuum resid and THQ were reacted in closed tubing bomb reactors at 400°C by immersing the reactors in a large sand bath at various times, followed by quenching the reactors in water at room temperature. THQ, quinoline, and distillable liquid product were distilled directly out of the tubing bomb. This distillation was done at 315°C and 1.4 mmHg using a small laboratory sand bath and a vacuum pump to simulate a vacuum resid cut-point. The remaining converted resid was separated into toluene insoluble coke, toluene soluble–n-heptane insoluble asphaltenes, and n-heptane solubles. The amount of distillable liquids were determined by subtracting the starting weight of THQ from the total weight of liquids distilled out of the tubing bomb.

The kinetic model for asphaltene conversion without THQ was more straight-forward because asphaltenes started in a separate phase and, therefore, all reacted asphaltenes formed coke and lighter products. The kinetic model for asphaltenes dissolved in THQ is similar to that of asphaltenes in the full resid in that unreacted asphaltenes can form asphaltene cores without forming coke. Also, in both cases, toluene-insoluble coke forms after an induction period, indicating that asphaltene cores are not soluble beyond a certain concentration in the mixture of THQ, quinoline, heptane solubles, and distillable liquids in the closed reactor. The data was not sufficient to determine a solubility limit, so the change in the sum of asphaltenes and toluene insolubles was modeled. Similarly, the formation of distillable liquids was unusual when asphaltenes were reacted in 50% THQ in that none was formed at 60 min, although considerable nonvolatile heptane solubles were formed. In contrast, without THQ, asphaltenes from Arabian Heavy resid form very little nonvolatile heptane solubles at 400°C (section 4.4.7). As a result, the sum of distillable liquids and heptane solubles were modeled. A comparison of the kinetic models for the thermolysis of Arabian Heavy asphaltenes with and without THQ are shown here:

No THQ: $A^+ \rightarrow 0.611$ TI + 0.389 (H + D); rate constant = 0.0204 min^{-1}

50% THQ: $A^+ \rightarrow 0.442$ A* + 0.558 (H + D); rate constant = 0.0103 min^{-1}

where A^+ represents unreacted asphaltenes, A* represents asphaltene cores, TI represents toluene-insoluble coke, H represents nonvolatile heptane solubles, and D represents distillable liquids. Based on the reaction in 50% THQ, the equations for the sum of heptane solubles and distillable liquids, and the sum of asphaltenes and coke are given as follows:

$$H + D = 100 \, [0.558][1 - \exp(-0.0103 \, t)]$$

$$A + TI = A^+ + A^* = 100 \exp(-0.0103 \, t) + 100 \, [0.442][1 - \exp(-0.0103 \, t)]$$

where t is the reaction time in minutes. The stoichiometric coefficient and rate constant were determined by fitting the data in Table 4.14 for the thermolysis of asphaltenes with 50% THQ in closed reactors at 400°C. The stoichiometric coefficient and rate constant for Arabian Heavy asphaltenes without THQ are part of

TABLE 4.14

Comparison of Experimental Data and Calculated Yields (wt%) for Arabian Heavy Asphaltenes Reacted in 50 wt% THQ in Closed Reactors at 400°C

Product	60 Min Data	60 Min Calculated	120 Min Data	120 Min Calculated	240 Min Data	240 Min Calculated
Distillable liquids	0.0	—	16.1	—	23.0	—
Heptane solubles	25.8	—	23.5	—	20.5	—
Dist liq + hept sol	25.8	25.7	39.6	39.6	49.2	51.1
Asphaltenes	74.6	—	53.9	—	37.8	—
Toluene insolubles	0.0	—	3.7	—	13.0	—
Asph + tol insol	74.6	74.3	57.6	60.4	50.8	48.9

Source: KA Gould and IA Wiehe. *Energy and Fuels* 21: 1199–1204, 2007. Reprinted with permission. Copyright 2007. American Chemical Society.

the series reaction kinetic model for Arabian Heavy vacuum resid given previously in section 4.4.7.

Therefore, the superreactive hydrogen donor, THQ, can reduce the fraction of asphaltenes that form asphaltene cores and increase the fraction of lower molecular weight products (heptane solubles and distillable liquids). This is probably a result of more effectively decreasing or eliminating molecular growth of aromatic fragments of cracking, but may include the thermal cracking of saturated rings that would otherwise aromatize. Although, 50% THQ delays the onset of coke formation from asphaltene thermolysis, coke still eventually forms when asphaltene cores become insoluble in the reaction medium. However, these improvements are at the expense of greatly reduced thermal reaction rate constant. Thermal cracking occurs primarily as a result of radical attack, not radical initiation. Therefore, when the superreactive hydrogen donor, THQ, rapidly terminates free radicals, the free radical concentration drops and thermal reactivity suffers.

The heptane-soluble fraction of Arabian Heavy vacuum resid was reacted in 50 wt% THQ for 120 min at 400°C in closed tubing bomb reactors and compared with the same reactant and conditions without THQ. The products were separated into distillable liquids, nonvolatile heptane solubles, asphaltenes, and toluene-insoluble coke with yields as shown in Table 4.15. It is clear that THQ greatly reduced the rate of conversion of heptane solubles because, with the same reaction conditions, 61.5% of the heptane solubles converted without THQ and 22.3% of the heptane solubles converted in the presence of 50% THQ. Of the heptane solubles that were converted, 25.0% formed asphaltenes and coke without THQ and 13.9% formed asphaltenes with 50% THQ. Although the superreactive hydrogen donor did not eliminate the formation of asphaltenes from heptane solubles, it did improve the selectivity to distillable liquids. Therefore, when pendants are cracked off polynuclear aromatic cores in the heptane solubles, they become *n*-heptane insoluble without requiring molecular weight growth reactions to form asphaltenes.

TABLE 4.15

Comparison of Experimental Yields for Arabian Heavy Heptane Solubles Reacted in Closed Reactors at 400°C for 120 Min, Neat and with 50 wt% THQ

Product	Neat Heptane-Soluble Yield (wt%)	50% THQ Yield (wt%)
Distillable liquids	46.4	18.4
Heptane solubles	37.2	77.7
Asphaltenes	14.9	3.1
Toluene insolubles	0.5	0.0

Source: KA Gould and IA Wiehe. *Energy and Fuels* 21: 1199–1204, 2007. Reprinted with permission. Copyright 2007. American Chemical Society.

4.7.5 IMPLICATIONS OF DONOR HYDROGEN IN THE CONVERSION OF RESIDS

It is now clear that natural hydrogen donors are very important in resid conversion. The addition of partially hydrogenated refinery aromatic oils to resids, to increase visbreaker conversion (donor diluent cracking), is well known.[15] The effect of greatly reducing the thermal reaction rate by adding high hydrogen donor concentrations is not well known. As discussed in section 4.6.7, the better solution may be to more effectively strip out the distillable liquid product during conversion. The distillable liquid product of resid thermal conversion is usually a nonsolvent for converted asphaltenes and is low in concentration of donor hydrogen. However, we also learned from the THQ experiment that reactive hydrogen donors do not just delay the onset of coke formation, but reduce the formation of asphaltenes from heptane solubles and reduce the formation of asphaltene cores from asphaltenes in favor of a greater selectivity to distillable liquids. Thus, one would like to add the reactive hydrogen donor slowly during the conversion to maintain a minimum concentration for better selectivity without greatly slowing the reaction rate. The saturate fraction or naphthenic oils do this function to a limited extent by generating donor hydrogen during the conversion. An alternative is to use hydrogen pressure and hydrogenation catalysts as is practiced by hydroconversion processes. This is why hydroconversion can attain much higher conversions of resids than visbreaking (≈30%) without forming coke. More will be discussed on this feature in chapter 10 on hydroconversion.

The ability of reactive hydrogen donors to inhibit thermal cracking reactions can be an advantage as a coprocessing oil when thermal cracking is undesirable. Examples are the furnace coils during vacuum distillation of reactive resids and the heaters of delayed cokers. However, one needs to measure the thermal reaction rate of the coprocessing oil. For instance, although Hondo contains the most reactive hydrogen donors of the resids reported here, we saw in section 4.4.8.3 that Hondo has a very high thermal cracking rate constant because of its high concentration of aliphatic sulfur. Therefore, coprocessing another resid with Hondo vacuum resid would probably increase the thermal cracking rate of the other one.

4.8 ACCOMPLISHMENTS, LIMITATIONS, AND FUTURE DEVELOPMENTS

4.8.1 ACCOMPLISHMENTS

The research described in this chapter set a new standard for the kinetic modeling of complex hydrocarbons. More importantly, kinetic modeling was used as a method to test hypotheses about resid thermal conversion mechanisms and arrive at new concepts for resid conversion. It should be clear to the reader that continuously testing and challenging a model or mechanism to determine its limitations is a good procedure not only to develop a more robust one, but also to acquire insights into making innovative process improvements.

Before this research, lumped kinetic models of complex hydrocarbons were usually parallel reactions from a single reactant to many products, each with a different reaction rate. There was strong suspicion that solubility fractions could not be used to define lumps for a chemical reaction model. It was believed that coke was a direct reaction product, but asphaltenes were relatively unreactive. Hydrogen donors could be added to resids in donor diluent cracking, but they were not present naturally. No one even dreamed of finding liquid crystalline particles in coke from resids. It was proved that reactions of complex hydrocarbons were closer to second order than first order and could not have an activation energy that was constant over more than a short range of temperatures. No one had even tried to input elemental balance in lumped models although every one did it for reactions of single compounds. Few, if any, tested their lumped model by changing the initial concentration or the reactor type from open to closed.

But in this chapter, it was demonstrated that solubility fractions can be used to define lumps, reactions of these lumps are first order, and TGA kinetics of Distact bottoms have activation energies over a greater temperature range than most single compounds. Each element of the phase-separation mechanism for coke formation and the Series Reaction Kinetic Model has been verified or modified to agree with independent evidence. The liquid crystalline coke (carbonaceous mesophase) is the key proof of liquid–liquid phase separation. Indeed, asphaltenes are the most reactive fraction in resids to an asphaltene core reaction limit. Similarly, resids contain high concentrations of reactive donor hydrogen that provides new insight, which has not been completely exploited. The Series Reaction Model was found to describe the kinetics when varying the initial asphaltene concentration and changing the reactor from open to closed. The use of stoichiometric coefficients not only enabled reducing the number of reaction rate constants, but also forced the model to allow for a lighter product by having a heavier by-product. Only then was it possible to evaluate all but one stoichiometric coefficient by hydrogen balance that enabled the solubility limit to be temperature-independent. If this was not simple enough, it was shown under a significant range of conditions that a simpler kinetic model could be used with a single independent reaction rate constant.

The insight gained from this research can lead to innovations in resid processing. Some techniques that have been mentioned include increasing the coke induction period in visbreaking by stripping out the distillable liquid product, or by adding a coprocessing oil that is a better solvent for converted asphaltenes and contains equal or greater concentration of hydrogen donors. Another technique is to use the carbonaceous mesophase in the heavy product of visbreaking or hydroconversion to detect and control coke formation. Reactive hydrogen donors can be used to suppress thermal reactions in furnace coils of vacuum distillation units and in heaters of delayed cokers. Other innovations will be discussed in subsequent chapters. Still others are limited only by the imagination of the reader.

4.8.2 LIMITATIONS AND FUTURE DEVELOPMENTS

The Series Kinetic Model is only the beginning; there is much more to be accomplished. The solubility limit in the kinetic model should be determined independently by solubility data. However, this requires the concepts on solubility of asphaltenes described in the next chapter and thus, will be covered in section 5.3.13. As described in section 4.4.11 the secondary cracking of distillable liquid products ought to be included in the kinetic model in order to apply it to higher temperatures. The kinetic model needs to be improved further to predict the boiling point distribution of the products. In order to accomplish this, one needs to go to a pendant-core model with a distribution of types and sizes of the pendants. High performance liquid chromatography (HPLC) data of the reactants and products will be invaluable to accomplish this. A pseudo free radical model is needed to describe the effect of donor hydrogen and subsequently make the transition to a hydroconversion kinetic model. Perhaps this will require a Monte Carlo simulation.[4] The kinetic model should include aliphatic sulfur as a way to predict kinetic rate constants of various resids without having to measure the reaction kinetics of each resid separately. At least we need to be able to predict the reaction kinetics of mixtures of resids from the kinetics of each resid separately. Of course, the kinetic model should be combined with models for vapor–liquid equilibria, fluid mechanics, and heat and mass transfer to form process models for each of the resid processes.

REFERENCES

1. RZ Magaril, EI Aksenora. Study of the mechanism of coke formation in the cracking of petroleum resins. *Int Chem Eng* 8: 727–729, 1968.
2. RZ Magaril, EI Aksenora. Investigation of the mechanism of coke formation during thermal decomposition of asphaltenes. *Khim Tekhnol Topl Masel* No.7: 22–24, 1970.
3. RZ Magaril, LF Ramazaeva, EI Aksenora. Kinetics of the formation of coke in thermal processing of crude oil. *Int Chem Eng* 11: 250–251, 1971.

4. PE Savage, MT Klein. Asphaltene reaction pathways—V. Chemical and mathematical modeling. *Chem Eng Sci* 44: 393–404, 1989.

5. RC Schucker, RC Keweshan. The reactivity of cold lake asphaltenes. Prep Paper—Am Chem Soc, *Div Fuel Chem* 25: 155–164, 1980.

6. IA Wiehe. A phase separation kinetic model for coke formation. Preprints ACS, *Div Pet Chem* 38: 428–433, 1993.

7. IA Wiehe. A phase-separation kinetic model for coke formation. *Ind Eng Chem Res.* 32: 2447–2454, 1993.

8. J. Sosnowski, DW Turner. J Eng. Upgrading heavy crudes to clean liquid products. Paper presented at 88th AIChE National Meeting, 1980.

9. ME Levinter. Mechanism of coke formation in the cracking of component groups in petroleum residues. *Khim Tekhnol Topl Masel* No. 9: 31–35, 1966.

10. ME Levinter. The mutual effect of group components during coking. *Khim Tekhnol Topl Masel* No. 4: 20–22, 1967.

11. GC Valyavin. Kinetics and mechanism of the macromolecular part of crude oil. *Khim Tekhnol Topl Masel* 8: 8–11, 1979.

12. T Takatsuka. A practical model of thermal cracking of residual oil. *J Chem Eng Japan* 22: 304–310, 1989.

13. TY Yan. Coke formation in visbreaking process. Prepr paper—Am Chem Soc, *Div Pet Chem* 32: 490–495, 1987.

14. T Takatsuka, Y Wada, S. Hirohama, YA Fukui. A prediction model for dry sludge formation in residue hydroconversion. *J Chem Eng Japan* 22: 298–299, 1989.

15. AW Langer. Thermal hydrogenation of crude residua. *Ind Eng Chem* 53: 27–30, 1961.

16. JD Brooks and GH Taylor. The formation of graphitizing carbons from the liquid phase. *Carbon* 2: 185–193, 1965.

17. A Oberlin. Carbonization and graphitization. *Carbon* 22: 521–541, 1982.

18. IA Wiehe. A series reaction, phase separation kinetic model for coke formation from the thermolysis of petroleum resid. AIChE Spring National Meeting, New Orleans, 1996.

19. GN George, ML Gorbaty. Sulfur k-edge x-ray adsorption spectroscopy of petroleum asphaltenes and model compounds. *J Am Chem Soc*, 111: 3182–3186, 1989.

20. SR Kelemen, GN George, ML Gorbaty. Direct determination and quantification of sulfur forms in heavy petroleum and coals. *Fuel*, 69: 939–944, 1990.

21. GN George, ML Gorbaty, SR Kelemen. Sulfur k-edge x-ray adsorption spectroscopy of petroleum asphaltenes and model compounds. In: WL Orr, CM White, Eds., *Geochemistry of Sulfur in Fossil Fuels*, Washington, D.C.: American Chemical Society, 1990, pp. 220–230.

22 IA Wiehe. Hydrogen balance constraints on pseudocomponent kinetic models. Proceedings of the 3rd International Conference on Refining Processing, AIChE Spring National Meeting, Atlanta, 2000, pp. 607–612.

23. RC Schucker. Thermal kinetics of residua by thermogravimetric analysis. preprints ACS, *Div Fuel Chem* 27: 314–320, 1982.

24. RC Schucker. Measurement dependent variations in asphaltene thermal cracking kinetics. ACS, *Div Pet Chem* 28: 683–690, 1983.

25. WN Olmstead, H Freund. Thermal conversion kinetics of petroleum residua. AIChE Spring National Meeting, New Orleans, 1998.

26. IA Wiehe. The Chemistry of Coke Formation from Petroleum: A Tutorial.

27. H Freund, SR Kelemen. Low-temperature pyrolysis of Green River kerogen. *Am Assoc Pet Geologists Bull* 73: 1011–1017, 1989.

28. H Freund. Application of a detailed chemical kinetic model to kerogen maturation. *Energy and Fuels* 6: 318–326, 1992.
29. JA Carberry. *Chemical and Catalytic Reaction Engineering.* New York: McGraw-Hill, 1976, p. 594.
30. D Luss, P Hutchinson. Lumping of mixtures with many parallel n-th order reactions. *Chem Eng J* 2: 172–177, 1971.
31. SV Golikeri, D Luss. Analysis of activation energy of grouped parallel reactions. *AIChE J* 18: 277–281, 1972.
32. KA Gould and IA Wiehe. Natural hydrogen donors in petroleum resids., *Energy and Fuels* 21: 1199–1204, 2007.
33. RT Morrison, RN Boyd. *Organic Chemistry.* Boston: Allyn and Bacon, 1959, p. 811.
34. PP Fu, RG Harvey. Dehydrogenation of polycyclic hydroaromatic compounds. *Chem Rev* 78: 317–361, 1978.

5 Phase Behavior

5.1 INTRODUCTION

This chapter addresses the reasons for the insolubility of asphaltenes, develops a tool, the Oil Compatibility Model, for predicting and avoiding asphaltene insolubility, applies this tool in a number of cases, reveals the limitations of this tool, and provides the direction for future improvements.

5.1.1 TARGET AND APPROACH

In chapter 4, we learned that during thermal conversion, coke formation is triggered by the phase separation of converted asphaltenes. As a result, the insolubility of asphaltenes is a clear limit to the thermal processing of petroleum macromolecules. In this chapter, we will learn that even blending of crude oils can precipitate unconverted asphaltenes and form coke when the insoluble asphaltenes are heated. In addition, during the production of petroleum, asphaltenes can precipitate and plug wells just by the decrease in pressure as the petroleum flows to the surface. Finally, converted asphaltenes that are soluble at conversion conditions can become insoluble on cooling after conversion to foul and plug process equipment downstream of conversion, as will be discussed in chapter 6. Thus, it is clear that to process petroleum macromolecules more effectively one needs to acquire the knowledge and the tools to predict when asphaltenes are soluble and insoluble. Such insight might be expected to lead to innovative ways to avoid asphaltene insolubility and result in step-out improvements in processing of petroleum macromolecules.

Although vapor–liquid equilibria might be included in a chapter on phase behavior as distillation is the most popular separation process in petroleum refining, it is not addressed in this chapter. This topic is well covered elsewhere. Petroleum macromolecules have little tendency to evaporate, and improved understanding offers little promise for process improvements. Instead, we continue the research approach, introduced in the preface, of directing the creation of knowledge along a priority path designed to give the biggest bang for the buck. For the processing of petroleum macromolecules, this approach clearly improves our ability to predict the solubility and insolubility of asphaltenes. Therefore, this chapter is devoted to this topic. In addition, this research was a good fit with the expertise of the author in solution thermodynamics and in polymer science. As will be seen, solution thermodynamics of polymers provides an excellent starting point for understanding the solubility of asphaltenes. In fact, as described in chapter 4, section 4.1.3, the author chose to do research on asphaltene solubility because of its strategic importance and because he was uniquely qualified in his organization to do it. As will be seen, the author follows another approach, described in the preface, which starts with simple models and only adds complexity when required by the experimental data. As in chapters 2 to 4, we will be surprised how well simple models describe petroleum macromolecules.

The phase behavior of waxes is not covered in this chapter. Wax phase behavior is dominated by its crystalline phases. However, as refining and upgrading of petroleum macromolecules for producing fuels is done at temperatures well above the wax melting points, wax phase behavior is not an issue in refining and upgrading for fuels. Conversely, wax deposition can be an important problem during production by plugging well bores and deep-sea pipelines. The ocean floor can be only a few degrees centigrade and often deep-sea pipelines can be many miles long. Therefore, it is important to determine in advance whether a crude oil will have a wax deposition problem. If a wax deposition problem is predicted, the mitigation strategy needs to be designed before the pipeline is built. Such mitigation methods include insulating the pipe, pigging the pipelines, or adding chemicals. Wax is removed from crude oils during the process of manufacturing lube oils by either a low temperature crystallization process or a catalytic process that isomerizes or cracks the wax, or both. Actually, wax is a low-volume, high-value product of refining. Edible (very pure) wax for crayons, for coating milk cartons, apples, etc., is usually the highest-price product of the refinery. Finally, wax can cause a problem during cold starting of diesel engines by plugging diesel fuel filters. Often chemicals are added to diesel fuels to modify wax crystallization to only make small crystals that pass through filters. Thus, the omission of wax phase behavior from this chapter is not done because of its low importance. On the contrary, wax phase behavior is so important and interesting scientifically that it would distract the reader from the emphasis on asphaltene phase behavior and require too much space to be described properly.

5.1.2 Causes of Insolubility

There are only a limited number of causes for insolubility in the liquid phase. In this section, we will discuss those that have been attributed to asphaltenes and other petroleum macromolecules.

5.1.2.1 Hydrogen Bonding or Other Electron Donor–Acceptor Interaction

Hydrogen bonding is when an electronegative atom, such as oxygen or nitrogen, pulls the electron cloud away from the covalent bonded hydrogen. This more positive and small hydrogen atom is attracted, at close approach, to the more negative electronegative atom of a different molecule to form the hydrogen bond. Hydrogen bonding of water is responsible for its insolubility with hydrocarbons, and hydrogen bonding of alcohols is responsible for its insolubility with alkanes.

Many researchers have attributed the insolubility of asphaltenes in paraffins and petroleum oils to hydrogen bonding. However, in section 5.2.3, we will learn that hydrogen bonding is not the principal cause for the insolubility of asphaltenes. On the other hand, hydrogen bonding is the major cause for the insolubility of coal-derived liquids with petroleum-derived liquids.

5.1.2.2 High Molecular Weight

Large molecules, such as organic polymers, have fewer possible arrangements than small molecules because of the covalent bonds between repeating units. This results in much lower entropy of mixing than the random mixing required for ideal solutions. In addition, the interaction energy between large molecules is greater than between small molecules just because of their greater surface of interaction. This results in greater heat of mixing for large molecules as opposed to small molecules. Because the Gibbs free energy of mixing is equal to the heat of mixing minus the absolute temperature times the entropy of mixing, the Gibbs free energy of mixing is greater for large molecules than small molecules and, thus, more susceptible to insolubility. Hence, organic polymers are more insoluble in a range of liquids than low-molecular-weight organic compounds. The most insoluble case is when two different organic polymers are blended, rarely resulting in solubility.

Asphaltenes are of much lower molecular weight than commercial organic polymers, which typically have molecular weights of more than 100,000. However, it will be shown in section 5.3.15 that the moderate molecular weights of asphaltenes contribute to their low solubility.

5.1.2.3 Difference in Dispersion Interactions between Like and Unlike Molecules

Insolubility can be caused when the dispersion interaction energy is much greater between like molecules than unlike molecules. Although this rarely causes insolubility between low-molecular-weight liquids, it can commonly cause a polymer to be insoluble in a low-molecular-weight liquid. We will find that this is the primary reason that large aromatic molecules, such as asphaltenes, are insoluble in *n*-alkanes and in some oils. The polynuclear aromatics in asphaltenes allow for a close parallel approach of large areas of interaction. This results in high dispersion interaction energies between asphaltene molecules, producing low solubility. In addition, the insolubility is promoted because of the asphaltenes' moderate molecular weight.

5.1.2.4 The Solute below Its Crystalline Melting Point

Since solid asphaltenes are amorphous (noncrystalline), crystallinity is not a cause for asphaltene insolubility. However, as mentioned earlier, crystallinity is the cause for insolubility of waxes in petroleum and in petroleum products at room temperature and below.

5.1.2.5 A Component of the Mixture Near or Above Its Critical Point

In this case the near-critical liquid or the gas component (above its critical point) expands much more rapidly than the solute and drops the solute out of solution. This can happen during production of petroleum when the pressure decreases as the oil with dissolved gases, such as methane, is brought to the surface, making asphaltenes precipitate and plug the well. This cause is used to advantage in the solvent recovery of the deasphalting process in refining. Instead of evaporating the solvent, such as propane or butane, the solvent and the deasphalted oil are heated above the critical point of the solvent causing the deasphalted oil to precipitate. As a result, considerable energy that would be required to supply the heat of vaporization is reduced.

5.1.2.6 Polarity

Asphaltenes are often said to be insoluble because of their polarity. This is a misconception. First, asphaltenes are not very polar, only more polar than the other fractions of petroleum. Second, polarity, as measured by dipole moment, is not a very significant influence on solubility (section 5.2.2.1). Electronegative atoms that cause dipole moment cause complex formation of the electron donor-electron acceptor type that can greatly affect solubility. However, even complex formation is a minor influence on asphaltene solubility compared with dispersion interactions (section 5.2.3).

5.2 TWO-DIMENSIONAL SOLUBILITY PARAMETERS

The author was introduced to solubility parameters in his doctoral research by his research advisor, Ed Bagley. Although the author's research at that time was to develop an alternative to solubility parameters for hydrogen bonding liquids, when the author needed to select solvents for polymer processes in his research at Xerox, he turned to solubility parameters. Not finding existing solubility parameter methods to be adequate, he developed the two-dimensional solubility parameter method described in this section. Then, when joining Exxon, he applied two-dimensional solubility parameters to fuel systems, described in section 5.2.3.

5.2.1 Previous Solvent Selection Methods

This section is devoted to the study of the range of solvents and nonsolvents for materials. It is an unusual application of solution thermodynamics, in which the dependency of chemical potential on concentration, temperature, and pressure has received the greatest attention. However, it has applications in the selection of liquids or liquid mixtures for processing materials in solution. In addition, the range of solvents and nonsolvents for a material can be used to elucidate the type and strength of the forces holding the material together. Although the procedure of solvent selection is still a trial-and-error process, one would like to minimize the number of trials. For instance, one would like to use a limited set of solvents and nonsolvents for a material to predict which of the remaining liquids are solvents and which are nonsolvents for that material. However, the major application is to predict what proportions of mixtures of liquids are solvents for a material based on the solubility of that material in pure liquids. It is clear that the purely trial-and-error procedure breaks down for mixtures of liquids because the possibilities are infinite. It will be shown that one cannot even rule out mixtures of liquids that are individually nonsolvents for the material because there are cases in which such mixtures are solvents. What is needed is a guide or a set of rules for making solvent selection. This is one of the applications of the two-dimensional solubility parameter. Often, controlling the "solvent power," the measure of the goodness of a solvent, is desirable. However, there is a scarcity of methods to predict, even qualitatively, whether a given solvent for a material barely dissolves the material or whether it is such a good solvent that only large dilutions with a nonsolvent produces insolubility. This is a second application of the two-dimensional solubility parameter.

Solvent selection science has been developed and applied in the paint and coating industry because of the limited solubility of polymers and because of the many environmental, safety, economic, and performance demands on solvent systems. Most utilize a variation of the solubility parameter from the regular solution theory of Hildebrand and Wood[1] and Scatchard[2]:

$$\delta^2 = \frac{E_v}{V} \tag{5.1}$$

where
δ = solubility parameter (positive root), (cal/cc)$^{1/2}$
E_v = energy of vaporization to an ideal gas, cal/g-mol
V = molar volume, cc/g-mol
E_v/V = cohesive energy density, cal/cc

The regular solution theory was derived for nonpolar liquids and not intended for hydrogen bonding and highly polar liquids. However, the energy of vaporization measures all forms of interaction energy. As a result, Hansen[3-5] divided the energy of vaporization into dispersion, polar, and hydrogen bonding contributions to form the three-dimensional solubility parameter:

$$\frac{E_v}{V} = \frac{E_v^d}{V} + \frac{E_v^p}{V} + \frac{E_v^h}{V} \tag{5.2}$$

$$\delta^2 = \delta_d^2 + \delta_p^2 + \delta_h^2 \tag{5.3}$$

This three-dimensional solubility parameter has found wide application. A strong advantage of Hansen's method is that he used the parameter from the regular solution model, the solubility parameter, but not the regular solution model for the chemical potential itself. Instead, Hansen mapped the solvents in three dimensions for each material using the three solubility parameter components as coordinates This took advantage of the work of Burrell,[6] Lieberman,[7] and Crowley et al.,[8] who made solubility maps with the solubility parameter as one coordinate and various measures of hydrogen bonding or polarity (or both) as other coordinates. Thus, the three-dimensional solubility parameter resulted in a graphical technique for predicting solubility in which the solvent power of each liquid is represented by a point in three-dimensional solubility parameter space, the solubility parameter vector. The solvent power of all mixtures of two liquids is on the line between the two points that represent the two liquids. The lever rule based on volume fraction is used to determine the point on that line for a given ratio of liquids in the mixture.

Even with modern computer graphics, it is not easy to determine if a given point is inside or outside an arbitrary three-dimensional volume. Therefore, Hansen[3-5] approximated the solubility volumes by spheres after multiplying the dispersion solubility parameters by two and, thus, greatly eased this graphical problem. However, as a result, the errors so introduced may have canceled out any advantage of three solubility parameter components over two components.

5.2.2 Development of the Two-Dimensional Solubility Parameter[9]

Two dimensions are much more convenient to use than three dimensions for graphical solutions, such as determining if a mixture of liquids is a solvent or a nonsolvent. Thus, the two-dimensional solubility parameter was developed with the objective of having equal, or better, predictive ability as the three-dimensional solubility parameter.

5.2.2.1 Relative Importance of Polar Interactions

Several early investigators of solubility parameters (Hildebrand and Scott,[10] Small,[11] and Burrell[6]) had already come to the conclusion that dipole interactions could be neglected for most liquids. This encouraged the author to determine if the polar solubility parameter component could be eliminated. Actually, the interaction energy between two permanent dipoles depends on their relative orientation as well as their separation distance, r. If they are lined up in their most favorable orientation, the interaction energy, ε, is given by

$$\varepsilon = -\frac{u_1 u_2}{r^3} \tag{5.4}$$

where u_1 and u_2 are the dipole moments of the two dipoles. However, usually this orientation is greatly disrupted at normal temperatures because of thermal agitation. Including this effect, Keesom[12,13] derived the expression for the average potential energy:

$$\bar{\varepsilon} = -\frac{2u_1^2 u_2^2}{3r^6 kT} \tag{5.5}$$

One can compare this equation with that for the interaction energy due to dispersion forces as predicted by London[14,15]:

$$\varepsilon = -\frac{3\alpha_1\alpha_2 I_1 I_2}{2r^6(I_1 + I_2)} \tag{5.6}$$

where I is the ionization potential and α is the polarizability. In most cases the dispersion interaction energy is much greater than the polar interaction energy. The exceptions include associated liquids, such as ammonia, alcohols, and water, where rotation is strongly hindered and the polar interaction energy is closer to equation (5.4) than equation (5.5).[6] Examination of the di-substituted benzene isomers offers further proof that dipole moment is not a significant parameter in solubility. The three isomers of dichlorobenzene and of dinitrobenzene have only slight differences in the ability to dissolve many solutes even though differences in dipole moment are large.[10] However, the number and type of substituents produce large differences in the ability to dissolve solutes.

The same electronegative atoms that cause dipole moment also provide molecules with the means to form electron donor–electron acceptor interactions. Thus, if an interaction requires a specific orientation of an atom of one molecule with a specific atom of another molecule, the energy of interaction can be large; here it is given the term "complexing interaction." Hydrogen bonds are just one type of electron donor–electron acceptor interaction. Compounds that have electronegative atoms and are spatially balanced, such as 1,4 dioxane or p-dichlorobenzene, have no dipole moment and yet have the capacity to form electron donor–electron acceptor complexes. Otherwise, polar interactions will be a small part of the overall interaction energy and can be grouped with the dispersion interaction that will also have an inverse sixth power dependency on the separation distance [equation (5.5) and equation (5.6)]. As these interactions act over a field and are not destroyed by orientational changes, they are grouped as "field force interactions."

5.2.2.2 Solubility Parameter Components

The two solubility parameter components are now identified as the field force solubility parameter component, δ_f, and the complexing solubility parameter component, δ_c. Obviously, dispersion interaction energy is part of the field force solubility parameter component and hydrogen bonding interaction energy is part of the complexing solubility parameter component. However, the interaction energy between randomized orientations of the molecules, caused by the dipole moment of the molecules, is part of the field force solubility parameter component. Conversely, the interaction energy between an electron-rich atom of one molecule and an electron-deficient atom of another molecule, requiring a specific orientation of the two molecules, is part of the complexing solubility parameter component. Therefore, a separate solubility parameter for interactions between polar molecules is no longer required, reducing the number of solubility parameter components from three to two.

5.2.2.3 Basic Postulates

The two-dimensional solubility parameter formalism is articulated by five postulates. Previous applications of three-dimensional solubility parameter suggested the first four postulates, but the fifth postulate was not used by previous applications of solubility parameters. Let us postulate that

1. The ability of a liquid or a liquid mixture to dissolve any other material is determined by two, and only two, variables:

$$\delta_f \geq 0 \quad \text{and} \quad \delta_c \geq 0$$

   Thus, the solvent power of a liquid or liquid mixture can be represented by a point in (δ_f, δ_c) space.

2. The two variables δ_f and δ_c for each liquid are related to the solubility parameter, δ, by

$$\delta^2 = \frac{E_V}{V} = \delta_f^2 + \delta_c^2 \tag{5.7}$$

   This, plus the addition rule below, allows δ to be represented as a vector with components δ_f and δ_c.

3. Mixtures of N liquids may be represented in (δ_f, δ_c) space using the mixing rules:

$$\delta_f^M = \sum_{i=1}^{N} \varphi_i \delta_{fi} \qquad \delta_c^M = \sum_{i=1}^{N} \varphi_i \delta_{ci} \tag{5.8}$$

   where superscript M refers to the mixture, subscript i refers to liquid i, and ϕ_i is the volume fractions of liquid i. These mixing rules indicate that the solubility parameter components of any mixture of two liquids 1 and 2 lie on a line between points $(\delta_{f1}, \delta_{c1})$, and $(\delta_{f2}, \delta_{c2})$.

4. The probability of a material, A, being dissolved by a given liquid or liquid mixture, B, increases as the absolute differences between their solubility parameter components,

$$|\delta_{fA} - \delta_{fB}| \quad \text{and} \quad |\delta_{cA} - \delta_{cB}|$$

   are decreased.

5. If each of two liquids individually dissolves a material up to a certain concentration, then any mixture of the two liquids also dissolves that material at least up to the given concentration.

Postulate 1 gives one the ability to represent the range of solvents for a given material at a given concentration by an area in (δ_f, δ_c) space.

Postulate 2 recognizes δ_f and δ_c as components of the vector δ. Therefore, if one of the solubility parameter components and its cohesive energy density are known, its point in (δ_f, δ_c) space is determined. Even if the individual solubility parameter components are not known, postulate 2 restricts its location in (δ_f, δ_c) space to a point on the positive quadrant of a circle of radius $|\delta|$ with its center at the origin.

Postulate 3 is a rule for predicting the solvent power for a mixture of liquids. This is the most practical application of solubility parameters because the number of combinations of liquids to form solvents, even at discrete ratios of liquids, is far too many for one to attempt in a given application.

Postulate 4 comes from the basic conclusion of the regular solution theory that solubility is more likely in a mixture as the solubility parameters of the components of the mixture become closer in magnitude. Postulate 4 just extends it to two solubility parameter components. However, the chemical potential derived from the regular solution theory generally does not provide a good description of the experimental solubility data. Although section 5.3.15 will indicate that the regular solution model for chemical potential for petroleum mixtures is actually a good approximation, it is not for most other liquid mixtures. Therefore, in the general case, the solubility or insolubility in a series of test liquids is used to determine the area in (δ_f, δ_c) space that defines solubility for each material. Then the solubility parameter components of the material can be estimated by the components at the center of the solubility area.

Exact matching of the solubility parameter components only assures solubility if the material is a liquid. A material may still be insoluble in a liquid of identical solubility parameter components if the material is a crystalline solid with a high heat of fusion and the temperature is well below the melting point. In addition, highly cross-linked polymers cannot be dissolved by any liquid. However, the liquid may dissolve into the polymer to form a swollen gel, a condition that is more likely achieved by more closely matching the solubility parameter components of the liquid to that of the polymer.

Postulate 5 is based on observations by the author and others as long as the liquids do not chemically react with each other or with the material to form a compound, salt, or complex that can be isolated as a stable species from the mixture. Although postulate 5 still applies with a chemical reaction, one needs to consider the stable species that forms additional components of the mixture and determine their solubility parameter components. A practical significance of postulate 5 is that it enables sharp boundaries of the area of solubility to be defined in (δ_f, δ_c) space. For example, all concave mappings of solubility areas are eliminated because all points on a line between two points known to represent liquids that dissolve a material are also predicted to dissolve the material. Therefore, the combination of postulates 3 and 5 predict solubility areas to be polygon shaped with linear borders between the external points of the areas.

5.2.2.4 Experiments

Solubility tests were run using 32 pure liquids on 38 materials (35 polymers and 3 dyes). A given amount of the material was weighed in a glass vial, a given volume of the liquid or liquid mixture was added, a cap with an aluminum insert was screwed on the top, and the vial was shaken and let stand at room temperature for three to four weeks. After each vial was shaken and checked once a day and no change for a week was noticed, the solubility results were recorded. As crystalline polymers have slower rate of dissolution, they were given at least six weeks before determining their solubility.

Later, it was determined that the time could be greatly reduced in solubility tests by putting them on a shaker rack. Nevertheless, the concentrations used for this work were 0.1 g of polymer per 25 cc of liquid and 0.5 g of dye per 25 cc of liquid.

The solubility results consisted in judging which of three groups to place each liquid–material combination:

Group	Symbol	Condition
1	•	Liquid completely dissolves the material.
2	■	Liquid swells or partially dissolves the material.
3	X	Liquid does not appear to affect the material.

This is all that is required for most applications because one usually wants either a liquid that completely dissolves a material or a liquid that has no noticeable effect on the material. In most cases the classification was obvious. However, in those cases in which there was any doubt, a drop of the mixture was placed between a glass slide and a cover slip and observed at 600× under an optical microscope. If particles could not be resolved (less than 0.5 μ), the material was called soluble.

5.2.2.5 Evaluation of Solubility Parameter Components

Hansen's dispersion solubility parameter components, δ_d, were used for initial estimates of δ_f. Where possible, the overall solubility parameters of Hoy[16] were used because of their high accuracy and consistency to calculate δ_c, using equation (5.7). Then the solubility data of Hansen[3] on 33 polymers and the present data on 38 materials were used to check for consistency in the two-dimensional solubility parameter diagrams for each material. The values of the solubility parameter components for each liquid were adjusted, when required, along their solubility parameter circles to best fit this large amount of data. After several trial-and-error cycles, the assignments grouped all the solvents for each material in an area that, except for a very few exceptions, excluded all nonsolvents. Fortunately, there are several features that confined the trial-and-error procedure. First, the location for each liquid is one point in (δ_f, δ_c) space for all 71 materials. Second, the location of each liquid point is limited to the positive quadrant of a circle, with a radius equal to the magnitude of the solubility parameter. Finally, the alkanes and carbon disulfide with no electronegative atoms or π electrons have zero complexing solubility parameter components. Thus, they are fixed on the δ_f axis with no adjustments. These serve as reference points when evaluating δ_f and δ_c values for the other liquids while defining a solubility area for each material.

The assignment of the solubility parameter components of 56 liquids are shown in Table 5.1. These points remain fixed for all materials. Liquids with similar chemical functionalities are generally grouped together at locations approximately expected for measuring complex and field force interactions. For instance, aromatic and halogenated compounds are low in δ_c and high in δ_f; ketones and esters have moderate δ_c and low δ_f, except aromatic and naphthenic ketones, which

TABLE 5.1
Solubility Parameter Components for Liquids in $(cal/cc)^{1/2}$

Liquid Name	Liquid Symbol	Overall Solubility Parameter	Field Force Solubility Parameter Component	Complexing Solubility Parameter Component
Acetone	Ac	9.62	7.31	6.25
Acetonitrile	AcN	12.11	7.50	9.51
Acetophenone	AcPH	10.58	8.55	6.23
Aniline	An	11.04	8.00	7.61
Benzene	B	9.16	8.95	1.95
1-Bromonaphthalene	BN	10.25	9.70	3.31
n-Butanol	ButA	11.60	9.42	6.78
n-Butyl acetate	BA	8.69	7.43	4.50
Carbon disulfide	CS2	9.92	9.92	0.00
Carbon tetrachloride	CCl4	8.55	8.55	0.25
Chlorobenzene	CB	9.67	9.39	2.30
Chloroform	C	9.16	8.65	3.01
Cyclohexane	CYH	8.19	8.19	0.00
Cyclohexanol	CYHOL	10.95	9.48	5.48
Cyclohexanone	CYHONE	10.42	8.50	6.03
Decalin	D	8.80	8.80	0.00
Dibromomethane	DBM	10.40	9.25	4.74
o-Dichlorobenzene	oDCB	10.04	9.66	2.74
Diethylamine	DEA	8.04	7.40	3.14
Diethylether	DEE	7.62	7.05	2.88
N,N-Dimethylacetamide	DMA	10.80	8.50	6.66
N,N-Dimethylformamide	DMF	11.79	8.52	8.15
Dimethyl sulfoxide	DMSO	12.93	9.00	9.28
1,4 Dioxane	1,4D	10.13	8.00	6.21
Dipropylamine	DPA	7.97	7.16	3.50
Ethanol	EtA	12.78	9.25	8.82
Ethyl acetate	EA	8.91	7.44	4.90
Ethylbenzene	ETB	8.84	8.73	1.40
Ethylene dichloride	EDC	9.86	9.39	3.00
Furan	FUR	9.09	8.82	2.20
n-Heptane	HEP	7.50	7.50	0.00
n-Hexane	HEX	7.27	7.27	0.00
Isophorone	ISOP	9.36	7.78	5.20
Methanol	MtA	14.50	9.04	11.35
Methyl acetate	MA	9.46	7.60	5.87
Methylcyclohexane	MCYH	7.80	7.80	0.00
Methylene chloride	MC	9.88	9.03	4.00
Methyl ethyl ketone	MEK	9.45	7.72	5.45
Nitrobenzene	NITROB	10.62	9.47	4.80

(Continued)

TABLE 5.1 (CONTINUED)
Solubility Parameter Components for Liquids in (cal/cc)$^{1/2}$

Liquid Name	Liquid Symbol	Overall Solubility Parameter	Field Force Solubility Parameter Component	Complexing Solubility Parameter Component
Nitroethane	NE	11.09	7.50	8.17
Nitromethane	NM	12.90	7.95	10.16
2-Nitropropane	2-NP	10.02	7.36	6.80
n-Pentanol	PentA	11.12	9.51	5.76
n-Propanol	PrA	12.18	9.34	7.82
Propylene carbonate	PC	13.30	8.20	10.47
Pyridine	PY	10.62	9.36	5.00
Quinoline	Q	10.80	9.93	4.25
Styrene	STY	9.35	9.13	2.00
Tetrahydrofuran	THF	9.52	8.22	4.80
Tetrahydroquinoline	THQ	11.06	9.60	5.50
Tetralin	Tet	9.50	9.35	1.72
Toluene	T	8.93	8.83	1.30
Trichlorobenzene	TCB	10.45	9.90	3.36
1,1,1-Trichloroethane	111TCE	8.57	8.25	2.33
Trichloroethylene	TCEY	9.16	8.88	2.25
p-Xylene	XYL	8.83	8.82	0.40

Source: IA Wiehe. *Fuel Sci Technol Int* 14: 289–312, 1996. With permission.

have moderate δ_f; and secondary amines and linear ethers have low δ_c and δ_f. The alkanols are located on a line at high δ_c and high δ_f. If water is on this line, it is expected to be located at $\delta_f = 8.12$ and $\delta_c = 22.08$, based on $\delta = 23.53$ at 25°C.[16]

5.2.2.6 Evaluation of Polygon Solubility Areas

The solubility area for each material is the minimum area in (δ_f, δ_c) space that includes all solvents while satisfying the requirements of the postulates, such as any mixture of solvents is also a solvent. These conditions are met by polygon solubility areas formed by connecting the external solvent points with line segments for each material. Figure 5.1 shows an example of a two-dimensional solubility parameter diagram for poly(phenylene oxide) copolymer. The area between the polygon and the nonsolvent points remains uncertain (whether to contain solvents or nonsolvents) until mixtures of solvents and nonsolvents are tested to obtain a sharper definition of the solubility area. When the solubility area extends in a direction not to exclude any nonsolvent points, such as the right side of the solubility area in Figure 5.1, the solubility area cannot be further defined by testing mixtures of solvents and nonsolvents. Even with these uncertain regions, the polygon

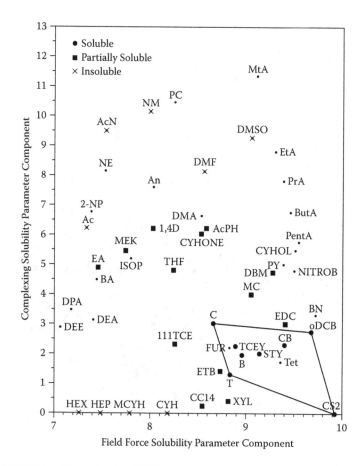

FIGURE 5.1 Two-dimensional solubility parameter diagram for poly(phenylene oxide) copolymer with a polymer concentration of 4 g/L. (From IA Wiehe. *Ind Eng Chem Res* 34: 661–673, 1995. Reprinted with permission. Copyright 1995. American Chemical Society.)

solubility areas usually can be used to predict solubility or insolubility of liquids or liquid mixtures not previously tested. For instance, using Figure 5.1, one can predict that furan and tetralin are solvents for poly(phenylene oxide) copolymer as well as mixtures of nonsolvents from one side of the solubility area with non-solvents on the opposite side, such as ethylene dichloride with ethyl benzene.

5.2.2.7 Mixtures of Solvents and Nonsolvents

By using liquid mixtures, the solubility areas can be better defined and the predictive ability of the areas can be tested. For instance, Figure 5.2, the solubility parameter diagram for poly(*N*-vinyl carbolzole), contains 13 mixtures (of solvents and nonsolvents) that are solvents (denoted by triangles). Three of these mixtures (carbon tetrachloride and carbon disulfide, methyl ethyl ketone and tetrahydrofuran,

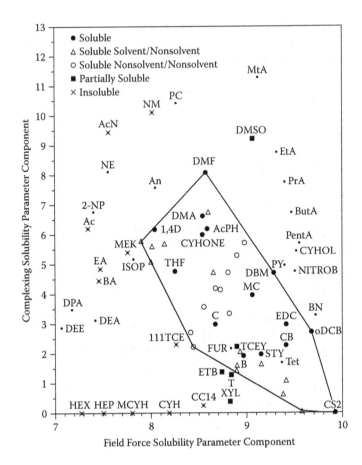

FIGURE 5.2 Two-dimensional solubility parameter diagram for poly(*N*-vinyl carbolzole) with a polymer concentration of 4 g/L. (From IA Wiehe. *Ind Eng Chem Res* 34: 661–673, 1995. Reprinted with permission. Copyright 1995. American Chemical Society.)

and methyl ethyl ketone and 1,4 dioxane) are external points, whereas the other 10 mixtures are correctly predicted by the solubility area to be solvents. Figure 5.3 for a polyamide provides an even better example, in which mixtures were used to better define the solubility region. Only two pure solvents, chloroform and tetrahydrofuran, were found to dissolve this polymer. However, the eight liquid mixtures found to be solvents enabled the solubility line to be extended into a solubility area.

5.2.2.8 Solvents from Mixtures of Nonsolvents

The prediction of solvents from mixtures of nonsolvents is one of the most useful applications of multidimensional solubility parameters. Figure 5.2 contains nine examples for poly(*N*-vinyl carbolzole), as denoted by the open circles, and numerous other cases were determined previously[6] for different materials.

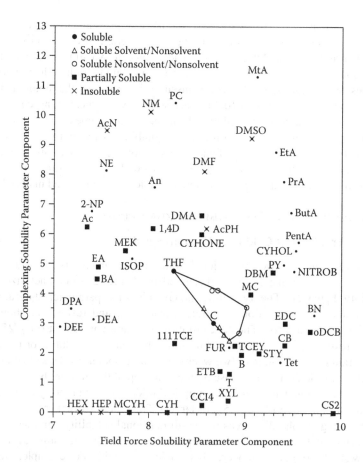

FIGURE 5.3 Two-dimensional solubility parameter diagram for a polyamide with a polymer concentration of 4 g/L. (From IA Wiehe. *Ind Eng Chem Res* 34: 661–673, 1995. Reprinted with permission. Copyright 1995. American Chemical Society.)

5.2.2.9 Degree of Success

As described by Hansen,[3] the three-dimensional solubility parameter method has problems predicting solubility near the borders. However, even discounting points near the border of the spherical solubility volume, the three-dimensional solubility parameter was only correct 95% of the time. On average 2.5% of the liquids lying at a distance greater than the spherical radius plus 0.5 actually dissolved the material even though they were predicted to be nonsolvents. Meanwhile, 2.5% of the liquids at a distance less than the radius minus 0.5 were actually nonsolvents even though they were predicted to be solvents. However, for the present two-dimensional solubility parameter, the overall accuracy in the placement of solvents and nonsolvents is 99.5%, including the data of Hansen[3] and of the author,[9] with no exclusion of points near the border.

There are at least three reasons for the large improvement in accuracy over previous techniques: (1) better accounting of interactions involving electronegative atoms and dipoles, (2) the perceptual advantage of two dimensions over three dimensions, and (3) the use of polygon solubility areas, rather than spherical volumes or circular areas. Because Hansen arbitrarily used the simplest three-dimensional shape, a sphere, to define each solubility volume, the inaccuracies so introduced more than eliminated the advantage of three dimensions.

The application of the two-dimensional solubility parameter to the selection of solvents for the spray drying of encapsulated polymers[9] was an important part of the innovation process. However, this is outside the scope of this book, which is limited to applications to petroleum macromolecules and related materials.

5.2.3 Application to Petroleum Macromolecules[17]

5.2.3.1 Fractions of Cold Lake Vacuum Resid

In this section, the two-dimensional solubility parameter is applied to the fractions of Cold Lake 1050°F+, discussed in chapter 2, to characterize their solubility behavior. The two-dimensional solubility parameter diagrams of the heptane-soluble fractions, saturates, aromatics, and resins, are shown in Figure 5.4 to Figure 5.6. The concentration was kept constant at 0.1 g/25 mL, and the resid fractions were judged to be soluble, partially soluble, or insoluble in each liquid. A minimum of two weeks at room temperature elapsed before the solubility was determined. In case of doubt, a drop of the mixture between glass slides was examined at 600× with an optical microscope. Solubility required no second phase to be resolved (0.5 μ or larger). The sets of solvents so determined defined polygon solubility areas in two-dimensional solubility parameter space as described in the previous section. Figure 5.4 shows that the saturates fraction has low solubility in even moderately complexing liquids. For example, this saturates fraction is insoluble in quinoline, one of the best solvents for asphaltenes. The solubility area of aromatics (Figure 5.5) is the largest among the fractions because it is the fraction of lowest molecular weight and because it has smaller aromatic rings than the other fractions with a high concentration of aromatics. As a result, the aromatics fraction is soluble in liquids from no complexing ability to those, such as aniline and dimethylformamide, with moderately high complexing ability.

All the solvents for resins are solvents for aromatics, but not all the solvents for aromatics are solvents for resins. The solubility of the resins is more limited because of its higher molecular weight and because it contains larger aromatic rings than the aromatics. As a result, acetophenone is the solvent for resins with the highest complexing solubility parameter component, and the lower alkanes, pentane and hexane, are nonsolvents for resins. To check the solubility areas, nonsolvents on one side of the solubility area can be mixed with nonsolvents on the other side to form solvents. Figure 5.6 provides an example in which an equal-volume mixture of the nonsolvents hexane and methyl ethyl ketone is shown to be a solvent for resins.

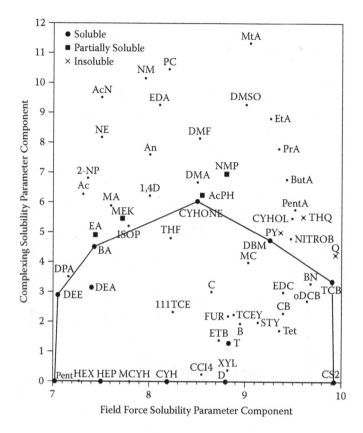

FIGURE 5.4 Two-dimensional solubility parameter diagram for the saturates fraction of Cold Lake vacuum resid at a concentration of 0.1 g/25 mL. (From IA Wiehe. *Fuel Sci Technol Int* 14: 289–312, 1996. With permission.)

Figure 5.7, the solubility parameter diagram for the asphaltenes from Cold Lake vacuum resid, shows that all the solvents for asphaltenes are solvents for resins, but the converse is not true. This is because the asphaltenes are higher in molecular weight and have more and larger aromatic rings than the resins. In fact, this only slightly lowered the maximum complexing solvent slightly from acetophenone to cyclohexanone. The largest difference is that only one liquid with zero-complexing solubility parameter component, carbon disulfide, is a solvent for asphaltenes, whereas all but only two were solvents for resins. Mixtures of nonsolvents were also correctly predicted to be solvents for asphaltenes. Equal-volume mixtures of tetrahydroquinoline with cyclohexane and with decalin are solvents for asphaltenes. Figure 5.7 shows that asphaltenes are insoluble in liquids of low field force solubility parameter component and in liquids of moderate and high complexing solubility parameter component. The large ring aromatics in asphaltenes (50% aromatic carbon) results in a high field force solubility parameter component and a preference for liquids with a low complexing solubility

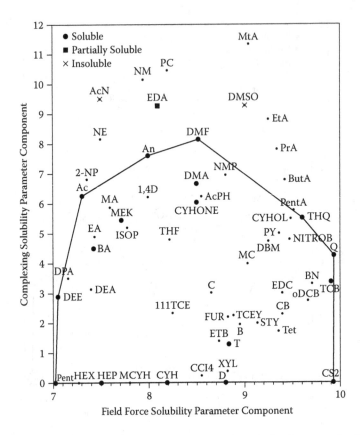

FIGURE 5.5 Two-dimensional solubility parameter diagram for the aromatics fraction of Cold Lake vacuum resid at a concentration of 0.1 g/25 mL. (From IA Wiehe. *Fuel Sci Technol Int* 14: 289–312, 1996. With permission.)

parameter component. On the other hand, it is the 50% aliphatic carbons and the high molecular weight that cause asphaltenes to be insoluble in even moderately complexing liquids.

Although asphaltenes are the least soluble fraction of Cold Lake vacuum resid, as seen in chapter 4, a lower soluble fraction, coke, can be formed by thermolysis. Most coke is insoluble in any solvent. The exception is coke formed by thermolysis of asphaltenes at short reaction times at moderate temperatures, such as at 30 min at 400°C, to form the coke used for the solubility parameter diagram in Figure 5.8. This coke has higher molecular weight, is more aromatic than asphaltenes, and has only three pure liquid solvents: quinoline, trichlorobenzene, and o-dichlorobenzene. In addition, the equal-volume mixture of the nonsolvents pyridine and carbon disulfide is also a solvent. These solvents are also asphaltene solvents and determine the best solvents for the large-ring aromatics in

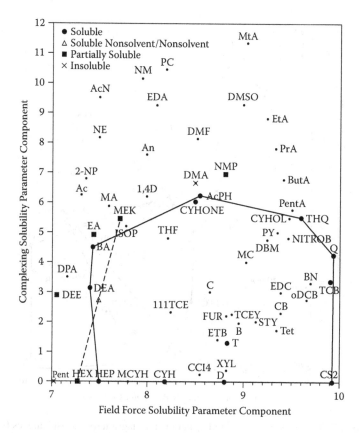

FIGURE 5.6 Two-dimensional solubility parameter diagram for the resins fraction of Cold Lake vacuum resid at a concentration of 0.1 g/25 mL. (From IA Wiehe. *Fuel Sci Technol Int* 14: 289–312, 1996. With permission.)

petroleum. This is a reason *o*-dichlorobenzene was selected for measuring the molecular weight of resid fractions by vapor pressure osmometry, as discussed in chapter 2. None of these coke solvents are hydrocarbons, but they contain nitrogen or chlorine atoms. Apparently, the aromatic interactions in coke are too strong to be broken up by single-ring aromatics alone. However, the addition of two or more chlorine atoms to the ring or a nitrogen atom within an aromatic ring structure increases the complexing ability of the solvent enough to break up aromatic interactions. An additional requirement for a liquid to be a solvent for coke is that the field force solubility parameter needs to be high.

All solvents for coke are solvents for asphaltenes. All solvents for asphaltenes are solvents for resins, and all solvents for resins are solvents for aromatics. Thus, the range of solvents gets larger as the molecular weight and aromaticity of the fractions are decreased. This causes the solubility areas of the more aromatic

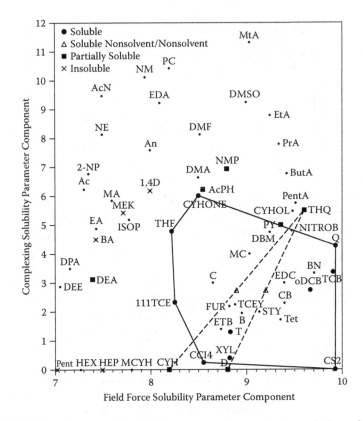

FIGURE 5.7 Two-dimensional solubility parameter diagram for the asphaltenes fraction of Cold Lake vacuum resid at a concentration of 0.1 g/25 mL. (From IA Wiehe. *Fuel Sci Technol Int* 14: 289–312, 1996. With permission.)

fractions to be concentric and demonstrates that the solubility of petroleum fractions is limited by molecular weight and aromaticity. As a result, there are no solvents that can extract, or dissolve, the most aromatic and highest-molecular-weight fraction without also dissolving the rest of the resid or resid reaction product. Thus, asphaltenes cannot be extracted from resid, crude oil, or their reaction products, but can only be separated by precipitation.

The resid fractions have the solubility characteristics of liquids of relatively low complexing ability. Thus, as the molecular weight of the aromatic fractions is increased, the highest complexing solubility parameter component of the solvents decreases to less than 5 $(cal/cc)^{1/2}$. This indicates that the primary interaction energy in petroleum fractions is caused by dispersion forces. In some cases, such as in asphaltenes and coke, these forces can be large because they are between large, flat rings of polynuclear aromatics. Nevertheless, this provides the justification for only using the overall solubility parameter to predict the solubility of resid fractions in purely hydrocarbon liquids. For low-molecular-weight liquids

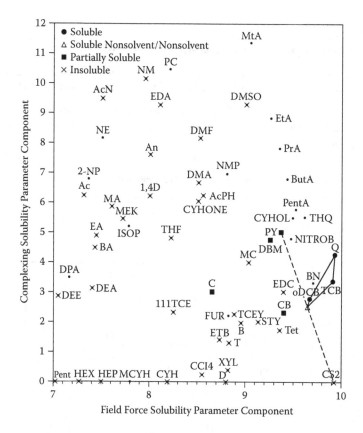

FIGURE 5.8 Two-dimensional solubility parameter diagram for the coke fraction formed from the thermolysis of Cold Lake asphaltenes at a concentration of 0.1 g/25 mL. (From IA Wiehe. *Fuel Sci Technol Int* 14: 289–312, 1996. With permission.)

that contain atoms other than carbon, hydrogen, and sulfur, the two-dimensional solubility parameter is still required.

5.2.3.2 Coal-Derived Liquids

Coal-derived liquids have much higher oxygen content, even in volatile fractions, than petroleum-derived liquids. As a result, coal-derived liquids are often found not to be miscible with petroleum liquids. Table 5.2 shows the properties of a fuel oil (177 to 538°C atmospheric boiling points) and its gas oil fraction (343 to 538°C) produced from the Exxon Donor Solvent Process operating on Wyodak coal. Their solubility parameter diagrams in Figure 5.9 and Figure 5.10 show that these volatile coal-derived fractions meet the definition of coke in that they are insoluble in toluene. Even two of the three best solvents for petroleum resids, trichlorobenzene and *o*-dichlorobenzene, fail to completely dissolve these coal derived fractions.

TABLE 5.2
Chemical Characteristics of Coal-Derived Liquids

Coal Liquid	C (wt%)	H (wt%)	O (wt%)	N (wt%)	S (wt%)	Atomic H/C	THF VPO MW
Wyodak VGO	88.93	7.09	3.33	0.90	0.11	0.950	385
Wyodak fuel oil	89.33	8.32	2.06	0.50	0.31	1.11	280
Hydrogenated Wyodak fuel oil	89.79	9.94	0.67	0.13	0.05	1.32	230

Source: IA Wiehe. *Fuel Sci Technol Int* 14: 289–312, 1996. With permission.

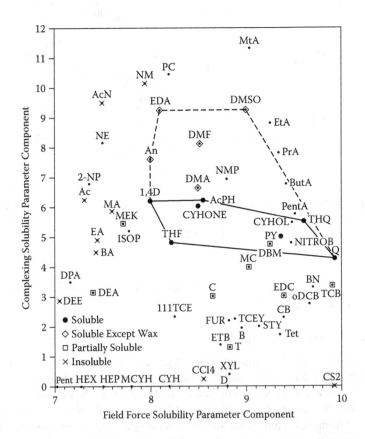

FIGURE 5.9 Two-dimensional solubility parameter diagram for the fuel oil fraction (177 to 538°C) formed from Wyodak coal by the Exxon Donor Solvent process at a concentration of 0.1 g/25 mL. (From IA Wiehe. *Fuel Sci Technol Int* 14: 289–312, 1996. With permission.)

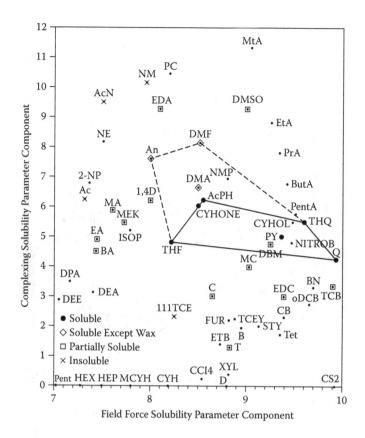

FIGURE 5.10 Two-dimensional solubility parameter diagram for the gas oil fraction (343 to 538°C) formed from Wyodak coal by the Exxon Donor Solvent process at a concentration of 0.1 g/25 mL. (From IA Wiehe. *Fuel Sci Technol Int* 14: 289–312, 1996. With permission.)

In contrast, these coal-derived liquids are soluble in moderately complexing liquids and would be soluble in highly complexing liquids, except for the precipitation of white, saturated waxes. The principal reason for this marked difference in solubility behavior of coal-derived liquids as compared with petroleum liquids is because oxygen functionality, such as phenolic oxygen, in the coal liquids allows for hydrogen bonding between the molecules.

Hydrotreating of Wyodak fuel oil removes two-thirds of the oxygen and reduces the aromaticity (higher H/C [hydrogen-to-carbon] atomic ratio), as shown in Table 5.2. The result is a greatly increased solubility area (Figure 5.11) to include toluene and carbon disulfide as solvents. The hydrotreated Wyodak fuel oil is also completely miscible with petroleum fuel oil. Again, this shows that hydrogen bonding caused by oxygen functionality limits the solubility of coal-derived liquids. This also demonstrates the contrast in solubility behavior of fractions containing

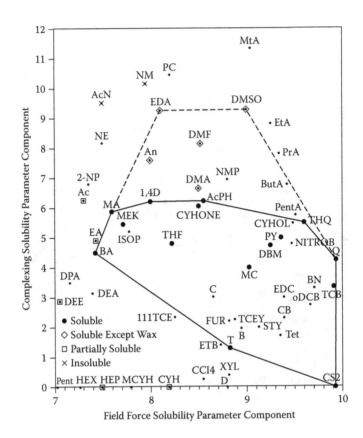

FIGURE 5.11 Two-dimensional solubility parameter diagram for the hydrogenated fuel oil fraction formed from Wyodak coal by the Exxon Donor Solvent process at a concentration of 0.1 g/25 mL. (From IA Wiehe. *Fuel Sci Technol Int* 14: 289–312, 1996. With permission.)

significant hydrogen bonding (coal-derived liquids) from fractions without significant hydrogen bonding (petroleum resid fractions).

5.2.3.3 Discussion of Results and Conclusions

This section shows that solvent mapping with two-dimensional solubility parameters applies as well for petroleum and coal-derived liquids as it does for organic polymers. In both cases, the groups of solvents form polygons in two-dimensional solubility parameter space, and solvents can be formed by mixing nonsolvents on opposite sides of the polygon.

The insolubility of petroleum asphaltenes has often been ascribed to polar and hydrogen bonding interactions. However, by examining the solvents and nonsolvents for petroleum fractions, one concludes that the solubility characteristics are dominated by aromaticity and molecular weight. Thus, solvents of the aromatic fractions form concentric solubility areas on the two-dimensional solubility parameter diagram. As the molecular weight and the aromaticity are increased, the solubility area shrinks to a region of high field force solubility parameter component and low, but nonzero, complexing solubility parameter component. Some complexing ability of a liquid is needed to break up aromatic interactions, but too much prevents the liquid from dissolving paraffinic side chains attached to the aromatics. However, this low range of complexing solubility parameter components of the best solvents, those that dissolve coke, is a strong indication that the dominant interaction between petroleum macromolecules is not hydrogen bonding. On the other hand, hydrogen bonding is the dominant interaction of coal-derived liquids as demonstrated by the high complexing solubility parameter components of solvents for coal-derived liquids. As a result, hydrogen bonding keeps the coal-derived liquids from being soluble in petroleum liquids. If petroleum liquids exhibited extensive hydrogen bonding, they would be compatible with coal-derived liquids. Instead, they are only compatible after much of the oxygen of coal-derived liquids is removed by hydrogenation.

Although the complexing solubility parameter is needed to characterize the solvent power of many liquids containing oxygen, nitrogen, or halogen functionalities, the solvent power of liquids containing only carbon, hydrogen, and sulfur atoms can be approximated with the one-dimensional solubility parameter. As a result, the solubility of petroleum liquids in hydrocarbons can be described by the overall solubility parameter, which is a linear function of the reciprocal of hydrogen content. This allows the solvent-resid phase diagram in chapter 2 to display the distribution of solubility parameters of the fractions of unconverted and converted resids. This again demonstrates that the solubility behavior of petroleum resids is determined by a combination of aromaticity and molecular weight.

5.3   THE OIL COMPATIBILITY MODEL[18]

As concluded in the previous section, the overall solubility parameter is sufficient to describe the solubility of oils in other oils. However, it was about 15 years after the author reached this conclusion that he applied it in a model. This is because there was not a defined need for such a model. Then, the group within Exxon that purchased crude oils wanted a method to predict fouling tendency of crude oils because some refineries turned down some of their purchases of crudes that were rumored to be high fouling. They needed a way to determine fouling tendency based on easily run laboratory tests on the crude so that it can be included in crude purchase decisions. They funded Pat Dennis, a retired Exxon "molecular planner," who asked the author to help after he had already promised the sponsors that he would arrive at a way to measure the "insolubility number" and "solubility blending number" of crude oils. Dennis suspected that crude oil incompatibility, asphaltenes precipitating on blending of different crudes, played a major role.

The author was dubious because the blending of unconverted resids in his laboratory never caused the precipitation of asphaltenes. Although it is well known that visbreaker tars and other processed oils can deposit asphaltenes when blended with paraffinic oils and that some suspected incidents of incompatible crude oils are mentioned in the patent literature,[19] there was no previous direct evidence of incompatible blends of crude oils in the scientific literature. However, the author decided to test the hypothesis by selecting two pairs of crude oils from the Exxon Crude Assay Laboratory that were most likely to be incompatible. These were blended and, amazingly, examination with an optical microscope showed that both pairs precipitated asphaltenes. Once the possibility of crude oil incompatibility was established, it took only about a month to derive and experimentally verify the Oil Compatibility Model because the research 15 years before had paved the way. Consequently, not only was it discovered that incompatible crudes are relatively common, but that the order of mixing is important. This provided the tests to measure the "insolubility number" and "solubility blending number" of crude oils and, based on these, to predict the fouling tendency as sought by those who purchase crude oil. Within six months after the Oil Compatibility Model was developed, it was applied to solve a refinery fouling problem caused by blending crude oils in the wrong order, to solve the plugging of the inlet of a hydrotreater at a second refinery, and to reject the purchase of a self-incompatible crude oil that would have severely fouled a third refinery.

5.3.1 PHYSICAL MODEL OF PETROLEUM

The phase behavior of petroleum is complex because of the large mixture of diverse molecules and because petroleum has some properties of a colloidal dispersion and some properties of a solution. Based on the class separation described in chapter 2, the million or so different molecules in petroleum are simplified to just four types: asphaltenes, resins, small ring aromatics, and saturates. Following the model of Pfeiffer and Saul[20] and, based on x-ray and neutron scattering data in chapter 2, the physical structure of petroleum, as shown in Figure 5.12, was obtained.[21] The largest, most aromatic molecules, the asphaltenes (A), are actually submicroscopic solids at room temperature dispersed in the oil by the resins (R), the next

```
            s s s
          s a a a s
        s a  R R R a s
      s a R A A R a s
      s a R A A R a s
        s a  R R R a s
          s a a a s
            s s s
```

A = Asphaltenes (Solute)
R = Resins (Dispersant)
a = Aromatics (Solvent)
s = Saturates (Nonsolvent)

FIGURE 5.12 More approximate physical model of petroleum. (From IA Wiehe. *Energy and Fuels* 14: 56–59, 2000. Reprinted with permission. Copyright 2000. American Chemical Society.)

```
         R s A
       a s a s R
      aa RaRas
     sRsAARas
     RsaAARsa
      asaRRas
       AsaaR
        s R a
```

A = Asphaltenes (Solute)
R = Resins (Dispersant)
a = Aromatics (Solvent)
s = Saturates (Nonsolvent)

FIGURE 5.13 More realistic physical model of petroleum.

largest, most aromatic group of molecules. This asphaltene–resin dispersion is dissolved into petroleum by small ring aromatics (a), which are solvents, but opposed by saturates (s), which are nonsolvents. Thus, asphaltenes are held in petroleum in a delicate balance, and this balance can be easily upset by adding saturates or by removing resins or aromatics. Because the blending of oils can greatly change the overall concentrations of these molecular types, it can upset this balance and precipitate asphaltenes. Probably, the physical structure shown in Figure 5.13 is closer to reality, with equilibrium between unassociated asphaltenes and asphaltenes associated with other asphaltenes and a separate equilibrium between unassociated resins and resins associated with asphaltenes. The aromatics and saturates are more randomly distributed. However, making the simplifying assumption that the asphaltenes are always associated with each other and all the resins are associated with asphaltenes, the phase behavior is based on solubility and on aromatics–saturates balance. This approximation provides both the power and the limitation of the resulting model. The power is that a relatively simple solubility parameter model is capable of describing a complex system, involving both colloidal dispersion and solubility. However, the limitation is that the predictions will not always be exact, but fortunately, as will be shown, it will err on the conservative side. As a result, a very practical tool was developed with a clear path for improvement in the future.

5.3.2 FLOCCULATION AND OIL SOLUBILITY PARAMETER

The square root definition of the solubility parameter (equation 5.1) results from the regular solution model. As the regular solution model is not used here, the cohesive energy density, the energy of vaporization divided by the molar volume, could be used instead. However, the rule for calculating the solubility parameter for a mixture is well established to be the volumetric average.[5,10,17] Although the solubility parameters at 25°C for pure liquids are well known because the energy of vaporization and the molar volume can be directly measured, they are not for asphaltenes that do not evaporate. The solubility parameter of an unconverted asphaltene has been estimated[17] to be about 9.5 $(cal/mL)^{1/2}$, the highest solubility parameter of any component in unprocessed oil. Nevertheless, what is more important is the solubility parameter at which asphaltenes just begin to precipitate, the flocculation solubility parameter.

The use of flocculation testing of oils by adding various amounts of a solvent, such as toluene, and titrating with a nonsolvent alkane to predict the compatibility of asphalts and of marine fuels that might contain processed oils, such as visbreaker tars, has been common in the petroleum industry.[22,23] However, the compatibility predictions are either empirical or based on the regular solution model. This work takes advantage of flocculation testing, but uses a model with a clear basis without the known limitations of the regular solution model. Although the resulting model also has limitations, the predictions will be conservative for well-understood reasons.

5.3.2.1 Basic Hypothesis

The basic hypothesis of the Oil Compatibility Model is that the flocculation solubility parameter for a given oil is the same whether the oil is blended with noncomplexing liquids or other oils. This hypothesis is supported, first, by the conclusion in section 5.2.3 that the primary interaction energy in petroleum fractions is caused by dispersion forces. Second, Buckley[24,25] observed that the onset of precipitation of asphaltenes occurs over a narrow range of solution refractive index, a measure of dispersion interaction energy.[10] By ordering the liquids that contain only hydrogen, carbon, and sulfur atoms by increasing the solubility parameter, the author has found that all liquids below a certain solubility parameter precipitate asphaltenes and all liquids of a higher solubility parameter dissolve the oil completely at the given concentration (see Ref. 17 and section 5.2.3). After developing the Oil Compatibility Model, the author discovered that this finding does not hold true for chain-type molecules, such as long-chain alkanes. For instance, n-hexadecane is a poorer solvent than n-heptane for oils in dilute mixtures even though n-hexadecane has a higher solubility parameter than n-heptane. The reasons for this will be discusses later in this chapter. Nevertheless, here we will neglect long-chain molecules. Thus, for Souedie crude (Figure 5.14), the flocculation solubility parameter lies between 7.80 (for methylcyclohexane, highest solubility parameter of the nonsolvents) and 8.19 (for cyclohexane, lowest solubility parameter of the solvents). Because solubility parameters for pure liquids only have discrete values, testing the oil with mixtures of solvents and nonsolvents is required to determine the value of the flocculation solubility parameter. However, one needs to account for the solubility parameter of the oil in the mixture to determine the exact flocculation solubility parameter because the oil contributes to the solvency of the mixture. Therefore, one needs to vary the oil to liquid volume ratio and extrapolate the measured flocculation solubility parameter to infinite dilution in oil. As a result, one can actually measure both the flocculation and the oil solubility parameters.

5.3.2.2 Insolubility Number and Solubility Blending Number

In determining asphaltene flocculation solubility parameters, one is focused on only a small portion of the solubility parameter scale. Because asphaltenes are

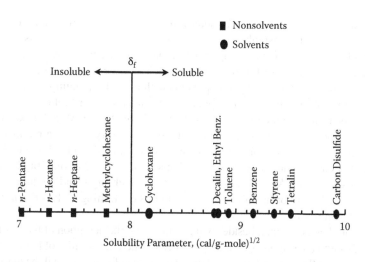

FIGURE 5.14 Solvents and nonsolvents for Souedie crude oil at a concentration of 0.1 g/25 mL of liquid. (From IA Wiehe. *Energy and Fuels* 14: 56–59, 2000. Reprinted with permission. Copyright 2000. American Chemical Society.)

defined as toluene soluble and *n*-heptane insoluble, it is convenient to stretch out this region of the solubility parameter scale by putting the solubility parameters on a scale referenced to *n*-heptane and toluene. It would also be helpful to avoid the cumbersome square root units of solubility parameters by using dimensionless solubility parameters. Therefore, two such dimensionless solubility parameters, the insolubility number, I_N, and the solubility blending number, S_{BN}, are defined as follows:

$$I_N \equiv 100 \frac{(\delta_f - \delta_H)}{(\delta_T - \delta_H)} \tag{5.9}$$

$$S_{BN} \equiv 100 \frac{(\delta_{oil} - \delta_H)}{(\delta_T - \delta_H)} \tag{5.10}$$

where
$\delta_f \equiv$ flocculation solubility parameter
$\delta_H \equiv$ solubility parameter of *n*-heptane
$\delta_T \equiv$ solubility parameter of toluene
$\delta_{oil} \equiv$ solubility parameter of the oil

If the oil is completely soluble in *n*-heptane, and thus, contains no asphaltenes, the insolubility number is assigned a value of 0. As resins can be precipitated out of oils by blending with liquids of solubility parameter below that of *n*-heptane, oils without asphaltenes can have negative insolubility numbers. However, because fouling of oils by the precipitation of resins without asphaltenes is very rare, it almost always is not necessary to determine negative insolubility numbers. This is the reason that the insolubility number for oils without *n*-heptane insolubles is almost always set at 0. On the other hand, if the oil is barely soluble in toluene, the insolubility number is 100. Processed oils, such as slurry oil from fluid catalytic cracking, often contain soluble components with insolubility numbers above 100. Although these are strictly not asphaltenes because they are toluene insoluble, they can be treated in the same way by the Oil Compatibility Model. Likewise, if an oil is as poor a solvent as *n*-heptane, it has a solubility blending number of 0, and if an oil is as good a solvent as toluene, it has a solubility blending number of 100. Both heavy crude oils and processed oils frequently have solubility blending numbers above 100. Although oils with negative solubility blending numbers are uncommon, the author has observed these in oils containing high concentrations of long-chain waxes.

It should be noted that, besides being convenient, the definition of the insolubility number and the solubility blending number satisfied the promise of Pat Dennis to the sponsors of this research that the fouling tendency of crude oils will be measured by two properties: the insolubility number and the solubility blending number.

5.3.3 Mixtures of Oil and Test Liquid

To measure the solubility parameters, one blends the oil with various volume ratios of toluene and *n*-heptane and determines if each dissolves or precipitates asphaltenes. Each mixture of toluene and *n*-heptane is called a "test liquid." As discussed earlier, the ratio of the volume of oil to the volume of test liquid also needs to be varied. The best method to detect insoluble asphaltenes is to observe a drop of the mixture between a cover slip and a glass slide in transmission light with an optical microscope at about 100 to 200×. One does not look for a single particle. Instead, if asphaltenes are insoluble, they are observed throughout the entire view in the form of brownish chain agglomerates, as shown in Figure 5.15.

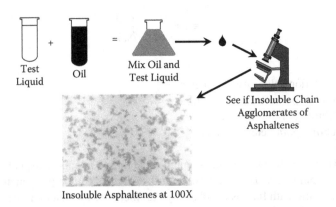

FIGURE 5.15 Schematic of the measurement of flocculation points for oils.

The minimum volume percent toluene in the test liquid required to keep the asphaltenes in solution is determined at various oil-to-test-liquid volume ratios. The solubility parameter of the mixture of oil, toluene, and n-heptane at each of these flocculation points is the flocculation solubility parameter, δ_f, determined with the volumetric mixing rule:

$$\delta_f = \frac{V_T \delta_T + V_H \delta_H + V_{oil} \delta_{oil}}{V_T + V_H + V_{oil}} \tag{5.11}$$

Cross-multiplying, rearranging, and substituting yields

$$\frac{100 \ V_T}{V_{TL}} = I_N + \frac{100 \ V_{oil}}{V_{TL}} \left[\frac{I_N - S_{BN}}{100} \right] \tag{5.12}$$

where V_{TL} is the volume of test liquid (toluene and n-heptane). Therefore, this model predicts that if the minimum percent toluene in the test liquid to keep asphaltenes soluble is plotted against 100 times the volume ratio of oil to test liquid, the data will fall on a line and the y-axis intercept will be equal to the insolubility number. If the x-axis intercept of the line is H_D, then,

$$S_{BN} = I_N \left[1 + \frac{100}{H_D} \right] = I_N \left[1 + \frac{V_H}{V_{oil}} \right]_{V_T = 0} \tag{5.13}$$

By determining the maximum n-heptane that can be added to the oil without precipitating asphaltenes, this x-axis intercept can be evaluated directly by what is called the "heptane dilution test." The volume ratio of n-heptane to oil determined by the heptane dilution test measures the reserve solvency of the oil beyond that required to dissolve the asphaltenes ($S_{BN} > I_N$).

The plot of the minimum percent toluene in the test liquid to keep asphaltenes soluble against 100 times the volume ratio of oil to test liquid has indeed been found to lie on straight lines for crude oils and for processed petroleum oils. Figure 5.16 for Arab Light crude is an example. However, at least two measurements are most commonly done for experimental efficiency. The heptane dilution with 5 mL of oil is usually one test, with the exception of very high viscosity oils, whereas the other test is typically the toluene equivalence test[23] at a concentration of 2 g of oil and 10 mL of the test liquid.

5.3.4 Mixtures of Oils

Once the insolubility numbers and the solubility blending numbers are evaluated for individual oils, they can be applied to the prediction of the compatibility of mixtures of oils. As the mixing rule for solubility parameters is the volumetric average, the solubility blending number of a mixture of oils can be calculated from

$$S_{BNmix} = \frac{V_1 S_{BN1} + V_2 S_{BN2} + V_3 S_{BN3} + \cdots}{V_1 + V_2 + V_3 + \cdots} \tag{5.14}$$

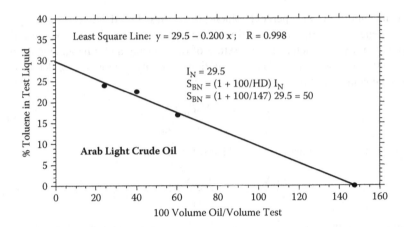

FIGURE 5.16 As predicted, a plot of toluene percentage in the test liquid versus volume ratio of oil to test liquid for Arab Light crude oil falls on a line. (From IA Wiehe. *Energy and Fuels* 14: 56–59, 2000. Reprinted with permission. Copyright 2000. American Chemical Society.)

By the definition of flocculation solubility parameter, if the solubility parameter of a mixture of oils (or oil with noncomplexing liquids) is above the flocculation solubility parameter of each oil in the mixture, the mixture is compatible. Therefore, the maximum insolubility parameter of all the oils in the mixture is the only one required. As a result, for a mixture of oils,

$$\text{Compatibility criterion: } S_{BNmix} > I_{Nmax} \tag{5.15}$$

Because the solubility number of an oil never exceeds the solubility number of its asphaltenes, the theoretically possible incompatibility condition when the solubility parameter of the mixture of oils is much higher than the solubility parameter of the asphaltenes does not happen in practice for oils. As a result, the compatibility criterion (5.15) avoids the need to evaluate the asphaltene solubility parameter and to determine the difference between the solubility parameters of the asphaltenes and the oil that causes incompatibility as required by applications of the regular solution model.

5.3.5 Blending of Souedie and Forties Crudes

Figure 5.17 shows the evaluation of the insolubility numbers and solubility blending numbers for Souedie and Forties crudes from the heptane dilution and toluene equivalence tests. Although the insolubility numbers are the y-axis intercepts, the insolubility numbers can also be calculated from

$$I_N = \frac{TE}{1 - \frac{V_H}{25 d}} \tag{5.16}$$

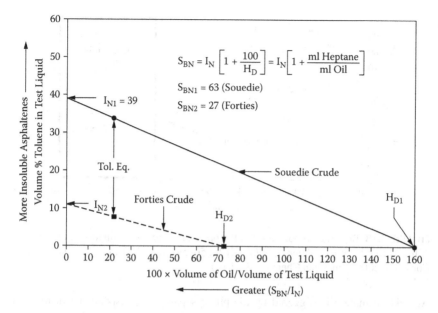

FIGURE 5.17 The evaluation of the compatibility numbers for Souedie and Forties crude oils. (From IA Wiehe. *Energy and Fuels* 14: 56–59, 2000. Reprinted with permission. Copyright 2000. American Chemical Society.)

where

TE = toluene equivalence

V_H = maximum amount (mL) of *n*-heptane that can be blended with 5 mL of oil without precipitating asphaltenes from the heptane dilution test

d = density of the oil, g/mL

The solubility blending number can be calculated from equation (5.13). As the insolubility number of each of these two crude oils is less than its own solubility blending number, each crude is compatible with itself. However, as the solubility blending number of Forties, 27, is less than the insolubility number of Souedie, 39, some mixtures of the two crudes are incompatible.

As shown in Figure 5.18, if one begins with a tank partially filled with Souedie crude and starts adding Forties crude, the blend remains compatible, but the solubility blending number of the mixture decreases until it reaches 39 at 67% Forties, when it equals the insolubility number of Souedie. Thereafter, any additional Forties that is added is predicted and found experimentally to precipitate asphaltenes. Now, if one begins with a tank partially filled with Forties and starts adding Souedie, from the first drop of Souedie until adding 33%, Souedie asphaltenes will continue to precipitate. Although precipitated asphaltenes can be redissolved when the proportions are brought in the compatibility region, it might take days or weeks because petroleum is both a dispersion and a solution. Meanwhile, such a blend has high potential for fouling and coking in the refinery timescale of minutes.

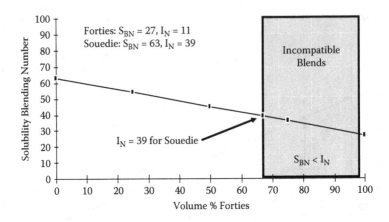

FIGURE 5.18 Both the proportions and the order of blending are important for compatibility. (From IA Wiehe. *Energy and Fuels* 14: 56–59, 2000. Reprinted with permission. Copyright 2000. American Chemical Society.)

As coke formation is triggered by the phase separation of asphaltenes at thermal cracking temperatures (chapter 4), insoluble asphaltenes that hit a metal surface above 350°C almost immediately form coke. Therefore, one should try to blend oils in the sequence of decreasing solubility blending number, from the highest to the lowest.

5.3.6 Refinery Processing of Souedie and Forties Crude Blends

A refinery had run blends of Souedie and Forties crudes for some time without fouling problems. Typically, these crudes were blended by adding Forties crude to a tank of Souedie crude. However, because of scheduling problems and not knowing the consequences, one day the refinery added Souedie crude oil to a tank containing Forties crude oil. As a result, although the crudes were blended in a tank in the correct proportions for compatibility (about equal volume), they were blended in the wrong order. The result was black sludge in the desalter, high fouling of preheat exchangers, and rapid coking of vacuum distillation furnace tubes. Once the problem was analyzed, this refinery continued to mix these crudes, but in the correct order and proportions without unusual fouling or coking. Therefore, blending order is important. One should blend potentially incompatible oils in the order of decreasing solubility blending number. The process of blending oils to ensure compatibility based on the Oil Compatibility Model is the subject of a patent[26] by the author, and is assigned to Exxon (now ExxonMobil).

5.3.7 The P-Test

Long after this research was completed and applied, the similarity to an old Shell proprietary test, the "P-Test,"[27] was pointed out to the author.* As the author

---

*Thanks to Frans van den Berg of Shell for providing the English translation of Ref. 27.

worked for a competitor and the test was not widely published, the derivation of the P-Test was not available to him. This test typically has been applied to residual fuel oils,[28] used to control visbreaker operation, and has been applied to asphalts,[22] vacuum resids used to construct roads after blending with gravel.

5.3.7.1 Derivation of the P-Test

Similar to the Oil Compatibility Model, in the P-Test one blends the oil with a diluent, composed of a mixture of an aromatic hydrocarbon and an aliphatic hydrocarbon, to determine flocculation points. At the flocculation point,

X = ratio of volume of diluent to the volume of oil
FR = flocculation ratio = volume fraction of aromatic hydrocarbon in the diluent

It was reasoned that there is some flocculation ratio, FR_{max}, above which the oil and diluent are compatible in all proportions and below which at some dilution they will precipitate asphaltenes. The peptizability, p_a, of the asphaltenes is defined as follows:

$$p_a = 1 - FR_{max} \qquad (5.17)$$

The peptizing power of the oil, p_o, is defined as the fraction of aromatic hydrocarbon in the diluent that has the same peptizing power as the oil. Finally, the state of peptization, P, of the asphaltenes is defined as

$$P = \frac{p_o}{1 - p_a} = \frac{p_o}{FR_{max}} \qquad (5.18)$$

It is reasoned that all the flocculation points of a given oil have equal peptizing power of the diluted oil medium. Two flocculation points are selected. One is for a diluent of only aliphatic hydrocarbon, $FR = 0$ and $X = X_{min}$. The second flocculation point selected is (FR, X). On the basis of the volume of oil = 1 and the volume of diluent = X, the volume average peptizing power of the oil and diluent at $X = X_{min}$ is

$$\frac{(1)p_o + X_{min}(0)}{1 + X_{min}} = \frac{p_o}{1 + X_{min}} \qquad (5.19)$$

and at X it is

$$\frac{(1)p_o + X\ (FR)}{1 + X} = \frac{p_o + X\ FR}{1 + X} \qquad (5.20)$$

Equating and solving for FR,

$$FR = \frac{p_o}{X_{min} + 1}\left[1 - \frac{X_{min}}{X}\right] \qquad (5.21)$$

At infinite dilution, FR approaches FR_{max} and thus,

$$FR_{max} = \frac{P_o}{X_{min} + 1} \tag{5.22}$$

$$FR = FR_{max}\left[1 - \frac{X_{min}}{X}\right] \tag{5.23}$$

This indicates that if one plots FR versus $1/X$, one obtains a line with a y-axis intercept of FR_{max} and an x-axis intercept of $1/X_{min}$. Finally, P may be calculated:

$$P = \frac{P_o}{FR_{max}} = \frac{FR_{max}[X_{min} + 1]}{FR_{max}} \tag{5.24}$$

$$P = X_{min} + 1 \tag{5.25}$$

The criterion to keep the asphaltenes peptized is for P to be greater than 1 (Ref. 27 incorrectly allows P = 1). To avoid being close to precipitating asphaltenes, it is more common to operate visbreakers so that P > 1.1.

5.3.7.2   Comparison of the P-Test and the Oil Compatibility Model for Single Oils

There are clearly a lot of similarities between the P-Test and the Oil Compatibility Model. If the aromatic hydrocarbon in the P-Test is toluene and the aliphatic hydrocarbon is n-heptane, then both techniques involve plotting the volume percent toluene in the test liquid or diluent (FR is defined as volume fraction, but sometimes plotted as volume percent) versus the ratio of the volume of oil to the volume of test liquid (the Oil Compatibility Model uses 100 times this ratio). The y-axis intercept is

$$I_N = 100\, FR_{max} \tag{5.26}$$

The x-axis intercept is

$$HD = 100/X_{min} \tag{5.27}$$

The P-Test assumes all flocculation points have equal peptizing power, whereas the Oil Compatibility Model assumes all flocculation points are at equal solubility parameter of the mixture. Both the peptizing power and the solubility parameter of a mixture are assumed to be the volumetric average of that of the components in the mixture. The value of P in terms of the Oil Compatibility Model is

$$P = X_{min} + 1 = 100/HD + 1 \tag{5.28}$$

From equation (5.13),

$$P = \frac{S_{BN}}{I_N} \quad \text{and} \quad p_o = \frac{S_{BN}}{100} \tag{5.29}$$

Thus, both agree that an oil is compatible with itself if $P > 1$ or $S_{BN} > I_N$. However, this is obvious because if one does not need to add any n-heptane to get an oil to precipitate asphaltenes, then clearly the oil is incompatible with itself. As the P-Test is designed for resids, the viscosity is often too high for the heptane dilution test, and one then needs to extrapolate the line to get $1/X_{min}$. In that case one should obtain positive values of $1/X_{min}$ (thus, of X_{min}) and values of $P > 1$ if the oil contained no insoluble asphaltenes. As will be described in section 5.3.9, if the oil contains insoluble asphaltenes, the flocculation points will lie on a line parallel to the x-axis so that $1/X_{min}$ will be infinite, X_{min} will be zero, and $P = 1$. Ref. 27 also allows for negative values of $1/X_{min}$ and, thus, $P < 1$. However, this is not allowed at thermodynamic equilibrium and shows that during the flocculation test the asphaltenes were not completely dissolved before adding the nonsolvent. This will be explained in section 5.3.10.

5.3.7.3   Comparison of the P-Test and the Oil Compatibility Model for Oil Mixtures

One significant difference between the P-Test and the Oil Compatibility Model is that the Oil Compatibility Model logically enables conservative prediction of the compatibility of mixtures of oils after making measurements on the individual oils, whereas the P-Test does not. In addition, the Oil Compatibility Model was the first to be applied to the compatibility of crude oils and, as will be shown in section 5.3.12, was the first to be applied to oils without asphaltenes. It is just this predictive ability for mixtures that is the greatest use of the Oil Compatibility Model. The P value does tell one if the oil one tested is compatible with itself or not and, if compatible, how far it is away from incompatibility (reserve solvency). However, that is the present state of the oil and only requires one flocculation test, the heptane dilution test. It does not naturally predict a future event, such as when the oil is blended with another oil. At the end of the original paper on the P-Test,[27] a method for predicting the compatibility of oils is promised. They did arrive at the value of p_o for a mixture to be the volume average, equivalent to the volume average solubility blending number in the Oil Compatibility Model. Although there is no logical mixing rule for p_a (equivalent to $1 - I_N/100$), they[27] empirically selected the weight average. This is equivalent to using a weight average insolubility number for a mixture, rather than the largest insolubility number of any component in the mixture. As a result, a later Shell paper* on fuel oils[28] recommended these mixing rules for predicting the compatibility of fuel oils. However, use of the weighted average leads to many cases when blends of incompatible oils are predicted to be compatible. As a result, after the publication of the Oil Compatibility Model, Shell now recommends using the equivalent maximum insolubility number (minimum p_a) for predicting the compatibility of crude oil mixtures.

---

* Thanks to Rudolph Kassinger of DNV Petroleum Services for providing this paper.

5.3.8 Solubility Parameter Model of Mertens Used by Andersen[29]

Another model similar to the Oil Compatibility Model was published much earlier, 1960, but also in an obscure part of the literature, an ASTM Bulletin on asphalt, by Mertens[30] of Chevron. The author still has not seen a copy of this paper, but Andersen[29] published an application of the method to the stability of crude oils and processed oils. According to Andersen, the flocculation point is determined for several oil/solvent ratios by flocculation titration with nonsolvent. The volume of nonsolvent at the flocculation point divided by the mass of oil (V_{NS}/m_o) is plotted versus the volume of solvent divided by the mass of oil (V_S/m_o), and a straight line is obtained. In order to relate this to the Oil Compatibility Model, let us assume the solvent is toluene and the nonsolvent is n-heptane. The mass of oil is equal to the volume of oil, V_{oil}, times the density of oil, d. The critical solubility parameter of Mertens is the same as what is called the flocculation solubility parameter, δ_f, in the derivation of the Oil Compatibility Model. Using the volume average mixing rule for the flocculation solubility parameter in a mixture of oil, toluene, and n-heptane,

$$\delta_f = \frac{V_{oil}\delta_{oil} + V_T\delta_T + V_H\delta_H}{V_{oil} + V_T + V_H} = \frac{(m_o/d)\delta_{oil} + V_T\delta_T + V_H\delta_H}{(m_o/d) + V_T + V_H} \qquad (5.30)$$

Cross-multiplying and rearranging, one obtains

$$\frac{V_H}{m_o} = \left[\frac{\delta_{oil} - \delta_f}{\delta_f - \delta_H}\right]\frac{1}{d} + \left[\frac{\delta_T - \delta_f}{\delta_f - \delta_H}\right]\frac{V_T}{m_o} \qquad (5.31)$$

This is an equation of a line when V_H/m_o is plotted versus V_T/m_o. Using the definition of insolubility number and solubility blending number (equation 5.9 and equation 5.10), one may put the slope and y-axis intercept of this line in terms of the insolubility number and the solubility blending number:

$$\text{Slope} = \frac{100}{I_N} - 1 \qquad (5.32)$$

$$y\text{-axis intercept} = \frac{\left[\frac{S_{BN}}{I_N} - 1\right]}{d} \qquad (5.33)$$

Andersen[29] used the criterion that the oil is unstable if the y-axis intercept is negative. From equation (5.33) this is clearly the same as when the insolubility number, I_N, is greater than the solubility blending number, S_{BN}. Nevertheless, as for the P-test, the Mertens model does not provide a logical method for predicting the compatibility on blending of oils. Thus, although at least two similar models were developed well before the Oil Compatibility Model, this model appears to be the first to provide a logical and conservative prediction of the compatibility on blending of oils, based on flocculation titration of the individual oils in the mixture.

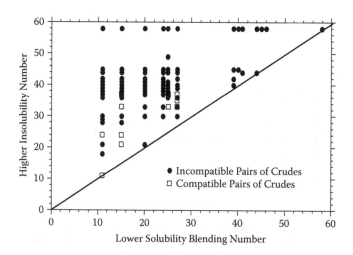

FIGURE 5.19 Over 100 pairs of crude oils have been found to be incompatible, but sometimes pairs predicted to be incompatible are found to be compatible.

5.3.9 INCOMPATIBLE PAIRS OF CRUDE OILS

Once the insolubility and the solubility blending numbers of a number of crude oils have been measured, the set of potentially incompatible pairs of crudes can be predicted and compared by experimentally blending a number of pairs of crudes in different proportions. Figure 5.19 is a plot of the higher insolubility number of a pair of crudes versus the lower solubility blending number of the pair of crudes. Incompatibility is predicted for some blends of each pair of crudes for points that fall on or above the diagonal line. Although over 100 pairs of crude oils have been experimentally found to be incompatible (solid circles in Figure 5.19), none that were predicted to be compatible in all proportions (below the diagonal in Figure 5.19) were experimentally found to be incompatible in some proportion. However, there are a number of crude pairs that were predicted to be incompatible (open squares in Figure 5.19), but found to be compatible. This happens when the lower insolubility number crude has an excess of resins and little or no asphaltenes and the high insolubility number crude has a low resin to asphaltene ratio. As a result, the maximum insolubility number decreases on blending because of the dispersant ability of resins.

To predict this effect is clearly outside the scope of the present model, but the subject of future investigations. Nevertheless, because reality is always equal to or better than the prediction, the present model provides a conservative prediction for preventing high rates of fouling and coking caused by incompatible oils. As the flocculation experiments are performed near room temperature and applied to elevated temperatures in the refinery, this also contributes to the conservative nature of the prediction. Asphaltene solubility in oils always increases with increasing temperature as long as no major fraction of a gas or other component at temperatures near its critical point is in the oil. Of course, a future direction is to

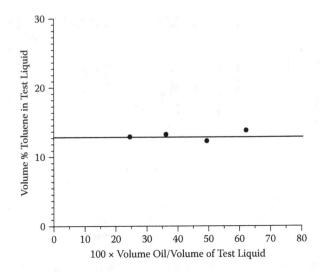

FIGURE 5.20 For the self-incompatible crude oil, Yme, the toluene percentage in the test liquid at incipient asphaltene precipitation is independent of oil concentration. (From IA Wiehe. *J Dispersion Sci Technol* 25: 333–339, 2004. With permission.)

improve the Oil Compatibility Model so that the predictions are less conservative, but an incompatible blend is never predicted as being compatible.

5.3.10 SELF-INCOMPATIBLE OILS[31,32]

Crude oils often contain salt, iron sulfide, rust, clays, and other inorganic solids, as well as wax, that can be observed with an optical microscope. Although these inorganic solids can be sources of fouling, waxes invariably dissolve on mild heating without fouling in the crude preheat train. However, the author has discovered that more than 20 commercial crude oils contain insoluble asphaltenes that can cause catastrophic fouling of crude preheat trains and coking of distillation furnace tubes. These can be observed with an optical microscope in a drop of the oil between a microscope slide and a cover slip. However, they can be verified by diluting the oil in toluene and seeing the insoluble asphaltenes disappear, but remain when diluting the oil in an equal volume of *n*-heptane. Another distinguishing feature of self-incompatible oils is that the volume percent toluene in the test liquid at the point of incipient asphaltene insolubility is independent of oil to test liquid volume ratio, as shown in Figure 5.20, for a commercial crude oil, Yme,* which contains less than 1 wt% asphaltenes. This is because the chemical potential of a solute remains constant when the soluble and insoluble solute are in phase equilibrium. As a result, for a self-incompatible oil the insolubility number, the solubility blending number, and the toluene equivalence are all equal.

---

* Thanks to Statoil for permission to publish the data on Yme crude oil.

One of the first applications of the Oil Compatibility Model early in 1995 was the discovery of a self-incompatible crude oil. This crude oil was being considered for purchase at a refinery that already had experienced furnace coking problems in their vacuum distillation unit and could not afford another. With the author's discovery and prediction of the consequences, the purchase was stopped within hours of signing the contract. Further probing revealed that this self-incompatible crude caused fouling and coking problems at other refineries. Thus, the refinery fouling problem was solved before it happened. This helped greatly to authenticate the Oil Compatibility Model as a tool for crude oil purchase.

It takes time for asphaltenes to precipitate during flocculation tests, depending on the viscosity of the medium. Thus, it is recommended that each bottle containing the oil and test liquid be held for at least 5 min at 60°C for the flocculation test before it is determined if the asphaltenes are in solution. There are automatic titrating instruments available for performing flocculation measurements. However, it is easy for these to titrate nonsolvent too quickly and overshoot because the time for precipitation is not taken into account. Nevertheless, the time to redissolve asphaltenes is much longer than the time to precipitate asphaltenes. As a result, for a self-incompatible oil if one tries to add toluene until the asphaltenes redissolve (toluene dilution test), one cannot help but overshoot the amount of toluene because of the very slow kinetics for redissolving. Therefore, it is recommended, for flocculation tests on incompatible oils, to add excess toluene, make sure the asphaltenes are in solution, and then add n-heptane to determine if the asphaltenes are in solution until reaching the flocculation point. Only after using this procedure were consistent results obtained and it was established that the percent toluene in the test liquid was independent of the volume ratio of oil to test liquid.

The existence of commercially traded crude oils that are self-incompatible means that the refiner even needs to be concerned of crude incompatibility when running only one crude at a time. It can be difficult to believe that a self-incompatible crude can be produced through the porous media of a petroleum reservoir. However, self-incompatible crudes may be the result of blending crudes or gas condensate with crudes at the production site or by asphaltenes precipitating on cooling and depressurizing after being produced from reservoirs at elevated temperatures and pressures. On the other hand, several self-incompatible crudes are known to have asphaltene plugging problems during production. The self-incompatible crude oil in the example, Yme (Figure 5.20), contains low amounts of very soluble asphaltenes as do many crudes that exhibit asphaltene production problems. In this case, the incompatibility is because the crude is quite paraffinic and, thus, is a poor solvent for asphaltenes. However, this is not always the case. Self-incompatible crudes can also have quite high insolubility numbers. The author has discovered self-incompatible crude oils that span insolubility numbers (thus, solubility blending numbers) from 12 to 58. However, the identity of these self-incompatible crude oils is quite proprietary because of the competitive advantage of refineries to avoid these highly fouling crude oils.

5.3.11 Nearly Incompatible Oils[33]

It is known that even compatible oils can undergo asphaltene fouling, albeit at a much slower rate than incompatible oils. Therefore, this section introduces the concept of fouling by the adsorption of asphaltenes at surfaces that depends on the distance from incompatibility, a quantity that can also be measured with the Oil Compatibility Model and tests.

5.3.11.1 Distance from Incompatibility

The ratio of the solubility blending number of a mixture of oils to the maximum insolubility number in the mixture is a measure of the distance from incompatibility. Coincidentally, this is equal to the P-value (section 5.3.7). When this ratio is 1 or less than 1, incompatibility is predicted. A value slightly greater than 1 predicts compatibility, but is nearly incompatible, and a value much greater than 1 is very compatible. This ratio may be directly measured by the heptane dilution test, in which one determines the maximum volume (mL) of n-heptane, V_H, that can be mixed with 5 mL of oil without precipitating asphaltenes:

$$\frac{S_{BN}}{I_N} = 1 + \frac{V_H}{5} \tag{5.34}$$

Alternatively, this ratio can be calculated for a blend of oils based on measurements of the compatibility numbers of the component oils. It is proposed that adsorption of asphaltenes on a metal surface of process equipment, such as a heat exchanger, is a mode of asphaltene fouling that is much slower than oil incompatibility. It is also proposed that the more soluble the asphaltenes are in the oil, the less tendency the asphaltenes will have to adsorb at a surface.

5.3.11.2 Fouling by Oil Blends

The fouling by Souedie and Forties crude oils as well as several blends were measured using accelerated fouling tests for the author in the laboratory of Ghaz Dickakian of Fouling and Coking Technology, formerly of Exxon. These tests were conducted using the Thermal Fouling/Coking Test Unit, now called a Hot Liquid Process Simulator and manufactured by Alcor Petroleum Instruments, Inc., as diagrammed in Figure 5.21. Oil (600 mL), at room temperature and under 700 psig nitrogen pressure to prevent volatilization, was pumped at 3 mL/min through an annulus in which a 318 carbon steel heater tube in the center was heated at a constant temperature of 760°F. As foulant built up on the heater tube surface, the foulant acted as an insulator and caused the temperature at the outlet of the annulus to decrease. Therefore, the fouling rate of the oil was measured by the decrease in temperature of the flowing oil at the annulus outlet over a three-hour period.

The variation in the fouling rate with volume percent Forties crude in blends with Souedie crude is shown in Figure 5.22. The fouling rate at 25% Forties is only slightly higher than expected from a line drawn through the points at 0%

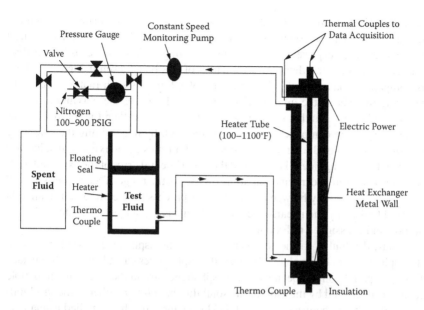

FIGURE 5.21 Schematic diagram of the thermal fouling/coking test unit. (From IA Wiehe, RJ Kennedy, G Dickakian. *Energy and Fuels* 15: 1057–1058, 2001. Reprinted with permission. Copyright 2001. American Chemical Society.)

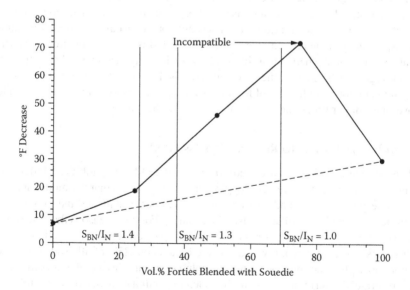

FIGURE 5.22 Thermal fouling data on Forties–Souedie crude oil blends show incompatible blends foul more, but nearly incompatible blends are also high fouling. (From IA Wiehe, RJ Kennedy, G Dickakian. *Energy and Fuels* 15: 1057–1058, 2001. Reprinted with permission. Copyright 2001. American Chemical Society.)

Forties (only Souedie) and 100% Forties. Because at 75% Forties the blend is incompatible, it is not surprising that the fouling rate is the highest measured for this set. The surprise is that the rate of fouling at 50% Forties is higher than the rate of fouling of either pure component of this blend, even though this blend is compatible. Based on this limited data, blending oils such that the solubility blending number of the mixture is greater than 1.3 times the maximum insolubility number assures that the fouling rate is not greater than the fouling rate of either of the pure components. On the other hand, by maintaining the solubility blending number of the mixture greater than 1.4 times the maximum insolubility number, the fouling rate is kept within the experimental error of the expected fouling rate. Therefore, nearly incompatible oils are in the range of S_{BN}/I_N between 1.0 and 1.4, but especially between 1.0 and 1.3. A process for blending petroleum oils to avoid fouling by such nearly incompatible oils is the subject of a patent[34] by the author, and is assigned to Exxon.

Thus, the fouling of heated metal surfaces by asphaltenes does not require the asphaltenes to be insoluble. Instead, the asphaltenes can adsorb on the surface from compatible oils, especially if the asphaltenes are in the nearly incompatible region as measured by the ratio of the solubility blending number to the insolubility number, being between 1.0 and 1.4. Mason and Lin[35] have studied asphaltene aggregation kinetics of blends of Forties and Souedie crudes in the incompatible and near incompatible regions using time-resolved small angle neutron scattering. Their aggregation kinetic model predicts that Forties and Souedie crudes are incompatible at volume fractions of Forties above 0.47 rather than the value of 0.67 reported here. This interpretation is that if one waited many days at room temperature, an equal volume mixture of Forties and Souedie would produce aggregates of asphaltenes in the micron-size range. Although this provides an interesting scientific question of whether near incompatibility is really incompatibility with very slow aggregation kinetics, the practical result is the same. This region does cause fouling but not as high a rate of fouling as that region observed to be incompatible over the smaller timescale (less than one day) in the laboratory.

5.3.12   Application to Refinery Process Oils[36]

This section describes a case study in which the Oil Compatibility Model was applied to solve a hydrotreater plugging problem. It shows that the model applies to refinery process oils and it extends to oils without asphaltenes and to oils containing oil-soluble, but toluene-insoluble, solutes. For these reasons the case study is included in this chapter, rather than the chapter on fouling mitigation. After this work was completed, it was discovered that about 25% of all crude oils have either no asphaltenes or asphaltenes too soluble (insolubility number less than 10) to accurately determine the solubility blending number using only the toluene equivalence and the heptane dilution tests. Therefore, the solvent oil equivalence and the nonsolvent oil dilution tests described in this section for process oils are also necessary for many crude oils. More details will be given later in section 5.3.14.

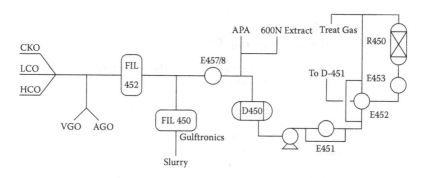

FIGURE 5.23 Schematic diagram of the fixed-bed, catalytic hydrotreater.

Actually, this application, mitigation of the hydrotreater plugging, took place only a month after the Oil Compatibility Model was developed, prior to solving any problem with crude oils, such as the first self-incompatible crude and the blending of Souedie and Forties crudes in the wrong order. Based on our analysis of the foulant and other evidence, the author suspected oil incompatibility was the cause of the problem at a European refinery. As a result, the author and colleague Ray Kennedy were asked to go to the refinery and help solve the problem. This was not a typical assignment for scientists in Corporate Research at Exxon. However, because new tests needed to be developed in the refinery using refinery process oils, the assignment turned out to be appropriate and quite successful.

5.3.12.1 Hydrotreater Plugging Problem

A refinery fixed-bed hydrotreater (schematic in Figure 5.23), was designed to desulfurize a wide mixture of feeds as shown in Table 5.3. This unit processed

---

TABLE 5.3
Components of Hydrotreater Feed

Virgin atmospheric gas oil
Virgin vacuum gas oil
Kerosene oil from fluid catalytic cracking
Light catalytic cycle oil
Heavy catalytic cycle oil
Departiculated fluid catalytic cracker bottoms
150 Neutral lube extract
600 Neutral lube extract
Propane asphalt

Source: IA Wiehe, RJ Kennedy. *Energy and Fuels* 14: 60–63, 2000. Reprinted with permission. Copyright 2000. American Chemical Society.

TABLE 5.4
Analysis of Scale Trap Foulant

Element	Before Plugging (wt%)	During Plugging (wt%)
Carbon	62.75	86.83
Hydrogen	2.71	3.82
Sulfur	4.07	3.42
Sodium	11.7	0.04
Iron	3.99	0.26
Aluminum	1.17	0.11
Magnesium	0.48	0.02
Calcium	0.48	0.03
Silicon	0.20	0.04
Potassium	0.15	
Zinc	0.14	

oil for several years without major problems. However, abruptly the hydrotreater began to build pressure drop that caused the unit to be shut down for cleaning. Only the top few inches of the bed were plugged with black solids. The foulant from the scale trap above the bed of the hydrotreater was washed with methylene chloride to remove the oil and analyzed for the elements. This is compared with the analysis of a similarly treated foulant from the scale trap before the hydrotreater had plugging problems in Table 5.4. Although the earlier foulant had more of the character of a true scale with high inorganic content, the foulant during times of plugging had the character of true coke, being mostly organic: carbon, sulfur, and hydrogen. Organic coke in the hydrotreater reactor inlet and the unusual wide range of feeds pointed to feed incompatibility as a possible cause. With further operation, the run lengths became increasingly shorter until with only two-week run lengths the unit was shut down, pending the resolution of the problem. Although hydrotreaters with fixed catalyst beds occasionally plug, this high frequency of plugging was intolerable. As the hydrotreater was a critical unit in the refinery, the cause and mitigation plan needed to be determined quickly. Although the Oil Compatibility Model had been developed for crudes, it had never been applied to oils without asphaltenes. Therefore, new tests had to be devised to diagnose and cure the plugging problem in the refinery laboratory.

5.3.12.2 Compatibility Testing

Feed compatibility was checked by blending the feed components in the laboratory in the proportions last run in the hydrotreater. Examination of this mixture with an optical microscope clearly showed insoluble asphaltenes (particles that were dissolved on addition of toluene, but not with n-heptane). Thus, feed incompatibility was ascertained to be the most likely cause. However, during the last few runs, incompatibility had already been suspected, though not proved. Thus, when pressure drop had increased, the propane asphalt, the feed component with the highest concentrations of asphaltenes, was removed from the feed. Nevertheless, the hydrotreater continued to plug rapidly. In addition, the insoluble asphaltenes in the simulated feed blend were less than 10 μ in size, too small to plug a fixed-bed hydrotreater with voids greater than 100 μ. Thus, it was difficult to explain these observations based on the incompatibility mechanism. Nevertheless, compatibility testing was initiated with the expectation of explaining these paradoxes once the feed phase behavior was understood.

Although the compatibility testing was at room temperature, the hydrotreater was at elevated temperature (above 350°C). As discussed in chapter 2, asphaltene solubility increases with increasing temperature as long as no component is near its critical point, such as with dissolved gases. Thus, if compatibility of the hydrotreater feed can be designed at room temperature, it will be definitely assured at elevated temperatures.

5.3.12.3 Compatibility Numbers for Components with Asphaltenes

Most of the feed components listed in Table 5.3 contained no asphaltenes. One volume of each feed component was mixed with 10 volumes of n-heptane and determined by optical microscopy whether or not asphaltenes precipitated (submicroscopic particles that cluster into chain- or shrimp-shaped agglomerates). Only propane asphalt and fluid catalytic cracker (FCC) bottoms were found to contain asphaltenes. The other feed components without asphaltenes by convention had insolubility numbers of zero. However, no tests were available to measure the solubility blending numbers of oils without asphaltenes. With the hydrotreater down, these tests had to be developed quickly. Nevertheless, the first step was to run the toluene equivalence and heptane dilution tests on samples of the feed components that contained asphaltenes. The result is shown in Table 5.5 along with the insolubility and solubility blending numbers, calculated with equation (5.16) and equation (5.13). It immediately became clear that FCC bottoms

---

TABLE 5.5
Compatibility Numbers for Feed Components Containing Asphaltenes

Feed Component	Toluene Equivalence (at 2 g oil/10 mL test liquid)	Heptane Dilution (mL n-heptane per 5 mL oil)	Oil Density (g/cc)	Insolubility Number	Solubility Blending Number
Propane asphalt	11	16.3	1.00	23	105
FCC bottoms	74	2.2	1.11	80	116

Source: IA Wiehe, RJ Kennedy. *Energy and Fuels* 14: 60–63, 2000. Reprinted with permission. Copyright 2000. American Chemical Society.

(also called slurry oil) with an insolubility number of 80 contained the asphaltenes that become insoluble on mixing and not the propane asphalt with an insolubility number of only 23.

The FCC bottoms became a reference oil to determine the solubility blending numbers of the feed components without asphaltenes. Ten volumes of each feed component were blended separately with one volume of FCC bottoms and determined if asphaltenes precipitated. Three feed components, atmospheric gas oil, vacuum gas oil, and kerosene oil, from fluid catalytic cracking were then determined to be nonsolvents for FCC bottoms, whereas the other feed components were solvents.

5.3.12.4 Nonsolvent Oil Dilution Test

The heptane dilution test was rerun on the FCC bottoms, but the n-heptane was replaced with each of the three nonsolvent oils. Thus, each of the nonsolvent oils was added to 5 mL of FCC bottoms and the maximum volume in mL of nonsolvent oil (V_{NSO}) that could be added without precipitating asphaltenes was determined. The mixture at that flocculation point must have the same solubility blending number as when n-heptane (V_H) was added to FCC bottoms in the heptane dilution test. From equation (5.14),

$$S_{BNmix} = \frac{V_H(0) + V_{FCCB}S_{FCCB}}{V_H + V_{FCCB}} = \frac{V_{NSO}S_{NSO} + V_{FCCB}S_{FCCB}}{V_{NSO} + V_{FCCB}} \qquad (5.35)$$

Rearranging and solving for the solubility blending number of the nonsolvent oil,

$$S_{NSO} = \frac{S_{FCCB}[V_{NSO} - V_H]}{V_{NSO}\left[1 + \frac{V_H}{5}\right]} \qquad (5.36)$$

The solubility blending numbers calculated with equation (5.36) for the nonsolvent oils are given in Table 5.6. Later in this chapter, the nonsolvent oil equivalence

---

TABLE 5.6
Compatibility Numbers for Nonsolvent Oils with FCC Bottoms as the Reference Oil

Nonsolvent Oil	Volume of N S Oil (mL)	Volume of n-Heptane (mL)	Insolubility Number	Solubility Blending Number
Atmospheric gas oil	5.5	2.2	0	48
Vacuum gas oil	5.5	2.2	0	48
FCC kerosene	16.25	2.2	0	69

Source: IA Wiehe, RJ Kennedy. Energy and Fuels 14: 60–63, 2000. Reprinted with permission. Copyright 2000. American Chemical Society.

---

test will be described also for determining the solubility blending number of non-solvent oils, the test that is currently preferred by the author.

5.3.12.5 Solvent Oil Equivalence Test

Four of the feed components contained no asphaltenes and are solvents for FCC bottoms: light and heavy catalytic cycle oil and 150 and 600 neutral lube extracts. For these, the toluene equivalence test (TE) was rerun on the FCC bottoms, but toluene was replaced by each of the solvent oils (SOE). Thus, the concentration of 2 g of FCC bottoms per 10 mL of test liquid was maintained. The minimum volume percent solvent oil in the test liquid (rest is n-heptane) to keep asphaltenes in solution was determined. The mixture at that flocculation point must have the same solubility blending number as the mixture of the toluene equivalence test on FCC bottoms:

$$S_{BNmix} = \frac{V_H^T(0) + V_T(100) + V_{FCCB}S_{FCCB}}{V_H^T + V_T + V_{FCCB}} = \frac{V_H^{SO}(0) + V_{SO}S_{SO} + V_{FCCB}S_{FCCB}}{V_H^{SO} + V_{SO} + V_{FCCB}} \quad (5.37)$$

$$S_{SO} = 100 \frac{\frac{V_T}{10}}{\frac{V_{SO}}{10}} = 100 \left[\frac{TE}{SOE} \right] \quad (5.38)$$

The solubility blending numbers calculated with equation (5.38) for solvent oils are given in Table 5.7.

5.3.12.6 Chlorobenzene Equivalence Test

As refinery oils can vary significantly with crude and process changes, samples were collected for a number of days of operation for each of the feed components. The compatibility numbers were measured as described earlier. One exception

TABLE 5.7
Compatibility Numbers for Solvent Oils with FCC Bottoms as the Reference Oil

Solvent Oil	Solvent Oil Equivalence (at 2g oil/10 mL T liq.)	Toluene Equivalence (at 2g oil/10 mL T liq.)	Insolubility Number	Solubility Blending Number
LCCO	78	74	0	95
HCCO	53	74	0	140
150 Neut. lube extr.	91	74	0	81
600 Neut. lube extr.	50	74	0	147

Source: IA Wiehe, RJ Kennedy. *Energy and Fuels* 14: 60–63, 2000. Reprinted with permission. Copyright 2000. American Chemical Society.

was for some samples of FCC bottoms that were either insoluble in toluene or close to it. Then chlorobenzene equivalence was used by replacing toluene in the toluene equivalence with chlorobenzene. Using the principle of equivalent flocculation solubility parameter, the toluene equivalence, TE, can be related to any other solvent equivalence, SE, by

$$\delta_f = \frac{V_H\delta_H + V_T\delta_T + V_{oil}\delta_{oil}}{V_H + V_T + V_{oil}} = \frac{V_H^S\delta_H + V_S\delta_S + V_{oil}\delta_{oil}}{V_H^S + V_S + V_{oil}} \tag{5.39}$$

$$V_H + V_T + V_{oil} = V_H^S + V_S + V_{oil} = 10 + V_{oil} \tag{5.40}$$

$$V_H\delta_H + V_T\delta_T = V_H^S\delta_H + V_S\delta_S \tag{5.41}$$

$$V_H = 10 - V_T; \quad V_H^S = 10 - V_S \tag{5.42}$$

$$10\delta_H + V_T[\delta_T - \delta_H] = 10\delta_H + V_S[\delta_S - \delta_H] \tag{5.43}$$

$$TE = 100\frac{V_T}{10}; \quad SE = 100\frac{V_S}{10} \tag{5.44}$$

$$TE = SE\left[\frac{\delta_S - \delta_H}{\delta_T - \delta_H}\right] \tag{5.45}$$

where
δ_S = solubility parameter for the solvent = 9.67 $(cal/cc)^{1/2}$ for chlorobenzene
δ_H = solubility parameter for n-heptane = 7.50 $(cal/cc)^{1/2}$
δ_T = solubility parameter for toluene = 8.93 $(cal/cc)^{1/2}$

This predicts that, when measured on the same oil, the toluene equivalence will be 1.52 times the chlorobenzene equivalence. This was checked on an FCC bottoms sample in which the toluene equivalence was measured to be 95 and the chlorobenzene equivalence, 65, giving a ratio of 1.46, within experimental error of the predicted ratio. Using equation (5.45), toluene equivalences now could be measured and insolubility numbers could be evaluated that were over 100 (toluene insoluble).

5.3.12.7 Overall Range of Compatibility Numbers

The full range of insolubility numbers and solubility numbers is shown in Figure 5.24. The range of propane asphalt insolubility numbers is lower than the range of solubility blending numbers for all feed components except for some atmospheric gas oils. As a result, propane asphalt was never insoluble in the feed mix. This is the reason why removing it from the feed did not reduce the rate of fouling. Actually, removing propane asphalt made the total feed a poorer solvent because of propane asphalt's high solubility blending number and high concentration

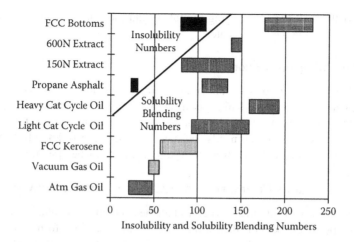

FIGURE 5.24 Range of insolubility numbers and solubility blending numbers for each feed component over days of operation. (From IA Wiehe, RJ Kennedy. *Energy and Fuels* 14: 60–63, 2000. Reprinted with permission. Copyright 2000. American Chemical Society.)

of resins (natural asphaltene dispersant). On the other hand, FCC bottoms are insoluble in the three nonsolvent oils in Table 5.6 and sometimes in 150 neutral lube extract as well. However, primarily the plugging was due to an incompatibility between the FCC bottoms and the virgin gas oils. With this knowledge, refinery records were reviewed, and it was discovered that the plugging started after the concentration of virgin gas oils in the hydrotreater feed was greatly increased. Without knowledge of the possibility of incompatibility, increasing the concentration of virgin gas oils was not flagged as a possible cause of fouling.

5.3.12.8 Root Cause Analysis

After obtaining the data in Figure 5.24, a root cause analysis was done with all the evidence. Although feed incompatibility was clearly the root cause, the high concentration of coke in the scale trap, the screen above the bed of catalyst, and in the first few inches of the bed indicated that the coke was formed upstream of the bed. The very small size (<10 μm) of the insoluble asphaltenes also showed that they could not have been captured in the first few inches of the bed with voids greater than 100 μm. Closer examination of the bed coke with an optical microscope showed that it was composed of flakes with some curvature. Therefore, the most likely location of the coke formation was the hottest heat exchanger (350°C), just before the reactor. After this analysis, the heat exchanger was opened and indeed, the tubes had a thick layer of coke, but not thick enough to plug. It was concluded that the insoluble asphaltenes formed coke, on contact with the hot walls, based on the knowledge that insoluble asphaltenes form coke without an induction period (chapter 4). Shear stress or upsets, or both, caused the coke to flake off the wall before plugging the tubes, and these large flakes of coke are what then plugged the hydrotreater.

5.3.12.9 Mitigating Action

The mitigating action was to control the feed composition so that the solubility blending number of the feed mix was always greater than the insolubility number of the FCC bottoms. With the data and the equations there was much flexibility for the refinery to meet their needs and stay within this constraint. After start-up, the hydrotreater operated for over three years without any bed plugging.

5.3.13 CHANGE OF COMPATIBILITY NUMBERS WITH THERMAL CONVERSION

In chapter 4, it was established that asphaltenes become less soluble when pendant groups are cracked off their aromatic cores and that coke formation is initiated when these converted asphaltenes become insoluble in the thermal reacting medium. Thus, it makes sense to determine how thermal conversion changes the insolubility number and the solubility blending number to devise a method to predict the coke induction period, based on data on thermally processed heavy oil within the induction period.

5.3.13.1 WRI Coking Indexes to Predict Coke Induction Period

Schabron,[37] using the P-Test (section 5.3.7) on thermally converted resids with isooctane as the aliphatic diluent, was the first to use solubility titration methods to predict coke induction periods based on the coking index:

$$C_{min} = \frac{d}{X_{min}} \tag{5.46}$$

$$\text{WRI Coking Index} = \frac{p_a}{C_{min}} = \frac{1 - FR_{max}}{C_{min}} \tag{5.47}$$

where
d = density of the oil, g/mL

When this WRI (Western Research Institute) Coking Index is less than 0.08, coke formation is predicted. More recently, Schabron[38] has also used other WRI Coking Indexes, including the portion of heptane-insoluble asphaltenes that are soluble in cyclohexane. As it turns out, WRI Coking Indexes describe the present status, but do not predict a future event. When WRI Coking Indexes are below certain values, coke is obtained. However, at that point it would be easier to observe mesophase in the oil by microscopy[39,40] to determine if coke is present than to make solubility or filtration measurements. On the more positive side, the WRI Coking Indexes do determine how far one is from the end of the coke induction period, which merely not finding mesophase in the thermally converted oil does not determine. Nevertheless, one cannot use this information to calculate quantitatively how much more the oil can be converted before coke formation will begin.

5.3.13.2 Oil Compatibility Model for Predicting Induction Period

The author worked on this problem as a consultant for the National Centre for Upgrading Technology of Canada, together with Parviz Rahimi and Dick Parker,[41,42] using the Oil Compatibility Model to predict the coke induction period. The insolubility number, I_N, and the solubility blending number, S_{BN}, always have been observed to vary linearly with themal severity. The thermal severity index, SI, is used to put data at different temperatures on the same base temperature for determining equivalent reaction times at that base temperature, 700°K:

$$SI = t\,e\left\{-\frac{E_A}{R}\left[\frac{1}{T} - \frac{1}{700}\right]\right\} \tag{5.48}$$

where
  t = actual reaction time, s
  E_A = activation energy, taken as 50.1 kcal/mol
  R = gas constant, 0.001987 kcal/(mol °K)
  T = reaction temperature, °K

An example is shown in Figure 5.25 for visbreaking of Athabasca bitumen in a pilot plant visbreaker. Whereas, more generally, the solubility blending number decreases with reaction time, in Figure 5.25 it remains constant. On the other

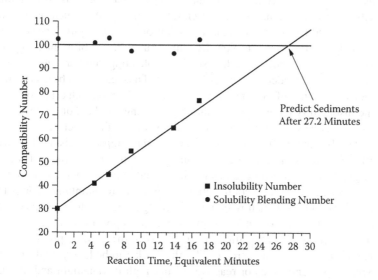

FIGURE 5.25 The oil compatibility model predicts the appearance of insoluble asphaltene sediments in visbroken Athabasca bitumen at a reaction time of 27.2 min and a temperature of 427°C when the insolubility number and the solubility blending number become equal. (From IA Wiehe. *J Dispersion Sci Technol* 25: 333–339, 2004. With permission.)

hand, the insolubility number of Athabasca bitumen, as well as of other heavy oils, always increases rapidly with reaction time. According to the Oil Compatibility Model, converted asphaltenes become insoluble to form coke when the insolubility number becomes equal to the solubility blending number. This enables the prediction of the end of the coke induction period at 27.2 equivalent min at 427°C for the data in Figure 5.25, based on flocculation experiments on the thermally converted oil within the induction period. Therefore, this is the only model that enables predicting the coke induction period from data on the feed and at least one point within the induction period. Of course, it is an advantage not to require actually forming coke, especially if the data are collected on a commercial visbreaker. Unfortunately, many visbreakers are operated in the same conditions for all feeds. This model enables the visbreaker conversion to be tuned to the maximum for each feed and still be operated safely within the coke induction period.

A model needs to be developed to explain why the compatibility numbers vary linearly with reaction time in order to elucidate the limitations to this approximation. To fit these measurements into the Phase Separation Kinetic Model (chapter 4), the solubility limit, S_L, needs to be related to the Oil Compatibility Model.

5.3.14 Status of the Application of the Oil Compatibility Model to Crude Oils

This section is inserted to review the current status of the Oil Compatibility Model and Tests as applied to crude oils. About 25% of the crude oils have been found to have either zero asphaltenes at the concentration range of the toluene equivalence test (2 g of oil and 10 mL of n-heptane) or too low of a toluene equivalence (below 5) to accurately determine the insolubility number and the solubility blending number. Many of these crudes have low solubility blending numbers and, so, act as nonsolvents when blended with other crude oils. Thus, care must be used to determine accurate values of the solubility blending numbers of such crude oils.

The insolubility numbers and solubility blending numbers of crude oils continue to be actively measured and applied. However, the exact values of various crude oils remain very proprietary because they impact decisions by various refineries on which crude oils they schedule and purchased. As crude oil accounts for the greatest expense in a refinery, which crude oils are purchased and how they are processed are kept confidential by refineries to maintain a competitive advantage over others in their region. It is a strong advantage for a refinery to process crude oils that are sold at lower prices (called "opportunity crudes") and still meet the target yields and quality specifications of the refinery products. These opportunity crudes may be perceived by the market to be heavy, high in sulfur, corrosive (naphthenic acids or reactive sulfur), high in vanadium and nickel, or high fouling. Depending on the experience, process capability, technical ability, and metallurgy of a refinery, a given "opportunity" crude may be a real economical opportunity or cause real problems and become a financial disaster. Thus, most refineries cannot afford either to pass up all opportunity crudes or to blindly purchase the lowest-priced crude oils. Because there are established tests for measuring all the potential problems of crude oils (crude assays) except fouling, the author developed the Oil Compatibility Model and Tests. Now we understand that fouling not only depends on the individual crude oil, but on which crude oils it is blended as well as the proportions and order that it is blended. This removes the mystery of asphaltene deposition and through the Oil Compatibility Model and Tests asphaltene fouling in crude preheat systems of the refinery can be predicatively prevented.

ExxonMobil has worldwide patents on using the Oil Compatibility Model to blend potentially incompatible crude oils to prevent incompatibility or near incompatibility.[26,34] This provides a strong advantage to ExxonMobil refineries—one they deserve for sponsoring the author's research that resulted in the patents. However, the Oil Compatibility Model and Tests could not be patented and, therefore, are available for anyone to use royalty free. Refineries not owned by ExxonMobil may also use the Oil Compatibility Model and Tests as long as their purchase and scheduling decisions based on these are not to process any mixture of crude oils that are incompatible or nearly incompatible in any proportions. Although the number of pairs of incompatible crude oils is significant, it is still a minor fraction of the total. Thus, this method of making crude oil purchase and scheduling decisions is not very restrictive, and it still is a significant economic advantage for a refinery in fouling reduction. Fouling in crude preheat systems alone commonly costs U.S. refineries 2 to 10 cents per barrel processed in energy, maintenance, and loss of process time.[43] Thus, for a refinery capacity of only 100,000 barrels per day, this amounts to $0.7 to $3.6 million per year. If avoiding fouling enables a refinery to process an opportunity crude oil that it would not normally process, the savings would be much more because crude oil is the greatest expense for a refinery whereras energy is second. Now, by avoiding incompatible and near-incompatible crude oil blends, asphaltene deposition can be greatly reduced, if not eliminated. More on the mitigation of asphaltene fouling as well as other types of fouling will be presented in the next chapter.

5.3.15 OIL COMPATIBILITY TEST PROCEDURES

Usually, refineries are only willing to pay for the minimum testing. Therefore, typically the tests run to determine the solubility blending number and the insolubility number of a crude oil are the heptane dilution and the toluene equivalence tests. In addition, the room temperature density of the oil needs to be measured.

5.3.15.1 Check Oil

The first step is to observe a drop of the oil in between a glass slide and a cover slip under an optical microscope in transmitted light at 100 to 200× magnification. This can enable the experienced eye to observe the solids present in crude oil. The occasional specs of clay, sand, iron sulfide, salt, etc., are common and not of concern. However, large numbers of solids or insoluble liquid are of concern as they may interfere with the tests or cause fouling themselves. If the solids are clear needles or platelets in normal transmitted light and white in cross-polarized light, they are probably wax and usually not of concern. As will be discussed later, part of the test procedure is to hold the oil and test solvent mixture at 60°C in an oven, which nearly always melts the wax and enables it to be dissolved into the mixture. If the insoluble solids are the yellow-to-brown curved agglomerates of asphaltenes, the oil is self-incompatible. This can be confirmed by blending the oil with toluene and finding the particles dissolved but equal dilution in n-heptane results in no dissolution. As discussed previously, self-incompatible oils require only the toluene equivalence test because both the solubility blending number and the insolubility number are equal to the toluene equivalence. If the insoluble phase is in the form of clear circles under the microscope, these are probably water drops. If only one or two drops are present in each view, these can usually be ignored. However, more water drops can interfere with the tests. In this case, they should be removed by centrifuging. If the particles are clear cubes, they are probably salt, which can be ignored. Toluene-insoluble black particles (often iron sulfide) can be difficult, but a trained eye can usually distinguish them from the curved chain agglomerates of insoluble asphaltenes. If this is the case, the tests can be continued and the black particles ignored. If not, the black particles need to be removed by centrifuging or by filtering before running the tests.

5.3.15.2 Check for Asphaltenes

The second step is to blend 5 mL of the oil with 25 mL of n-heptane. As oil is more accurately measured by weight rather than by volume, the density of the oil in grams per milliliter is multiplied by 5 and this amount in grams is weighed into a bottle. After 25 mL of n-heptane is added to the bottle, a cap is screwed on, and the bottle is shaken and placed in a 60°C oven for 5 min. The temperature is designed to melt any wax in the oil and to lower the viscosity of the oil–test liquid mixture. To detect insoluble asphaltenes, they need time to agglomerate and the 5 min period at 60°C is designed to provide this. If the oil has an unusually high viscosity, the time period and the temperature should be increased. However, n-heptane and toluene boil near 100°C, so that the temperature should not exceed 90°C to avoid exploding bottles. It is recommended that two methods be used simultaneously to detect insoluble asphaltenes: the spot test and the microscope method. In the spot test, the oil–test liquid mixture is shaken and a drop of the mixture is placed on filter paper. The drop spreads out laterally and any insoluble asphaltenes are filtered. Thus, when dried, any insoluble asphaltenes leave a dark ring or a circle within the oil-stained spot on the filter paper. If no insoluble asphaltenes are present, a uniform, circular, oil-colored spot results. Of course, if the oil contains insoluble black particles at the start, this method cannot be used because black circles or rings will appear even when asphaltenes are soluble. Viewing a drop of the oil–test solvent mixture between a glass slide and a cover slip at 100 to 200× in transmitted light under an optical microscope is more sensitive than the spot test and should be used as the standard. With experience one can distinguish insoluble asphaltenes from other particles, unlike the spot test. However, one can also be fooled by not focusing the microscope in the plane where the insoluble asphaltenes are located. Thus, the spot test provides a check and produces a permanent record. Nevertheless, with the microscope one looks for many particles across the entire view and not for the occasional isolated particle. If blending 5 mL of the oil with 25 mL of n-heptane produces insoluble asphaltene particles, the heptane dilution and toluene equivalence tests are run. If no insoluble asphaltene particles are detected, the insolubility number is set equal to zero and a different test is run to measure the solubility blending that will be described later.

5.3.15.3 Heptane Dilution Test

For the heptane dilution test one needs to weigh the equivalent in grams of 5 mL of oil into three bottles. In the first bottle is added 5 mL of chromographic grade n-heptane. The bottle is capped, shaken, and heated 5 min in a 60°C oven. The mixture in the bottle is checked with the spot test and microscope methods. If no insoluble asphaltenes are detected, increments of 5 mL of n-heptane are added until insoluble asphaltenes are detected. To the second bottle a volume of n-heptane of 4 mL less than the total amount added to the first bottle is added. The bottle is capped, shaken, and heated 5 min in a 60°C oven. The mixture in the bottle is checked with the spot test and microscope methods. This procedure is continued with 1 mL increments of n-heptane until insoluble asphaltenes are detected. To the third bottle a volume of n-heptane of 0.5 mL less than the total amount added to the second bottle is added. The third bottle is capped, shaken, and heated 5 min in a 60°C oven, and the mixture in the bottle is checked with the spot test and microscope methods. The heptane dilution is reported as the average of the lowest total (in mL) of n-heptane added when insoluble asphaltenes were detected and the highest total (in mL) of n-heptane added without detecting asphaltenes.

5.3.15.4 Toluene Equivalence Test

Into each of four bottles, weigh 2 g of oil. In the first bottle, add 2 mL of chromographic grade toluene. After swirling the bottle to mix oil and toluene, add 8 mL of chromographic grade n-heptane, 2 mL at a time with swirling after each addition. A cap is screwed on, and the bottle is shaken and placed in a 60°C oven for 5 min, after which the bottle is shaken and the oil–test solvent mixture is checked with the spot test and the microscope method for insoluble asphaltenes. If the asphaltenes are found to be insoluble, the total test liquid, n-heptane and toluene, are kept at 10 mL for the second bottle, but the toluene is increased to 40% (4 mL of toluene followed by 6 mL of n-heptane, 2 mL at a time). If the asphaltenes are found to be soluble in the first bottle, the toluene in the test liquid for the second bottle is reduced to 10% (1 mL of toluene followed by 9 mL of n-heptane, no more than 2 mL at a time). Continue this procedure until the difference between the highest level of toluene in the test liquid that dissolves asphaltenes is 2.5% (absolute) less than the lowest percentage of toluene that precipitates asphaltenes. Average these two percentages to obtain the final toluene equivalence. The insolubility number is calculated from equation (5.16), and the solubility blending number is calculated from equation (5.13). However, if at a test liquid containing 5% toluene, the asphaltenes remain to be soluble, the toluene equivalence cannot be used to determine accurate compatibility numbers. Instead, the solvent oil equivalence, described in the next section, should be used to determine the solubility blending number and equation (5.13) can be used to calculate the insolibility number.

5.3.15.5 Solvent Oil Equivalence Test

If no insoluble asphaltenes were detected when blending 5 mL of the oil with 25 mL of n-heptane or if the toluene equivalence is less than 5, one needs to use a reference oil containing asphaltenes to measure the solubility blending number of the sample oil. Ideally, the reference oil should have been recently measured with the toluene equivalence test and have a TE above 20. First, the two oils should be checked for compatibility by blending 25 mL of the sample oil and 5 mL of the reference oil in a bottle, capping and shaking the bottle, and putting it in a 60°C oven for 10 min, shaking the bottle, and checking a drop from it for insoluble asphaltenes using the microscope method. The filter paper method also can be tried, but usually does not work because of the high viscosity of the blend of oils. Better success with the filter paper method can be obtained if it is done inside the 60°C oven. Nevertheless, if insoluble asphaltenes are detected in the blend of oils, the nonsolvent oil equivalence test, described in the next section, should be used. Otherwise, the solvent oil equivalence is used.

As described previously, the solvent oil equivalence is just like the toluene equivalence run on the reference oil except that, instead of toluene, the sample oil is used. Instead of measuring the oil by volume, the sample oil is measured by weight, determined by multiplying the desired volume by the density of the sample oil. Also, the starting point for the test liquid is typically 50% sample oil, rather than 20% toluene for the toluene equivalence. Otherwise, follow the same procedure to determine the SOE as was done for the toluene equivalence and use equation (5.38) to calculate the solubility blending number of the sample oil.

5.3.15.6 Nonsolvent Oil Equivalence

This test is applied to oils with no asphaltenes or with toluene equivalence below 5 and found to be a nonsolvent for the reference oil. Previously (section 5.3.12.4), the nonsolvent oil dilution test was applied in this case, but the high viscosity of oil blends greatly slows down the kinetics of asphaltene phase separation. Therefore, the nonsolvent oil dilution test has been replaced by the nonsolvent oil equivalence test. In this test the toluene equivalence test is repeated on the reference oil, however, n-heptane is replaced by the nonsolvent oil. A typical starting point for the test liquid is 2 mL of toluene and 8 mL of sample oil (NSOE = 20) and the sample oil is weighed, rather than measured by volume. The final percentage of toluene in the test liquid (NSOE) is compared with the TE of the reference oil, and the solubility blending number of the sample oil (S_{SO}) is calculated from

$$S_{SO} = \frac{100[TE - NSOE]}{100 - NSOE}$$

(5.49)

5.3.16 Effect of Changing the Normal Paraffin

The research related to this section was the result of a collaboration of the author with Harvey Yarranton and Kamran Akbarzadeh of University of Calgary, and Parviz Rahimi and Alem Teclemariam of the National Centre for Upgrading Technology.[44] This is another example in which a model is pushed beyond the boundaries that it describes. Despite the large number of successes of the Oil Compatibility Model, the data in this section show that the basic assumption of the Oil Compatibility Model is not always true. As previously, this forces one to increase knowledge about the phenomena to account for the discrepancies. However, this case is not just an extension, but threatens the basic core of the model. Why has this not caused false predictions by the Oil Compatibility Model? What corrections are required? These questions need to be answered in this section.

Previously, equation (5.45) was shown to successfully convert solvent equivalence data collected using a different solvent than toluene (such as chlorobenzene) into the toluene equivalence reference based on the solubility parameter of the solvent. Table 5.8 shows that replacing toluene with cyclohexane in the toluene equivalence test for three crude oils produces an average ratio of the toluene equivalence to the cyclohexane equivalence of 0.49. This is in good agreement with the ratio of 0.48 predicted by equation (5.45) based on the solubility parameters of toluene, cyclohexane, and n-heptane. However, replacing n-heptane in the heptane dilution test with a different normal paraffin has not been successful. For normal paraffins, the solubility parameter increases with increasing number of carbons. As a result, normal paraffins are expected to become better solvents

TABLE 5.8
Cyclohexane and Toluene Equivalence

Crude Oil	Toluene Equivalence	Cyclohexane Equivalence	TE/Cyhex Equivalence
A	30	64	0.47
B	12	23	0.52
C	27	58	0.47
Average			0.49
Predicted by OCM			0.48

Source: IA Wiehe, HW Yarranton, K Akbarzadeh, PM Rahimi, A Teclemariam. *Energy and Fuels* 19: 1261–1267, 2005. Reprinted with permission. Copyright 2005. American Chemical Society.

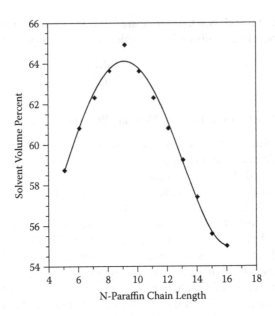

FIGURE 5.26 The volume percentage of normal paraffin at the flocculation point for Cold Lake bitumen 1 is maximum for nonane. (From IA Wiehe, HW Yarranton, K Akbarzadeh, PM Rahimi, A Teclemariam. *Energy and Fuels* 19: 1261–1267, 2005. Reprinted with permission. Copyright 2005. American Chemical Society.)

(poorer nonsolvents) for an oil with increasing number of carbons. As expected, at high dilutions of an oil in normal paraffins as one increases the carbon number of the normal paraffin, the amount of precipitate decreases.[45] Therefore, it is surprising to find that the volume percentage of normal paraffin at the point of incipient precipitation does not increase monotonically with increasing carbon number (chain length) of the normal paraffin. Instead, as shown in Figure 5.26 for Cold Lake bitumen 1 and in Table 5.9 for two different samples of Cold Lake bitumen, Athabasca bitumen, Maya crude, and an Alberta crude at room temperature, the volume percentage of normal paraffin at the point of incipient precipitation increases, reaches a maximum at about 8 to 10 carbons, and decreases with increasing carbon number of the normal paraffin.[44] This indicates that after the maximum, normal paraffins become a poorer solvent for asphaltenes with increasing carbon number. Previously, Hotier and Robin[46] found that the maximum in volume of n-paraffins at the onset of asphaltene precipitation occurred at a carbon number of 7 (n-heptane). Although the carbon number at the maximum volume varies with the oil, it is clear that for n-paraffins that the onset of asphaltene precipitation does not occur at the same solubility parameter of the mixture as required by the basic assumption of the Oil Compatibility Model. What is different about increasing molecular size, such as the paraffin series, that decreases the solvency below that expected from their solubility parameters? How can long-chain normal paraffins be a better solvent for asphaltenes at high

TABLE 5.9
Volume Fractions of *n*-Paraffin at the Onset of Asphaltene Precipitation

n-Paraffin Carbon No.	Cold Lake Bitumen 1	Cold Lake Bitumen 2	Athabasca Bitumen	Maya Crude	Alberta Crude
5	0.587	0.645	0.650	0.412	0.587
6	0.608	0.669	0.669	0.438	0.618
7	0.623	0.669	0.671	0.462	0.640
8	0.636	0.669	0.669	0.474	0.645
9	0.649	0.669	0.673	0.474	0.653
10	0.636	0.669	0.669	0.462	0.667
11	0.623	0.669	0.669		0.645
12	0.608	0.645	0.669	0.438	0.640
13	0.592	0.632	0.660	0.438	0.624
14	0.574	0.621	0.645		0.618
15	0.556	0.612	0.630		0.600
16	0.550	0.590	0.618	0.412	0.590

Source: IA Wiehe, HW Yarranton, K Akbarzadeh, PM Rahimi, A Teclemariam. *Energy and Fuels* 19: 1261–1267, 2005. Reprinted with permission. Copyright 2005. American Chemical Society.

dilutions, but a poorer solvent for asphaltenes at the flocculation point? These questions express the paradox that needs be explained. Actually the paradox is not necessarily about the maximum because the appearance of a maximum depends on how the data are plotted. As is shown in Figure 5.27, if the data are plotted as g-moles of *n*-paraffin added per 5 mL oil versus *n*-paraffin carbon number, no maximum appears. Although this data is for Cold Lake bitumen 1, the data for the other oils also do not show maximums when plotted this way. Nevertheless, the data still show that *n*-paraffins are poorer solvents for flocculation data with increasing *n*-paraffin carbon number, contrary to what is expected.

5.3.16.1 OCM Approximate Method for Normal Paraffin Nonsolvents

The approximate method treats normal paraffins, except *n*-heptane, as nonsolvent oils and uses the data on incipient precipitation of the five oils in Table 5.9 to determine an average effective solubility blending number, or solubility parameter, for the normal paraffins. Thereafter, the effective solubility blending numbers can be used to calculate the incipient precipitation values for other oils. Equation (5.36) can be used to calculate effective solubility blending numbers once the solubility blending number of Cold Lake bitumen is determined. Equation (5.13) and equation (5.16) were used to calculate the insolubility blending number and the solubility blending number of the five oils using the measured density, toluene equivalence,

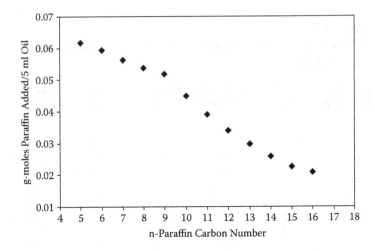

FIGURE 5.27 The g-moles of normal paraffin per 5 mL of Cold Lake bitumen 1 at the flocculation point decreases with increasing normal paraffin carbon number without a maximum. (From IA Wiehe, HW Yarranton, K Akbarzadeh, PM Rahimi, A Teclemariam. *Energy and Fuels* 19: 1261–1267, 2005. Reprinted with permission. Copyright 2005. American Chemical Society.)

and heptane dilution as shown in Table 5.10 along with other properties. Rather than using equation (5.36) directly, it was combined with equation (5.13) to obtain

$$S_{\text{Paraffin}} = I_N \left[1 - \frac{V_H}{V_{\text{Paraffin}}} \right] \tag{5.50}$$

The average effective solubility blending numbers of the *n*-paraffins so calculated are in Table 5.11. Now these average effective solubility blending numbers of the *n*-paraffins can be used to calculate the *n*-paraffin dilution volumes at the flocculation point by rearranging equation (5.50):

$$V_{\text{paraffin}} = \frac{V_H}{1 - \frac{S_{\text{Paraffin}}}{I_N}} \tag{5.51}$$

The result will be compared in section 5.3.15.8 with experimental values along with the regular Flory–Huggins model.

5.3.16.2   Mixing Molecules of Different Sizes

One difference about large molecules, such as polymers, is that they have restricted motion compared with molecules made up of about the size of the repeating unit because of the covalent bonds between units. As a result, chain molecules have much fewer possible arrangements and, thus, much lower entropy of mixing (ΔS^M) than the random mixing required for ideal entropy of mixing.

TABLE 5.10
Properties of the Five Oils in This Study

Property	Cold Lake Bitumen 1	Cold Lake Bitumen 2	Athabasca Bitumen	Maya Crude	Alberta Crude
Saturates					
Yield (wt%)	19.5[a]	19.5	16.4		38.4
Density (g/mL)	0.882[a]	0.882	0.885		0.882[a]
Molecular weight	508[a]	508	524		508[a]
Aromatics					
Yield (wt%)	38.2[a]	38.2	40.1		37.9
Density (g/mL)	0.995[a]	0.995	1.003		0.995[a]
Molecular weight	522[a]	522	550		522[a]
Resins					
Yield (wt%)	26.8[a]	26.8	28.7		12.1
Density (g/mL)	1.037[a]	1.037	1.04		1.037[a]
Molecular weight	930[a]	930	976		930[a]
C_5.Asphaltenes					
Yield (wt%)	15.5[a]	15.5	14.8		11.6
Density (g/mL)	1.203[a]	1.203	1.203[a]		1.203[a]
Molecular weight	2850[a]	2850	2910		2600
Density (g/mL)	0.991	1.001	0.9952	0.876	0.9327
Toluene equivalence	19	19.5	18.5	38	21.0
Insolubility number	28.5	33	31.4	47	34
Solubility blending number	76	99.7	95.5	71	94.5

[a] Not measured, but assumed to be same as Cold Lake Bitumen 2.

Source: IA Wiehe, HW Yarranton, K Akbarzadeh, PM Rahimi, A Teclemariam. *Energy and Fuels* 19: 1261–1267, 2005. Reprinted with permission. Copyright 2005. American Chemical Society.

The Oil Compatibility Model uses only the solubility parameter as the measure of solvent quality among liquids for a given oil. The solubility parameter measures energy of interaction that can be used to estimate the enthalpy of mixing (ΔH^M) if one neglects any volume of mixing. However, phase equilibria at constant temperature and pressure is determined by the minimum in the Gibbs free energy. If the Gibbs free energy of mixing (ΔG^M) two liquids is positive, the two liquids will be completely immiscible. Even though the Gibbs free energy of mixing is negative, two liquids may be insoluble if by splitting into two phases containing both liquids in which the Gibbs free energy is lower than if they were in one phase. Nevertheless, the simplified view is that, for solubility, the Gibbs free energy of mixing should be negative.

$$\Delta G^M = \Delta H^M - T\Delta S^M \qquad (5.52)$$

The entropy of mixing is always positive with the ideal entropy of mixing being its maximum value. If we neglect complexing interactions as implied by using

TABLE 5.11
Effective Solubility Blending Numbers for *n*-paraffins and Five Oils

n-Paraffin Carbon No.	Cold Lake Bitumen 1	Cold Lake Bitumen 2	Athabasca Bitumen	Maya Crude	Alberta Crude	Average Eff. S_{BN}
5	−4.62	−3.63	−3.04	−10.74	−8.62	**−6.13**
6	−1.84	0.00	−0.31	−4.82	−3.36	**−2.07**
7	0.00	0.00	0.00	0.00	0.00	**0.00**
8	1.63	0.00	−0.31	2.09	0.75	**0.83**
9	3.08	0.00	0.30	2.09	1.81	**1.46**
10	1.63	0.00	−0.31	0.00	3.74	**1.01**
11	0.00	0.00	−0.31		0.75	**0.11**
12	−1.84	−3.63	−0.31	−4.82	0.00	**−2.12**
13	−3.93	−5.76	−1.62	−4.82	−2.46	**−3.72**
14	−6.33	−7.65	−3.80		−3.36	**−5.28**
15	−9.12	−9.19	−6.26		−6.35	**−7.73**
16	−10.05	−13.29	−8.14	−10.74	−8.03	**−10.05**

Source: IA Wiehe, HW Yarranton, K Akbarzadeh, PM Rahimi, A Teclemariam. *Energy and Fuels* 19: 1261–1267, 2005. Reprinted with permission. Copyright 2005. American Chemical Society.

the one-dimensional solubility parameter, the enthalpy of mixing can be zero or positive, but not negative. In order to maintain solubility of two liquids, one needs to minimize the enthalpy of mixing and maximize the entropy of mixing. The Oil Compatibility Model focuses on minimizing the enthalpy of mixing by requiring the solubility blending number (or solubility parameter) of the mixture to be below the insolubility number (solubility parameter at which asphaltenes precipitate) of either component. As long as we compare the solubility of oils in liquids of about the same size, the entropy of mixing will be about the same, and its influence can be neglected. Although the entropy of mixing helps determine the solubility parameter at which asphaltenes precipitate, as long as the liquids are about the same size, this flocculation solubility parameter remains the same. However, once we vary the size of the liquid molecules that we blend with oils, the entropy of mixing will vary, and a method will be required to account for its influence on the flocculation point. However, because all the liquids that we consider blending with oils have lower molecular weight than asphaltenes, the size difference between asphaltenes and these liquids becomes less as the carbon number of the paraffin is increased. Therefore, this predicts that increasing paraffin carbon number will increase the entropy of mixing with asphaltenes and make longer paraffins better solvents for asphaltenes than expected, not poorer solvents as observed. There must be some other reason for what is observed.

Although the lower entropy of mixing is usually the only reason that is discussed for the low solubility of polymers, there is another reason. Actually, the solubility is lowest when two polymers are blended. Very few mixtures of two polymers are compatible, requiring the formation of complexes between the polymers for compatibility.[47] Even mixing two different polymers separately dissolved in the same solvent will usually result in precipitation of one of the polymers. Mixing two different polymers of the same molecular weight is predicted to have almost ideal entropy of mixing, but they are likely to be incompatible. Although the repeating units of polymers have similar interaction energies (solubility parameters) as low-molecular-weight liquids, when this is added up, the interaction energies per molecule (or mole) is huge. Thus, mixing of two polymers would result in extremely high heat of mixing per mole causing the polymers not to mix (positive free energy of mixing). Therefore, when an oil containing asphaltenes is mixed with a paraffin, the heat of mixing per mole will increase with increasing paraffin carbon number even though the difference between the solubility parameter of the paraffin and the asphaltenes becomes less because the molecular weight (and molar volume) of the paraffin increases. Thus, we conclude that increasing the carbon number of the paraffin can make the paraffin a better solvent for asphaltenes because of higher entropy of mixing, but a poorer solvent for asphaltenes because of higher enthalpy of mixing. As solubility is governed by Gibbs free energy, the result is the solvency will go through a maximum with increasing paraffin carbon number. Now, a model will be discussed that includes both effects.

5.3.16.3 The Regular Flory–Huggins Model

One of the advantages of the Oil Compatibility Model is that solubility parameters were used without the requirement of formulating how the chemical potential, or other Gibbs free energy function, varies with composition. Although solubility parameters were devised as part of the regular solution model, very few liquid mixtures are found to be described by the regular solution model. The regular solution model is only expected to produce a semiquantitative representation for mixtures of nonpolar compounds.[48] However, to quantitatively account for solubility one needs to model free energy functions that include contributions from the entropy and the enthalpy (equation 5.52). The simplest formulation of the enthalpy of mixing with solubility parameters is the regular solution model. If the volume of mixing is assumed to be zero, the enthalpy of mixing is equal to the energy of mixing. For the mixing of two liquids, 1 and 2, the regular solution model[49] gives

$$\Delta H^M = (V_1 x_1 + V_2 x_2)(\delta_1 - \delta_2)^2 \phi_1 \phi_2 \qquad (5.53)$$

where
V_i = molar volume of component i
x_i = mole fraction of component i
δ_i = solubility parameter of component i
ϕ_i = volume fraction of component i, given by

$$\phi_i = \frac{V_i x_i}{\sum V_i x_i}$$

or for two components,

$$\phi_1 = \frac{V_1 x_1}{V_1 x_1 + V_2 x_2}; \quad \phi_2 = \frac{V_2 x_2}{V_1 x_1 + V_2 x_2} \qquad (5.54)$$

The entropy of mixing is ideal for a regular solution:

$$\Delta S_{\text{Ideal}}^M = -R[x_1 \ln(x_1) + x_2 \ln(x_2)] \qquad (5.55)$$

The simplest model for the entropy of mixing of chain molecules is the Flory–Huggins equation:

$$\Delta S_{\text{F-H}}^M = -R[x_1 \ln(\phi_1) + x_2 \ln(\phi_2)] \qquad (5.56)$$

Although the first system, benzene and rubber, compared with the Flory–Huggins equation was in fair agreement with the data, other systems in general do not conform to the model.[50] In reality the Flory–Huggins equation is a lower limit on the entropy of mixing of noncomplexing chain liquids, whereas the ideal entropy of mixing is an upper limit.[51] Although the Flory–Huggins equation explained why polymer solutions have low entropy of mixing, it should not be expected to be more than semiquantitative.[50] More modern models use the Wertheim theory[52,53], such as the SAFT (statistical associating fluid theory) equation of state.[54–56] The SAFT equation of state was developed by Chapman and Gubbins at Cornell University with collaboration and financial support of Exxon through the author's former colleague Mac Radosz. Although the SAFT equation of state was initially derived to account for asphaltene association in petroleum mixtures, its greatest success has been the application to polymer solutions. Nevertheless, the SAFT equation of state has been applied to several petroleum systems.[57–59] However, Ting, Chapman, and Hirasaki[58] found it was not necessary to include asphaltene association and verified that the onset of asphaltene precipitation occurs at a nearly constant solubility parameter (or refractive index). These are assumptions in the Oil Compatibility Model. However, the SAFT equation of state is a very mathematically complex model. Thus, following the theme in this book, it makes sense to start with a simple model, such as the regular solution model with the Flory–Huggins entropy of mixing, and only add complexity when needed. As will be seen, even a simple model can become complex when used to predict phase equilibria.

Although applications to petroleum systems of the Flory–Huggins equation combined with regular solution have been relatively recent, it has been applied to polymer systems for over 50 years.[60] Polymers with typical molecular weights of over 100,000 are a much more extreme case than asphaltenes of molecular weight of 3000. Nevertheless, one can learn from the experiences and follow the example of pioneers in polymer solution thermodynamics, most of which was recorded in a book by Nobel Laureate Paul Flory.[60] He combined equation (5.52), equation (5.53), and equation (5.56) for the Gibbs free energy of mixing and differentiated with respect to the number of moles of each component at constant pressure and temperature to determine chemical potentials. We express the result in terms of the chemical potential of asphaltenes, μ_A, and the chemical potential of the maltenes (the rest of the oil after removing asphaltenes), μ_M:

$$\mu_A = \mu_A^* + RT\left[\ln(\phi_A) + \left(1 - \frac{v_A}{v_M}\right)\phi_M + \phi_M^2\frac{v_A(\delta_A - \delta_M)^2}{RT}\right] \quad (5.57)$$

$$\mu_M = \mu_M^* + RT\left[\ln(\phi_M) + \left(\frac{v_M}{v_A} - 1\right)\phi_A + \phi_A^2\frac{v_M(\delta_A - \delta_M)^2}{RT}\right] \quad (5.58)$$

where
μ_A^* = reference chemical potential of the pure asphaltenes
μ_M^* = reference chemical potential of the maltenes alone
v_A = molar volume of the asphaltenes
v_M = molar volume of the maltenes

At equilibrium the chemical potentials are equal in both phases (I and II):

$$\frac{\mu_A^I - \mu_A^*}{RT} = \frac{\mu_A^{II} - \mu_A^*}{RT} \quad (5.59)$$

$$\frac{\mu_M^I - \mu_M^*}{RT} = \frac{\mu_M^{II} - \mu_M^*}{RT} \quad (5.60)$$

Flory[60] discusses the case of a polymer mixed with a solvent and a nonsolvent, which is similar to asphaltenes mixed with maltenes and a nonsolvent. A three-component mixture requires three chemical potentials:

$$\frac{\mu_A - \mu_A^*}{RT} = \ln(\phi_A) + (1 - \phi_A) - \phi_M\left(\frac{v_A}{v_M}\right) - \phi_L\left(\frac{v_A}{v_L}\right)$$
$$+ (\phi_M\chi_{AM} + \phi_L\chi_{AL})(\phi_M + \phi_L) - \phi_M\phi_L\left(\frac{v_A}{v_M}\right)\chi_{ML} \quad (5.61)$$

$$\frac{\mu_M - \mu_M^*}{RT} = \ln(\phi_M) + (1 - \phi_M) - \phi_A\left(\frac{v_M}{v_A}\right) - \phi_L\left(\frac{v_M}{v_L}\right)$$
$$+ (\phi_A\chi_{MA} + \phi_L\chi_{ML})(\phi_A + \phi_L) - \phi_A\phi_L\left(\frac{v_M}{v_A}\right)\chi_{AL} \quad (5.62)$$

$$\frac{\mu_L - \mu_L^*}{RT} = \ln(\phi_L) + (1 - \phi_L) - \phi_A\left(\frac{v_L}{v_A}\right) - \phi_M\left(\frac{v_L}{v_M}\right)$$
$$+ (\phi_A\chi_{LA} + \phi_M\chi_{LM})(\phi_A + \phi_M) - \phi_A\phi_M\left(\frac{v_L}{v_A}\right)\chi_{AM} \quad (5.63)$$

where

$$\chi_{AM} = \frac{v_A(\delta_A - \delta_M)^2}{RT}; \quad \chi_{MA} = \frac{v_M(\delta_A - \delta_M)^2}{RT} = \frac{v_M \chi_{AM}}{v_A}$$

$$\chi_{AL} = \frac{v_A(\delta_A - \delta_L)^2}{RT}; \quad \chi_{LA} = \frac{v_L(\delta_A - \delta_L)^2}{RT} = \frac{v_L \chi_{AL}}{v_A} \qquad (5.64)$$

$$\chi_{ML} = \frac{v_M(\delta_M - \delta_L)^2}{RT}; \quad \chi_{LM} = \frac{v_L(\delta_M - \delta_L)^2}{RT} = \frac{v_L \chi_{ML}}{v_M}$$

The condition for equilibrium requires that chemical potential of each component is equal in each phase:

$$\frac{\mu_A^I - \mu_A^*}{RT} = \frac{\mu_A^{II} - \mu_A^*}{RT} \qquad (5.65)$$

$$\frac{\mu_M^I - \mu_M^*}{RT} = \frac{\mu_M^{II} - \mu_M^*}{RT} \qquad (5.66)$$

$$\frac{\mu_L^I - \mu_L^*}{RT} = \frac{\mu_L^{II} - \mu_L^*}{RT} \qquad (5.67)$$

As the three volume fractions sum to one in each phase, there are four independent compositional variables and three equations, which need to be solved by trial and error. The fourth condition is usually the overall composition. However, solving three equations simultaneously by trial and error is still tedious in this computer age. As a result, simplifying approximations are warranted.

5.3.16.4 Hirschberg Approximations

Hirschberg et al.[61] were the first to apply the regular Flory–Huggins model to petroleum mixtures. They made the simplifying assumption that mixtures of asphaltenes, maltenes, and a liquid could be treated as pseudobinary by considering the solvent to be a mixture of maltenes and the liquid, and applying the following:

$$v_S = x_M v_M + x_L v_L \qquad (5.68)$$

$$\delta_S = \phi_M \delta_M + \phi_L \delta_L \qquad (5.69)$$

This approximation would not account for the fact that the maltenes and liquid will partition differently between the two phases. However, as Hirschberg et al. also made the simplifying, but questionable, assumption that the asphaltene-rich phase only contained asphaltenes, the partition of the maltenes and liquid between phases is not the issue. Nevertheless, from equation (5.65),

$$\mu_A^{II} = \mu_A^* = \mu_A^I \tag{5.70}$$

$$\frac{\mu_A^I - \mu_A^*}{RT} = 0 = \ln(\phi_A) + \left(1 - \frac{v_A}{v_S}\right)\phi_S + \phi_S^2 \frac{v_A(\delta_A - \delta_S)^2}{RT} \tag{5.71}$$

They made still another simplifying assumption that the asphaltenes are dilute at the flocculation point:

$$\phi_A \ll 1$$

$$\phi_S \cong 1 \tag{5.72}$$

Therefore, they arrive at the maximum volume fraction of asphaltenes that is soluble in the oil–solvent mixture:

$$\phi_A^{Max} = \exp\left\{\frac{v_A}{v_S} - 1 - \frac{v_A}{RT}(\delta_A - \delta_S)^2\right\} \tag{5.73}$$

5.3.16.5 Cimino Approximations

Cimino et al.[62] made a different simplifying assumption at the point of phase separation that all the asphaltenes precipitate along with the solvent, leaving a pure solvent phase. This is only a good approximation for very-high-molecular-weight polymers but not for the molecular weight range of asphaltenes.[60] Nevertheless, they obtained:

$$\mu_S^I = \mu_S^* = \mu_S^{II} \tag{5.74}$$

$$\ln(1 - \phi_A^{II}) + \left(\frac{v_S}{v_A} - 1\right)\phi_A^{II} + (\phi_A^{II})^2 \frac{v_S(\delta_A - \delta_S)^2}{RT} = 0 \tag{5.75}$$

Cimino et al.[62] approximated values for asphaltene molar volume (1500 mL/g-mol) and solubility parameter (11.5 {cal/mL}$^{1/2}$) and for oil molar volume (150 mL/g-mol) and solubility parameter (9.98 {cal/mL}$^{1/2}$). As a result, the maximum volume of n-paraffin that could be added without precipitating asphaltenes was predicted to be at n-octane. They also showed that their approximation better described flocculation data using aromatic solvents and n-paraffin nonsolvents with an oil than the approximation of Hirschberg et al.[61] and the steric-stabilization model of Leontaritis and Mansoori.[63] However, when flocculation is

observed with an optical microscope, an asphaltene-rich phase precipitate
is seen, leaving an asphaltene-lean phase behind. At no time is a solvent–oil
phase observed without asphaltenes. It is well known that to precipitate all the
asphaltenes in an oil one needs to dilute the oil with at least 20 times the volume
of the oil in n-paraffin. This is not surprising because only polymers of infinite
molecular weight completely precipitate just beyond the flocculation point.[60]
Thus, the approximation of Cimino et al. does not, and should not, match physical reality.

5.3.16.6   Wang and Buckley Approximations

Wang and Buckley[64] approximated the regular Flory–Huggins model by assuming that asphaltenes, maltenes, and n-paraffin mixtures are a binary mixture with
maltenes and n-paraffin acting as a single solvent in the same proportion in both
the light and heavy phases. Although much better than the other approximations,
this pseudobinary approximation has been criticized for polymer solutions.[65] Nevertheless, the approximation of Wang and Buckley has successfully been applied
to describe quantitatively the change of refractive index of crude oils in n-paraf-
fins at the onset of asphaltene precipitation. Wang and Buckley calculated the
onset condition when the Gibbs free energy of mixing as a function of asphaltene
mole fraction curve has an inflection point with a horizontal tangent line at the
low asphaltene concentration end. Although the variation of refractive index at
the onset of asphaltene precipitation with n-paraffins indicates that the solubility parameter of the mixture also varies, Wang and Buckley did not mention the
maximum in n-paraffin volume.

5.3.16.7   Yarranton et al. Approximations

Yarranton et al.[66–68] have used similar approximations as Hirschberg et al.[61] for
the regular Flory–Huggins (RF-H) model except that Yarranton et al. greatly
increased the number of pseudocomponents to include n-paraffin, saturates, aromatics, resins, and associated asphaltenes with a molecular weight distribution, and
included resins with asphaltenes in the precipitated phase. As a result, their model
is capable of quantitatively describing the fractional precipitation of asphaltenes
from Athabasca and Cold Lake bitumens with volume fraction of n-paraffin from
n-pentane through n-hexadecane. Figure 5.28 provides an example of the result of
the dilution of Athabasca bitumen, separately, in n-pentane, n-hexane, and n-hep-
tane. Because the aromaticity of the precipitated phase varies significantly with
yield, it is necessary to have a number of asphaltene pseudocomponents.

The multicomponent (n-components) regular Flory–Huggins model[60,69] for
the chemical potential of component i, μ_i, is given by

$$\frac{\mu_i - \mu_i^*}{RT} = \ln(\phi_i) + 1 - \sum_{j=1}^{j=n} r_{ij}\phi_j + \frac{v_i(\delta_i - \delta_M)^2}{RT} = \ln(x_i\gamma_i) \qquad (5.76)$$

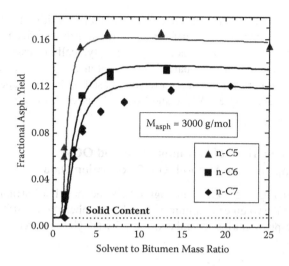

FIGURE 5.28 The yield of insolubles as a function of dilution of Athabasca bitumen in *n*-pentane, *n*-hexane, and *n*-heptane at 23°C as compared with predictions of the regular Flory–Huggins model with the Yarranton approximations. (From H Alboudwarej, K Akbarzadeh, J Beck, WY Svrcwk, HW Yarranton. *AIChE J*, 49: 2948–2956, 2003. With permission.)

where

μ_i^* = standard chemical potential of pure component i

R = gas constant

T = absolute temperature

x_i = mole fraction of component i

v_i = molar volume of component i

ϕ_i = volume fraction of component i = $x_i v_i/(\Sigma x_i v_i)$

$r_{ij} = v_i/v_j$

δ_i = solubility parameter of component i

δ_M = solubility parameter of the mixture = $\Sigma \phi_i \delta_i$

γ_i = activity coefficient of component i

With the assumptions discussed previously, Yarranton et al.[66–68] determined that the ratio of the equilibrium mole fractions of each of the asphaltene and resin species in light (L) and heavy (H) phases is given by

$$K_i = \gamma_i^L = \exp\left\{\ln\frac{v_i^L}{v_M^L} + 1 - \frac{v_i^L}{v_M^L} + \frac{v_i^L}{RT}(\delta_i - \delta_M^L)^2\right\} = \frac{x_i^H}{x_i^L} \qquad (5.77)$$

where

v_M^L = molar volume of the mixture in the light phase = $\Sigma x_i v_i^L$

The regular Flory–Huggins model is extremely sensitive to the values of the solubility parameters with differences in the fourth significant figure being important. This requires higher accuracy than is typically measured for solubility parameters, including for n-paraffins. Yarranton et al.[68] calculated solubility parameters for n-paraffins from a correlation based on heat of vaporization and molar volume data, whereas Wang and Buckley[64] used a correlation with refractive index data.

5.3.16.8　Comparison of Yarranton RF-H and OCM Approximate Method with Flocculation Data

As is shown in Figure 5.29 through Figure 5.33, the model of Yarranton et al. correctly predicts the volume maximum with carbon number of n-paraffins at the onset of asphaltene precipitation (calculated for 0.5 wt% yield of precipitate) for four

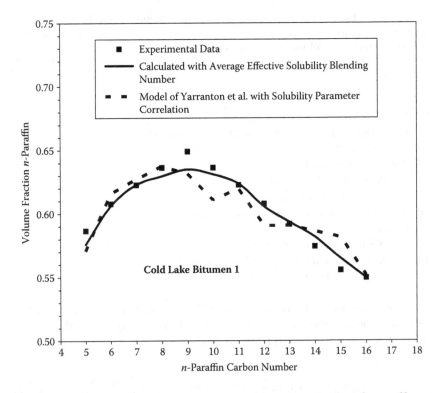

FIGURE 5.29 Comparison of experimental data of the volume fraction of n-paraffins at the onset of asphaltene precipitation with two methods of calculation for Cold Lake bitumen 1. (From IA Wiehe, HW Yarranton, K Akbarzadeh, PM Rahimi, A Teclemariam. *Energy and Fuels* 19: 1261–1267, 2005. Reprinted with permission. Copyright 2005. American Chemical Society.)

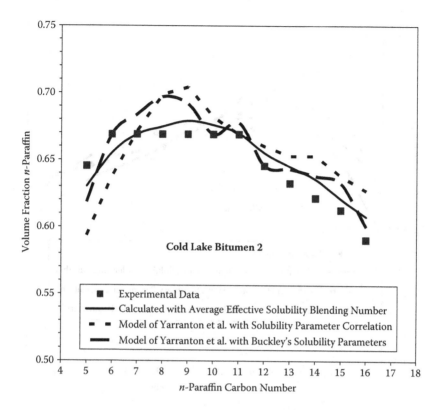

FIGURE 5.30 Comparison of experimental data of the volume fraction of *n*-paraffins at the onset of asphaltene precipitation with three methods of calculation for Cold Lake bitumen 2. (From IA Wiehe, HW Yarranton, K Akbarzadeh, PM Rahimi, A Teclemariam. *Energy and Fuels* 19: 1261–1267, 2005. Reprinted with permission. Copyright 2005. American Chemical Society.)

different oils with either the solubility parameters of Yarranton et al. and of Wang and Buckley. Therefore, the cause of this volume maximum is consistent with mixing molecules of much different sizes as predicted by the regular Flory–Huggins model.

As shown in Figure 5.29 through Figure 5.33, the Oil Compatibility Model with average effective solubility blending numbers can describe well the asphaltene onset data for all the oils. This demonstrates why the Oil Compatibility Model with compatibility numbers measured by asphaltene onset titrations has been successful at predicting oil compatibility even though the onset does not strictly occur at the same solubility parameter of the mixture. The effective solubility parameter (solubility blending number) of the oil measured with toluene and *n*-heptane is a good approximation of the effective solubility parameter of the oil when blended with other oils.

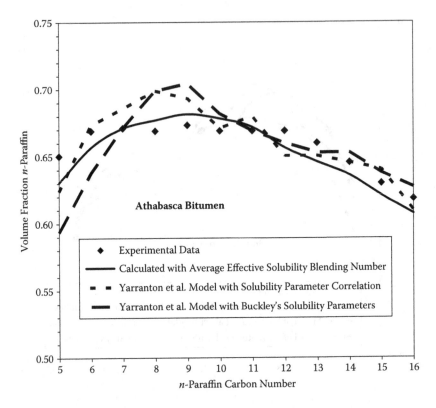

FIGURE 5.31 Comparison of experimental data of the volume fraction of *n*-paraffins at the onset of asphaltene precipitation with three methods of calculation for Athabasca bitumen. (From IA Wiehe, HW Yarranton, K Akbarzadeh, PM Rahimi, A Teclemariam. *Energy and Fuels* 19: 1261–1267, 2005. Reprinted with permission. Copyright 2005. American Chemical Society.)

5.3.16.9　Conclusions and Discussion of Results

The maximum in volume of *n*-paraffin as a function of carbon number of the *n*-paraffin at the onset of asphaltene precipitation is general for crude oils and bitumens. This phenomenon is a result of mixing liquids of greatly different molecular sizes and, as a result, the precipitation of asphaltenes from the onset to large excesses of nonsolvent can be quantitatively described by the Yarranton et al. approximations to the regular Flory–Huggins model. This model is even more successful for petroleum oils than for organic polymers. This is yet another case in which a simple model works better for petroleum oils than for other less complex systems. Part of the reason this model succeeds is that only dispersion interactions are important for petroleum oils (section 5.2.3), but the complete reason is unknown. Nevertheless, it confirms the conclusion of Wang and Buckley[64] that the onset of asphaltene precipitation does not necessarily happen at the same

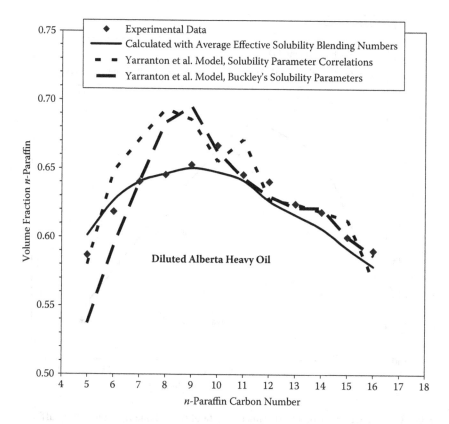

FIGURE 5.32 Comparison of experimental data of the volume fraction of *n*-paraffins at the onset of asphaltene precipitation with three methods of calculation for a diluted Alberta heavy crude oil. (From IA Wiehe, HW Yarranton, K Akbarzadeh, PM Rahimi, A Teclemariam. *Energy and Fuels* 19: 1261–1267, 2005. Reprinted with permission. Copyright 2005. American Chemical Society.)

solubility parameter of the mixture for a given petroleum oil, but can be affected by the molar volumes of the components in the mixture. When the solvent and nonsolvent blended with oil is kept at the same order of magnitude of molar volume, the solubility parameter of the mixture at the flocculation point is constant. However, if the solvent and nonsolvent blended with the oil varies significantly in molar volume, the solubility parameter at the flocculation point varies. Although this contradicts the basic assumption of the Oil Compatibility Model, the practice of determining effective solubility parameters (solubility blending numbers) by asphaltene onset titrations directly on petroleum oils accounts for the size effect.

The model of Yarranton et al. and the Oil Compatibility Model are directed to different applications. The model of Yarranton et al. is more scientifically satisfying and predicts both the maximum in the onset volume and the greater precipitation at large dilutions with increasing carbon number of the *n*-paraffin nonsolvent.

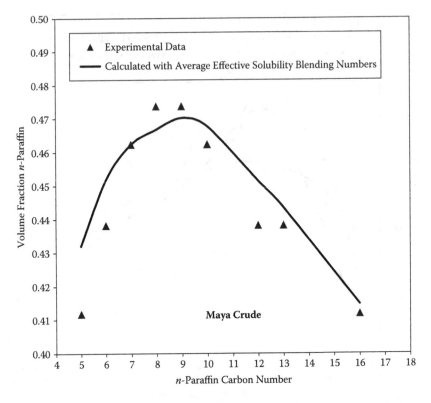

FIGURE 5.33 Comparison of experimental data of the volume fraction of *n*-paraffins at the onset of asphaltene precipitation with the approximate oil compatibility model for Maya crude oil. (From IA Wiehe, HW Yarranton, K Akbarzadeh, PM Rahimi, A Teclemariam. *Energy and Fuels* 19: 1261–1267, 2005. Reprinted with permission. Copyright 2005. American Chemical Society.)

This ability to describe the complete precipitation of asphaltenes with increasing dilution should find applications to deasphalting processes. The main application of the Oil Compatibility Model is in cases in which there are a minimum of characterization data and in fouling mitigation where no asphaltene precipitation, or a tendency for asphaltene precipitation, is desired. Maya crude is an example in which the Oil Compatibility Model can describe the asphaltene onset data with only density and toluene equivalence data.

5.4   EFFECT OF PRESSURE ON ASPHALTENE SOLUBILITY FOR LIVE OILS

Petroleum in the reservoir commonly contains dissolved methane and other gases, called "live oil." Under high pressure the dissolved hydrocarbon gas can be a solvent for asphaltenes. However, at moderate pressures the lower density gas

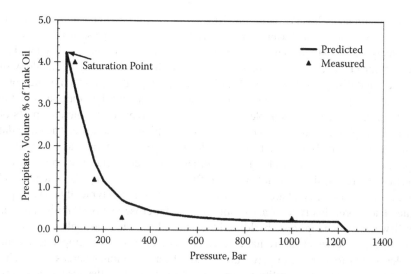

FIGURE 5.34 The equation of state model of Firoozabadi describes the precipitation of asphaltenes from a live oil as the pressure is decreased to the bubble (saturation) point. (From A. Firoozabadi, *Thermodynamics of Hydrocarbon Reservoirs*, McGraw-Hill, New York, 1999, p. 328. With permission.)

can be a nonsolvent for asphaltenes, causing precipitation of asphaltenes in the well bore. At low pressures, below the saturation point, the hydrocarbon gas is no longer soluble in the oil and bubbles out. This can make the asphaltenes soluble in the oil again. An equation of state model is required to quantitatively describe the change in solvency of the mixture as a function of pressure. An example in Figure 5.34 shows how well an equation of state model of Firoozabadi[70] describes the precipitation of asphaltenes of an oil mixed with *n*-propane as a function of pressure based on properties of the oil. Gonzalez et al.[71] have used the SAFT equation of state to predict the onset of asphaltene precipitation on addition of light gases, such as methane, ethane, carbon dioxide, and nitrogen under pressure. Wang, Buckley, and Creek[72] have combined a single flocculation titration on the dead (or tank) oil with routine PVT data, and composition analysis to predict if asphaltenes can be expected to precipitate during oil production.

5.5 FUTURE NEEDS FOR ASPHALTENE PHASE BEHAVIOR

Although significant progress has been made recently, many important scientific questions remain that need to be answered. The most basic question remaining is about asphaltene association and the role of resins in breaking up this association.

5.5.1 ASPHALTENE ASSOCIATION

Evidence presented in this chapter shows that the dominant attractive interaction between asphaltene molecules is caused by dispersion forces. How can one then explain asphaltene association if the dominant interaction is not directional? How can one best model asphaltene association? These are questions that need to be answered to make progress in understanding petroleum phase behavior. The author currently favors the explanation built on the block copolymer analogy of Sirota[73] and discussed in section 2.7.3. Part of the asphaltene molecule, the poly-nuclear aromatics, is insoluble in the oil and solvents, and the other parts of the asphaltene molecule are soluble. This forces the polynuclear aromatics parts of different asphaltene molecules to cluster on a submicroscopic scale. The large flat surface of polynuclear aromatics allows for large dispersion interaction energies when they are stacked parallel to each other. However, because the polynuclear aromatics are covalently bonded to paraffin chains, small ring naphthenes, and small ring aromatics that are soluble, the asphaltenes do not typically phase sepa-rate. Therefore, this type of association is a submicroscopic insolubility and fits well with the pendant-core building block model, in which asphaltenes contain two or more polynuclear aromatic cores per molecule. Higher temperatures and higher solubility parameter solvents decrease the association, whereas lower tem-peratures and lower solubility parameter solvents increase the association until phase separation results. Thermal cracking the pendant groups off the cores then clearly increases association and asphaltene insolubility as is observed. The chal-lenge remains as how to put this concept into a relative simple model of asphal-tene association and solubility.

5.5.2 ROLE OF RESINS IN ASPHALTENE SOLUBILITY

Both the Oil Compatibility Model and the regular Flory–Huggins model neglect the role of resins as a dispersant. However, the author used the dispersant nature of resins to explain why the Oil Compatibility Model predicts oil blends to be incompatible, but sometimes experimentally they are found to be compatible. The next chapter on fouling mitigation will provide evidence that the selective conversion of resins relative to asphaltenes in hydrotreating can cause asphal-tene insolubility. Wang and Buckley[74] titrated 1 vol% solutions of asphaltenes in toluene with several n-paraffins and repeated on addition of 1 to 5 vol% resins. The refractive index at the flocculation point decreased linearly with vol% res-ins, showing that resins made asphaltenes more soluble. Wang and Buckley con-cluded that resins increase the solubility of asphaltenes more than as a solvent, but that there is no evidence of specific interactions between asphaltenes and resins. Rahimi et al.[75] added 5 to 20 wt% Athabasca resins to Athabasca bitumen (resin to asphaltene weight ratio of 1.42) and its visbroken product (resin to asphaltene weight ratio of 0.79) and applied the heptane dilution and toluene equivalence tests. The calculated insolubility numbers and solubility blending numbers are given in Table 5.12 along with P-values (ratio of solubility blending number to insolubility number). The decrease in insolubility number measures the disper-sant effect of resins, whereas the increase in solubility number measures the sol-vent effect of resins. Although the combination of dispersant and solvent effects of resins increase stability as measured by P-values, the solvent effect is greater

TABLE 5.12

Effect of Resins on Compatibility of Athabasca Bitumen (AB) and Visbroken (VB) Athabasca Bitumen

	I_N	S_{BN}	P-Value
VB	89.3	104	1.17
VB + 5 wt% resin	88.9	114	1.28
VB + 10 wt% resin	84.2	119	1.41
VB + 20 wt% resin	82.4	131	1.59
AB	34.2	93.0	2.72
AB + 5 wt% resin	32.9	104	3.17
AB + 10 wt% resin	31.9	110	3.45
AB + 20 wt% resin	31.1	117	3.75

than the dispersant effect. Thus, based on this data, the dispersant effect of Athabasca resins is less than expected. As a method of comparison, these experiments were repeated with a synthetic dispersant, dodecylbenzene sulfonic acid (DBSA), instead of resins. The results in Table 5.13 show that DBSA, which has a strong specific interaction with asphaltenes, decreases the insolubility number much greater than resins. The effect of DBSA on the solubility blending number is odd in that it greatly increases the solubility blending number when added to visbroken Athabasca bitumen, but greatly decreases the solubility blending number when added to Athabasca bitumen. As a result, DBSA increases the stability, as measured by P-value, of visbroken Athabasca bitumen more than resins, but not

TABLE 5.13

Effect of Dodecylbenzene Sulfonic Acid (DBSA) on Compatibility of Athabasca Bitumen (AB) and Visbroken (VB) Athabasca Bitumen

	I_N	S_{BN}	P-Value
VB	89.3	104	1.17
VB + 5 wt% DBSA	71.8	101	1.40
VB + 10 wt% DBSA	62.6	122	1.94
VB + 20 wt% DBSA	43.0	214	4.97
AB	34.2	93.0	2.72
AB + 0.5 wt% DBSA	30.5	94.5	3.10
AB + 5 wt% DBSA	22.7	71.5	3.15
AB + 10 wt% DBSA	14.4	45.0	3.13

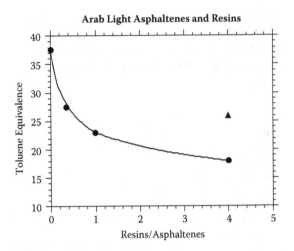

FIGURE 5.35 Arab Light resins decrease the toluene equivalence of Arab Light asphaltenes (Data: ●) up to a resin/asphaltene weight ratio of 4, the ratio in Arab Light crude oil (Data: ▲).

as much as resins for Athabasca bitumen. More will be discussed about synthetic dispersants in the next chapter.

Different data by the author indicate that resins can play a greater role in solubilizing asphaltenes. In Figure 5.35, the toluene equivalence is plotted versus resin-to-asphaltene weight ratio for Arab Light resins and asphaltenes without the rest of the Arab Light crude oil (saturates and aromatics). Thus, up to equal weight resins and asphaltenes, the resins greatly decrease the toluene equivalence, but in going from a resin to asphaltene ratio of 1 to 4 (the ratio in the crude oil) there is only a gradual decrease in toluene equivalence. This is more like the expected behavior if resins acted like an asphaltene dispersant.

Another method to probe asphaltene–resin interactions is to study mixtures that are very dilute in oil. As asphaltene self-association is expected to be stronger than the interaction of asphaltenes with resins, mixtures that are very dilute in oil should break resin–asphaltene interactions and make asphaltenes less soluble than expected. This is what was observed by the author for Cold Lake bitumen. When the percentage of toluene in the test liquid (rest is n-heptane) is plotted versus the volume ratio of oil to test liquid, in the dilute oil range the data curves above (less soluble asphaltenes) the linear behavior predicted by the Oil Compatibility Model. However, the experiments in this very dilute oil region are very tedious by optical microscopy and subject to much greater error than more typical oil-to-test-solvent volume ratios (0.2 and higher). In contrast, Wang and Buckley[76] have observed for a different oil that asphaltenes become more soluble when very dilute. Therefore, the study of asphaltene solubility in this very dilute range deserves much more attention, including more accurate methods for detecting insoluble asphaltenes.

It is clear that much more experimental work needs to be directed to the effect of resins on asphaltene solubility. The dispersant ability of resins for asphaltenes appears to be very feed-dependent. The author currently favors a mechanism that builds on the block copolymer analogy of asphaltene association. Resins are hypothesized to have one polynuclear aromatic core per molecule. As a result, they have a low tendency to self-associate. However, by this one core associating with asphaltene cores, the rest of the resin molecules can reduce asphaltene association by steric interference while increasing solubility by having a larger soluble part of the resin molecule than asphaltenes.

NOMENCLATURE

E_v Energy of vaporization, cal/g-mol
I   Ionization potential
k   Boltzmann constant
N   Number of liquids in a mixture
r   Molecular separation distance
T   Absolute temperature, °K
u   Dipole moment
V   Molar volume, cc/g-mol
d   Solubility parameter, $(cal/cc)^{1/2}$
d_c Complexing solubility parameter component, $(cal/cc)^{1/2}$
d_d Dispersion solubility parameter component, $(cal/cc)^{1/2}$
d_f Field force solubility parameter component, $(cal/cc)^{1/2}$
d_h Hydrogen bonding solubility parameter component, $(cal/cc)^{1/2}$
d_p Polar solubility parameter component, $(cal/cc)^{1/2}$
d_v Volume dependent solubility parameter component, $(cal/cc)^{1/2}$
e   Interaction energy
f   Volume fraction

SUPERSCRIPTS

d   Dispersion
h   Hydrogen bonding
M   Mixture
p   Polar

SUBSCRIPTS

A   A material
B   A liquid or liquid mixture
c   Complexing
d   Dispersion
f   Field force
h   Hydrogen bonding

i   Any of the liquids in a mixture
p   Polar
v   Volume dependent
1   Component 1
2   Component 2
12   Mixture of components 1 and 2

REFERENCES

1. JH Hildebrand, SE Wood, The derivation of equations for regular solutions. *J Chem Phys* 1: 817–822, 1933.
2. G Scatchard, Equilibria in non-electrolyte solutions in relation to the vapor pressures and densities of the components. *Chem Rev* 8: 321–333, 1931.
3. CM Hansen, *The Three Dimensional Solubility Parameter and Solvent Diffusion Coefficient*, Copenhagen: Danish Technical Press, 1967, pp. 1–46.
4. CM Hansen, The three dimensional solubility parameter–key to paint component affinities: I. Solvents, plasticizers, polymers, and resins. *J Paint Technol* 39: 104–117, 1967.
5. CM Hansen, The universality of the solubility parameter. *I & EC Prod Res Dev* 8: 2–11, 1969.
6. H Burrell, Solubility Parameters for Film Formers. *Off Digest Fed Soc Paint Technol* 27: 726–758, 1955.
7. EP Lieberman, Quantification of the hydrogen bonding parameter. *Off Digest Fed Soc Paint Technol* 34: 444–453, 1962.
8. JD Crowley, GS Teague, JW Lowe, A three-dimensional approach to solubility. *J. Paint Technol* 38: 269–280, 1966.
9. IA Wiehe, Polygon mapping with two-dimensional solubility parameters. *I & EC Res* 34: 661–673, 1995.
10. JH Hildebrand, RL Scott, *The Solubility of Nonelectroytes*, 3rd ed., New York: Dover Publications, 1964, pp. 160–162.
11. PA Small, Some factors affecting the solubility of polymers. *J Appl Chem* 3: 71–80, 1953.
12. WH Keesom, Die van der Waalsschen Kohasionskrafte. *Z Physik* 22: 126–141, 1921.
13. WH Keesom, Die Berechnung der Molekularen Quadrupolmomente aus der Zustandsgleichung. *Z Physik*, 23: 225–228, 1922.
14. F London, Zur Theorie und Systematik der Molekularkrafte. *Z Physik* 63: 245–279, 1930.
15. F London, The general theory of molecular forces. *Trans Faraday Soc* 33: 8–26, 1937.
16. KL Hoy, New values of the solubility parameters from vapor pressure data. *J Paint Technol* 42: 76–118, 1970.
17. IA Wiehe, Two-dimensional solubility parameter mapping of heavy oils. *Fuel Sci Technol Int* 14: 289–312, 1996.
18. IA Wiehe, The oil compatibility model and crude oil compatibility. *Energy and Fuels* 14: 56–59, 2000.
19. GB Dickakian, Blending of Hydrocarbon Liquids. U.S. Patent 4,853,337. Assigned to Exxon, 1989.
20. JP Pfeiffer, RN Saul, Asphaltic bitumen as colloid system. *J Phys Chem* 44: 139–145, 1940.

21. IA Wiehe, KS Liang, Asphaltenes, resins, and other petroleum macromolecules. *Fluid Phase Eq* 117: 201–210, 1996.
22. JJ Heithaus, Measurement and significance of asphaltene peptization. *J Inst Pet* 48: 45–53, 1962.
23. MG Griffith, CW Sigmund, Controlling compatibility of residual fuel oils. In *Marine Fuels*, CH Jones, Ed, Philadelphia: ASTM, 1985, pp. 227–247.
24. JS Buckley, Microscopic investigation of the onset of asphaltene precipitation. *Fuel Sci Technol Int* 14: 55–74, 1996.
25. JS Buckley, Predicting the onset of asphaltene precipitation from refractive index measurements. *Energy and Fuels* 13: 328–332, 1999.
26. IA Wiehe, RJ Kennedy, Process for Blending Potentially Incompatible Petroleum Oils. U.S. Patent 5,871,634. Assigned to Exxon, 1999.
27. WJ van Kerkvoort, AJJ Nieuwstad, M van der Waarden, Le comportement des syn\stemes hydrocarbures-asphaltenes. Un facteur important pour la stabilite et compatibilite das fuel-oils residuels. IV Congr Intern Chauffage Ind, Paris, 1952, Preprint # 220.
28. CWG Martin, DR Bailey, The stability and compatibility of fuel oil and diesel fuel. *J Inst Pet* 40: 138–150, 1954.
29. SI Andersen, Flocculation onset titration of petroleum asphaltenes. *Energy and Fuels* 13: 315–322, 1999.
30. EW Mertens, Predicting weatherability of coating grade asphalts from asphaltene characteristics. *ASTM Bull*: 40–44 (TP 218–222), 1960.
31. IA Wiehe, Are you processing incompatible crude oils? *World Refining* 12: 24–28, October 2001.
32. IA Wiehe, Self-incompatible crude oils and converted petroleum resids. *J Dispersion Sci Technol* 25: 333–339, 2004.
33. IA Wiehe, RJ Kennedy, G Dickakian, Fouling of nearly incompatible oils. *Energy and Fuels* 15: 1057–1058, 2001.
34. IA Wiehe, RJ Kennedy, Process for Blending Petroleum Oils to Avoid Being Nearly Incompatible. U.S. Patent 5,997,723. Assigned to Exxon, 1999.
35. TG Mason, MY Lin, Time-resolved small angle neutron scattering measurements of asphaltene nanoparticle aggregation kinetics in incompatible crude oil mixtures. *J Chem Phys* 119: 565–571, 2003.
36. IA Wiehe, RJ Kennedy, Application of the oil compatibility model to refinery streams. *Energy and Fuels* 14: 60–63, 2000.
37. JF Schabron, AT Pauli, JF Rovani, FP Miknis, Predicting coke formation tendencies. *Fuel* 80: 1435–1446, 2001.
38. JF Schabron, AT Pauli, JF Rovani, Free solvent volume correlation with pyrolytic coke formation. Preprints, ACS, *Div Pet Chem* 46: 99–103, 2001.
39. IA Wiehe, A phase-separation kinetic model for coke formation. *Ind Eng Chem Res* 32: 2447–2454, 1993.
40. T Gentzis, P Rahimi, Processability and thermal behaviour of Athabasca bitumen with varying asphaltenes concentration. Preprints, *ACS Fuel Chem* 44: 810–816, 1999.
41. PM Rahimi, RJ Parker, R Knoblauch, IA Wiehe, Stability and compatibility of partially upgraded bitumen for pipeline transportation. Proceedings of the 7th International Conference on Stability and Handling Fuels, Graz, Austria, 177–192, 2000.
42. PM Rahimi, A Teclemariam, E Taylor, T deBruijn, IA Wiehe, Determination of coking onset of petroleum feedstocks using solubility parameters. preprints, *ACS Fuel Chem* 48: 103–105, 2003.

43. AR De Jong, Determining fouling mechanisms by monitoring and chemical analysis. In the Proceedings of the 4th International Conference on Refinery Processing; Podder, S.K, Boock, L.T., Eds.; New York: AIChE 2001, 193–203.

44. IA Wiehe, HW Yarranton, K Akbarzadeh, PM Rahimi, A Teclemariam, The paradox of asphaltene precipitation with normal paraffins, *Energy and Fuels* 19: 1261–1267, 2005.

45. JG Speight, *The Chemistry and Technology of Petroleum*, 3rd ed., New York: Marcel Dekker, 1998, p. 416.

46. G Hotier, M Robin, Effects of different diluents on heavy oil products: measurement, interpretation, and a forecast of asphaltene flocculation, *Revue de l'IFP* 38: 101, 1983.

47. PJ Flory, *Principles of Polymer Chemistry*, Ithaca, NY: Cornell University Press, 1953, pp. 554–555.

48. JM Prausnitz, *Molecular Thermodynamics of Fluid-Phase Equilibria*, Englewood Cliffs, Prentice-Hall, NJ: 1969, p. 273.

49. JH Hildebrand, RL Scott, *The Solubility of Nonelectroytes*, 3rd ed., New York: Dover Publications, 1964, p. 134.

50. PJ Flory, *Principles of Polymer Chemistry*, Ithaca, NY: Cornell University Press, 1953, pp. 515–516.

51. JH Hildebrand, RL Scott, *The Solubility of Nonelectroytes*, 3rd ed., New York: Dover Publications, 1964, p. 109.

52. MS Wertheim, *J Stat Phys* 42, 459, 477, 1986.

53. MS Wertheim, *J Chem Phys* 85, 2929, 1986.

54. W.G. Chapman, K.E. Gubbins, G. Jackson, M. Radosz, *I & EC Res* 29, 1709, 1990.

55. S.H. Huang, M. Radosz, Equation of state for small, large, polydisperse, and associating molecules, *I & EC Res* 29, 2284–2294, 1990.

56. S.H. Huang, M. Radosz, Equation of state for small, large, polydisperse, and associating molecules: Extension to fluid mixtures, *I & EC Res* 30, 1994–2005, 1991.

57. S.H. Huang, M. Radosz, *Fluid Phase Equilibria*, 70, 33–54, 1991.

58. PD Tang, WG Chapman, GJ Hirasaki, Asphaltene phase behavior: Experiment and modeling using the SAFT equation of state, in the Proceedings of the 3rd International Conference on Petroleum Phase Behavior and Fouling; Wiehe, I.A., Ed.; New York: AIChE, 204–210, 2002.

59. MS Zhuang, MC Thies, Extraction of petroleum pitch with supercritical toluene: Experiment and prediction, *Energy and Fuels,* 14, 70–75, 2000.

60. P.J. Flory, *Principles of Polymer Chemistry*, Ithaca, Cornell University Press, New York: 1953, pp. 495–594.

61. A Hirschberg, LNJ. de Jong, BA. Schipper, JG Meijers, Influence of temperature and pressure on asphaltene flocculation, *Soc Petrol Eng*, 6, 283–293, 1984.

62. R Cimino, S Correra, A Del Bianco, TP Lockhart, Solubility and phase behavior of asphaltenes in hydrocarbon media, In *Asphaltenes Fundamentals and Applications*, E.Y. Sheu and O.C. Mullins, Eds., New York: Plenum Press, pp. 97–130 (1995).

63. KJ Leontaritis and GA Mansoori, Aggregation flocculation during oil recovery and processing: A thermodynamic-colloidal model, Paper SPE 16258, 1987 SPE International Symposium On Oilfield Chemistry, Society of Petroleum Engineers: Richardson, TX.

64. JX Wang, JS Buckley, A two-component model of the onset of asphaltene flocculation in crude oils. *Energy and Fuels*, 15: 1004–1012, 2001.

65. JH Hildebrand, RL Scott, *The Solubility of Nonelectroytes*, 3rd ed., New York: Dover Publications, 1964, p. 380–381.

66. HW Yarranton, JH Masliyah, Molar mass distribution and solubility modeling of asphaltenes, *AIChE J*, 42: 3533–3543, 1996.
67. KD Mannistu, HW Yarranton, JH Masliyah, Energy and fuels, solubility modeling of asphaltenes in organic solvents, *Energy and Fuels*, 11: 615–622, 1997.
68. H Alboudwarej, K Akbarzadeh, J Beck, WY Svrcwk, HW Yarranton, Regular solution model for asphaltene precipitation from bitumens and Solvents, *AIChE J*, 49: 2948–2956, 2003.
69. JH Hildebrand, RL Scott, *The Solubility of Nonelectroytes*, 3rd ed., New York: Dover Publications, 1964, pp. 198–202.
70. A. Firoozabadi, *Thermodynamics of Hydrocarbon Reservoirs*, New York: McGraw-Hill, 1999, p. 328.
71. DL Gonzalez, PD Ting, GJ Hirasaki, WG Chapman, Prediction of asphaltene instability under gas injection with the PC-SAFT equation of state, *Energy and Fuels* 19: 1230–1234, 2005.
72. J Wang, J Buckley, J Creek, Rapid screening for asphaltene precipitation during oil production, Proceedings on 7th International Conference on Petroleum Phase Behavior and Fouling, Paper No. 6, 2006.
73. EB Sirota, Physical structure of asphaltenes. *Energy and Fuels* 19: 1290–1296, 2005.
74. J Wang, JS Buckley, Asphaltene stability in crude oil and aromatic solvents—The influence of oil composition. *Energy and Fuels* 17: 1445–1451, 2003.
75. PM Rahimi, IA Wiehe, D Patmore, A Teclemariam, T de Brujin, Resin-asphaltene interactions in virgin and cracked bitumen. Preprints, *ACS Fuel Chem*, 49: 545–546, 2004.
76. J Wang, JS Buckley, Effect of dilution ratio on amount of asphaltenes separated from stock-tank oil. Presentation at 6th International Conference on Petroleum Phase Behavior and Fouling, Amsterdam, Paper No. 33, 2005.

6 Fouling Mitigation

6.1 INTRODUCTION

From the well through the refinery, condensed phases can form in petroleum and petroleum products that interfere with flow and insulate walls from heat transfer. These fouling phases may be asphaltenes, coke, wax, water emulsions, or inorganic solids. In the past, the petroleum industry accepted such fouling as a price of doing business. However, significant progress has been made recently in elucidating the chemical and physical mechanisms of petroleum phase behavior and fouling. As a result, new methods of diagnosis, treatment, and cure of petroleum fouling are now available. The determination of the cause of fouling is based on the combination of process history and analysis of the foulant, and the oil flowing through the process unit. Tracing the foulant precursors to their source usually suggests a number of alternative mitigation actions. Although most organic fouling in refineries by petroleum is caused by insoluble asphaltenes, including coke formation, inorganic solids, such as iron sulfide and salts, cause a significant fraction of refinery fouling. Thus, general methods to distinguish between these fouling causes and develop mitigation methods as well as specific case studies are discussed.

In section 6.2, a general fouling mitigation strategy will be examined. This grew out of a review and analysis of a number of refinery fouling mitigation case studies by the author to develop a more systematic approach. Some case studies have already been discussed or implied. In section 4.3.7.4, the presence of carbonaceous mesophase in the heavy product of a resid conversion process was used to indicate that coke was forming at thermal cracking temperature by the phase separation of converted asphaltenes. This will be analyzed in more detail in this chapter along with mitigation methods that take advantage of the phase-separation mechanism for coke formation. In section 5.3, the case study on blending the incompatible crude oils, Soudie and Forties, in the wrong order was discussed. Also, the case study of rejecting a self-incompatible crude oil before it fouled a refinery, as well as the possibility of fouling by nearly incompatible crude oils, was covered. In addition, solving the plugging of the inlet of a hydrotreater using the Oil Compatibility Model (OCM) was discussed. Clearly, the carbonaceous mesophase and the OCM are useful tools for fouling mitigation that will be further applied in this chapter.

Although more fouling mitigation case studies will be given in this chapter, there are many more by the author that cannot be mentioned because they remain proprietary, whereas others occurred in refinery units that do not process petroleum macromolecules, such as reformer fouling.[1] One of the cases that is covered in this chapter is popcorn coke caused by the polymerization of conjugated olefins where coking can be mitigated by raising temperature. In another case, fouling of a heat exchanger after a hydrotreater is caused by preferentially converting resins relative to asphaltenes. The fouling by stable oil–water emulsions most commonly happens in desalters and, therefore, will be discussed in section 7.1 where desalting is covered. Although stable oil–water emulsions can cause fouling anywhere in the refinery downstream where water and oil are supposed to be separated, these cases solved by the author are proprietary and, thus, cannot be discussed. On the other hand, in this chapter, mitigation methods will also be given. An example is the testing and optimization of dispersants for mitigating insoluble asphaltene sediments. Finally, engineering methods, and design and physical process changes to mitigate fouling, will be briefly discussed.

6.1.1 DEFINITION OF FOULING

Fouling is defined here as the formation and accumulation of an unexpected phase that interferes with processing. Although the fouling phase is often a solid, it could be a liquid when a gas is expected, or it could be an emulsion when separate liquid phases are expected. Coke formation on a surface of a furnace tube is fouling because it is not expected and it insulates the furnace tube from heat transfer. However, coke formation inside the drum of a delayed coker is not fouling because it is expected and means for removal of the coke are provided.

Fouling during petroleum processing has been so common that petroleum processors expect that units need to be shut down periodically for cleaning. However, such fouling produces an unneeded expense because usually it can be greatly reduced or eliminated by a systematic procedure that determines the cause and suggests alternatives for its elimination.

6.1.2 INCENTIVES

There are large incentives to mitigate fouling in refining. Most only consider the maintenance cost of cleaning. However, in these times of high crude and energy costs, maintenance costs may only rank fourth among the costs of fouling. The lost production when units are shut down for cleaning is a significant cost when there is a shortage of refinery capacity. The insulating effect of layers of foulant on heat exchange surfaces can cost refineries large amounts of energy without it being realized. In addition, foulants reduce the efficiency of fractionators and reduce the reactivity in catalytic reactors. Finally, to lower cost, opportunity crudes are often not purchased because of the fear of fouling, or they are purchased and greater fouling results. In either case, fouling mitigation would reduce the cost of crude oil. One estimate, not including opportunity crudes, is that fouling costs refineries more than $2 billion per year in the United States.[2] In 1999, Lemke[3] estimated that two-thirds of refinery fouling expense is in the crude unit, and 76% of this expense is due to increased energy consumption. Today, these numbers are probably even higher with the higher cost of energy and the widespread use of opportunity crudes that can be half the cost of light, sweet crudes. Refining is second only to electrical power generation among industries in the production of carbon dioxide. Thus, fouling mitigation may be the best method to reduce both greenhouse gas emissions and refining costs. To obtain the right properties of refinery products, it is common to blend light crudes with heavier opportunity crudes. This can cause incompatibility and asphaltene deposition on heat exchangers and coke formation in furnace tubes of atmospheric and vacuum distillation units. However, the foulant may also contain high amounts of inorganics because of stable emulsion formation in the desalter or because of corrosion. Therefore, each refinery should carefully estimate these costs in determining their incentives for fouling mitigation. Most will conclude that they need not wait for a large fouling incident to justify a significant program on fouling mitigation.

6.2 FOULING MITIGATION STRATEGY[4]

The best strategy to mitigate fouling is to elucidate the foulant chemistry and to use this basic knowledge to determine how and where to eliminate its formation. Although there are methods to reduce the rate of fouling without knowing the cause, many of these merely pass the fouling problem on to the next unit. However, by stopping the foulant precursors from forming, the unexpected foulant phase can be completely eliminated from the refinery. A systematic procedure will be given for determining the cause of refinery fouling, tracing the precursors of fouling to the source, devising ways to stop the precursor formation, and selecting among alternatives for mitigation action. These are based on understanding the chemical and physical mechanisms of the six most common causes of refinery fouling and the use of key indicators that enable the quick diagnosis of the fouling and focuses on the correct precursors.

The four basic steps of this fouling mitigation strategy are

1. Diagnosis: Determine the cause of fouling.
2. Investigation: Trace the cause to the source.
3. Innovation: Device ways to interrupt/reduce foulant precursors at each step.
4. Mitigation: Select the best of these ways for the particular refinery to implement.

6.2.1 MOST COMMON CAUSES

Although the general fouling mitigation strategy will solve any refinery fouling problem, it makes sense to keep in mind the most common causes. As 90% of refinery fouling is caused by only six basic causes, most fouling causes can be quickly identified by using tests or indicators for the diagnosis and investigation steps. Although some of these tests and indicators are proprietary, examples will be discussed among those that are not proprietary. The most common causes of refinery fouling are

1. Oil incompatibility on mixing
2. Coke from over-thermal treating crude oil or resid
3. Insoluble asphaltenes on cooling after conversion
4. Inorganics
5. Polymerization of olefins after thermal conversion
6. Oil–water emulsions

Of these, causes 1 to 3 are related as they each involve insoluble asphaltenes.

6.2.2 DIAGNOSIS

The diagnosis is arrived from three sources of evidence:

1. Process conditions/history
2. Analysis of foulant
3. Analysis of oil flowing through fouling unit

6.2.2.1 Process Conditions/History

The process conditions/history should include the range and average conditions (temperature, pressure, flow rate, and feeds) and should include upstream units as well as the fouling unit. What is most revealing is determining the difference in conditions when the unit was not fouling and when the unit was fouling. Can the initiation of fouling be correlated with a particular incident, such as a process upset, a contaminant (slop or sludge) addition, or a new feed? The case study in section 5.3.12 on the plugging of the inlet of a hydrotreater revealed that evidence of the plugging occurring when the percentage of virgin gas oils was increased was available from the beginning of the problem. However, it was not noticed because a careful enough comparison before and after fouling was not made and incompatibility was not considered.

6.2.2.2 Analysis of the Foulant

The analysis of the foulant usually provides maximum information about the cause. As a result, the foulant sample should be taken with care while recording information as to its location, amount, and exposure to cleaning liquids before sampling. Pictures of the foulant in the unit before sampling are very revealing. If the foulant is a solid, it should be ground and mixed to homogenize to obtain an average sample. The sample should be washed with a volatile solvent, such as methylene chloride or toluene, to remove the oil and then dried. If the foulant is soluble in the solvent, as in the case of an asphaltene sediment, a different sample should be washed with heptane.

The first foulant analysis should be to determine how much is inorganic. An ash test, thermal gravimetric analysis (TGA), or elemental analysis may be used. If it is determined to be over 10 wt% inorganic, this cause should be included, traced to the source, and mitigation action implemented. Inorganics are easy to detect and traced to the source because, with the exception of nickel and vanadium, they are not a natural part of petroleum. Often the inorganic foulant adsorbs organics out of the oil, and elimination of the inorganic deposit also eliminates the organic portion of the foulant. Likewise, if inorganic solids are present, they will be pulled into the deposit by organic deposition. Common inorganic foulants are iron sulfide, rust, sea salts, catalyst fines, clays or dirt, and ammonium chloride. Iron sulfide and rust are corrosion products. Therefore, they should be traced to the corrosion source or determined if they arrived in the crude oil. Often, corrosion can be detected as a foulant before it can be measured by decreasing wall thickness. Different metallurgy of the fouled and upstream units will enable determining the location of the corrosion by metals other than iron detected in the foulant. Iron sulfide, one of the most common refinery foulants, is a black, granulated, insoluble solid that is often misidentified as coke. Iron sulfide can form directly from hydrogen sulfide reacting with iron or steel surfaces. Alternatively, iron naphthenate can form as an oil-soluble salt by naphthenic acids in the oil reacting with iron or steel surfaces and then reacting with hydrogen sulfide that is released when oil is thermally cracked. Panchal[2] has shown that iron naphthenate decomposes to iron carbonate at 299°C and then to iron oxide at 474°C. Iron oxide can then react with hydrogen sulfide to form iron sulfide. As will be discussed later in this chapter, iron sulfide on a tube metal surface can promote the adhesion of coke to the surface. In addition, iron sulfide can act as a dehydrogenation catalyst to promote the formation of coke by asphaltenes. In chapter 3, we learned that asphaltenes have high concentrations of donor hydrogen, which when removed, increase the size of the polyaromatic core and encourage coke formation. Therefore, there are considerable ways asphaltenes and iron sulfide can interact to form foulants.

If sea salts, such as sodium, calcium, and magnesium chlorides, are not removed in the desalter, they may deposit wherever the water is evaporated in the preheat train. As calcium and magnesium chloride are not thermally stable, they can hydrolyze in resid conversion units, hydroconversion, or coking to form hydrogen chloride and react with ammonia released by the conversion to form the solid ammonium chloride. This salt can be washed out of the unit, such as a coker fractionator, with water. Also, caustic can be injected after the desalter to convert calcium and magnesium chloride to sodium chloride. However, the preferred solution is to correct the desalter operation to more effectively remove the sea salts from the crude oil.

Oil incompatibility was discussed in section 5.3, and the phase-separation mechanism for coke formation from insoluble asphaltenes was discussed in chapter 4. Therefore, the presence of the carbonaceous mesophase in a foulant is used as an indicator to show that the foulant is coke that was formed by the asphaltene phase-separation mechanism during thermal cracking. If the coke was formed by oil incompatibility on mixing and the insoluble asphaltenes were later thermally cracked, no carbonaceous mesophase would be observed. The presence of carbonaceous mesophase in the heavy product of resid hydroconversion has been used[5] to detect that coke was forming in the reactor so that process conditions could be modified to eliminate further coke formation and avoided shutting the reactor down due to coke plugging. In addition, other indicators have been determined to detect the other common causes that enable quick diagnosis based upon foulant analysis. One of the simplest indicators[6] is to determine the hydrogen-to-carbon (H/C) atomic ratio of the foulant (divide the wt% hydrogen by wt% carbon and multiply by 11.91):

H/C Atomic	Foulant
1.4–1.9	Wax
1.0–1.2	Unconverted asphaltenes
0.7–1.0	Thermally converted asphaltenes
0.3–0.7	Coke

6.2.2.3 Analysis of the Oil

The analysis of the oil flowing through the fouled unit should be based on what was learned from the foulant analysis. If the foulant analysis indicated the cause of the foulant, the precursors should be sought in the oil to confirm the diagnosis. For example, if the foulant is over 10 wt% iron sulfide, the incoming oil should be analyzed for soluble and insoluble iron and the total acid number for naphthenic acids[7] and for reactive sulfur.[8] This should determine if the corrosion was in the unit or upstream of the unit, and the type of corrosion if in the unit. On the other hand, if the foulant analysis is not able to distinguish among several causes, the oil should be analyzed for the complete range of possible precursors. Examples are shown below:

Common Cause	Precursor Analysis
1. Oil incompatibility on mixing	Oil compatibility tests
2. Coke	Conradson carbon, asphaltenes, carbonaceous mesophase
3. Insoluble asphaltenes on cooling	Asphaltenes with H/C atomic <1.0, oil compatibility tests
4. Inorganics	Metals found in foulant (IC Plasma Emission Spect.)
5. Polymerization of olefins	Dienes
6. Oil–water emulsion	Presence of emulsion, surface active contaminants

6.2.3 INVESTIGATION

At this point the cause has been determined from the analysis of the foulant, identification of the precursor in the oil, and matching with process conditions and history. Now use the same analyses for the precursors in the oil to trace them in upstream units to the source. Where is the corrosion happening? Is the precursor coming into the refinery in one particular crude oil? Is the temperature too high in an upstream unit? Are the precursors coming from the recycle of slop or sludge? Basically, determine each step in the progression from incoming crude oil through upstream units until the foulant is deposited in the fouled unit, and write this down along with evidence for each step.

6.2.4 INNOVATION

Now one needs a range of ideas on how to stop or interrupt each step in the progression from precursor in the crude oil to deposited foulant. Sometimes the possible ways are obvious. Do not process one particular crude oil, lower the temperature of one process unit, do not recycle one kind of slop, or mix oils in different proportions or order. However, often most of the possible ways seem to have negative implications or large costs. These could be direct costs, such as for additives or for testing each oil flowing in a particular unit. On the other hand, it might require capital costs, such as using more expensive alloys or some other modification of a unit. For these, it is a good idea to meet with a group with diverse technical backgrounds, including those responsible for the affected upstream and downstream units. Then the group should brainstorm, flush out, and record all possible ways to stop or interrupt each step along with advantages and disadvantages of doing each, including projected costs.

6.2.5 MITIGATION

The mitigation phase focuses in on what is the best mitigation action for the refinery. The same group doing the innovation phase or a different group needs to evaluate each possible mitigation action that comes from the innovation and put them in priority order. This group should establish a quantitative measure of success to evaluate and test the preferred mitigation action and make a recommendation to refinery management. Usually, the highest priority mitigation action is tested, and if it meets the criteria of success, no others are tried. Of course, if the criteria are not satisfied, the next highest priority mitigation action is then tested.

The fouling mitigation strategy is a suggested procedure to be followed, but should not be restrictive when there are good reasons to deviate from the procedures. Each fouling case has its own characteristics and, with experience, one can often jump quickly to the cause and mitigation action based on similarities to other fouling cases. For instance, if the fouling is at a similar process location to a fouling case previously solved, it makes sense to immediately analyze for the same indicator and precursor to confirm that it has the same cause. If it is the same cause, often the refinery will want to try first the mitigation action that was successful in the previous fouling case. However, because each refinery is different, it is still worthwhile to cover the innovation and mitigation phases as there may be mitigation actions that are better suited to the more recent refinery or not available to the previous refinery.

Some examples of fouling problems and mitigation solutions are given in the following text.

6.3 CASE OF FOULING CAUSED BY POLYMERIZATION OF CONJUGATED OLEFINS[9]

Refinery process equipment downstream of processes that thermally crack petroleum resids, such as cokers or visbreakers, often become coated or plugged with a puffed carbonaceous solid called popcorn coke. Examples of such downstream process equipment include fractionators, heat exchangers, reboilers, and hydrotreater reactors. Unlike the coke formed from thermal cracking, popcorn coke can be crushed with the fingers or by stepping on it. However, the mechanism for the formation of popcorn coke and its mitigation were not available in the literature until the research reported here. This study began by investigating the buildup of popcorn coke in fractionators after Fluid Cokers and Flexicokers. These processes licensed by Exxon had frequent problems with fractionator fouling that did not respond favorably to design changes. The amount and location of the coke in such fractionators varied significantly from unit to unit and typically took six months to a year to affect fractionation. However, some units experienced no significant formation of popcorn coke over the multiyear run length of the coker.

A multifunctional team was organized to elucidate the mechanism and to devise mitigation solutions. This was composed of the author and Glen Brons of Corporate Research, Linda Cronin of Engineering, and Leo Brown of Development. Cronin surveyed the operators of each of the units to determine the fouling history of each of the coker fractionators, and arranged for samples of the foulant and oil to be sent to Development and to Corporate Research. Brown visited some of the units during times when they were shut down to obtain first-hand knowledge and to collect samples. Brown, trained as a geologist, was particularly adept at obtaining foulant morphology from optical microscopy. Brons, a chemist, worked out the procedures for foulant sample preparation and analysis. He also studied the procedure for measuring conjugated olefins in detail and confirmed its validity. The author used his background in polymer science to propose and test the polymerization of conjugated olefin hypothesis. This is an example that a multitalented team is much better than a single individual at solving a problem.

6.3.1 ANALYSIS OF POPCORN COKE

Popcorn coke and typical oils from many Fluid Coker and Flexicoker fractionators were analyzed. Popcorn coke was found to be isotropic (no carbonaceous mesophase) and to have less than 1 ppm of nickel plus vanadium, and the hydrogen-to-carbon ratio was in the range of 0.65 to 0.75. This can be compared with thermal coke from resids that is anisotropic (contains carbonaceous mesophase), has well over 100 ppm of nickel plus vanadium, and has hydrogen-to-carbon atomic ratios in the range of 0.20 to 0.45. Thus, it can be concluded that popcorn coke foulants in these coker fractionators are not caused by either entrained coke or entrained resid from the coker.

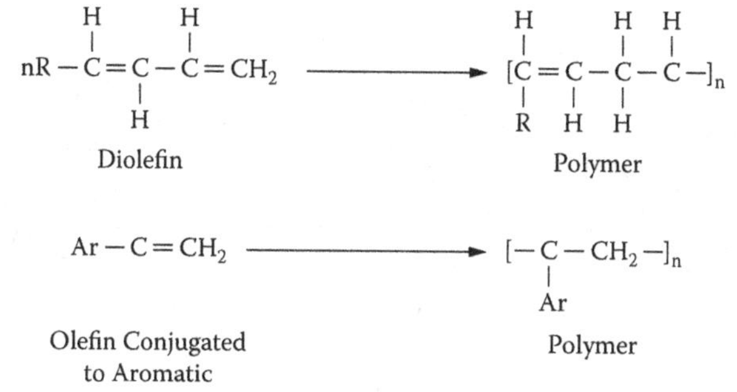

FIGURE 6.1 Solids can form by the polymerization of olefins conjugated to olefins or to aromatics. (From IA Wiehe. *Pet Sci Technol* 2: 673–680, 2003. With permission.)

6.3.2 POLYMERIZATION HYPOTHESIS

Although popcorn coke can form on various trays, depending on the coker units, the oil sampled from the trays fouled with popcorn coke all contained measurable amounts of compounds with conjugated olefins. Conjugated olefins are olefins one carbon away from another olefin (called diolefins) or from an aromatic (Ar) containing one or more fused rings. The importance of conjugated olefins is that they easily form free radicals and undergo thermal polymerization as represented in Figure 6.1. Because coking of highly aromatic resids involves the thermal cracking of aliphatic side chains off aromatics, nearly all the conjugated olefins in volatile coker liquids are expected to be conjugated to aromatics. This can be compared with the predominately diolefin-type conjugated olefins formed by steam cracking of highly aliphatic feeds. As a result, fractionator deposits after steam crackers tend to be clear, elastic polymers, whereas fractionator deposits after cokers tend to be black, brittle popcorn coke. Therefore, the hypothesis was made that popcorn coke is caused by the polymerization of olefins conjugated to aromatics. Laboratory reaction tests with model compound conjugated olefins, vinyl naphthalene, and acenaphthalene verified that they polymerize in the temperature range (450°F to 580°F) of coker fractionators without adding free-radical initiators.

6.3.3 MEASUREMENT OF CONJUGATED OLEFIN CONCENTRATION

Several methods were investigated to measure the concentration of conjugated olefins in oils. Although none was completely satisfactory, measurement of the diene value by UOP method number 326-82 was the best method tried, as represented in Figure 6.2 for the conjugated olefin, vinyl naphthalene. Of course, in practice, the structures of the conjugated olefins are unknown. Nevertheless, the procedure is to reflux the oil with a known excess of maleic anhydride in toluene solution. The maleic anhydride forms a water-insoluble adduct with conjugated olefins in

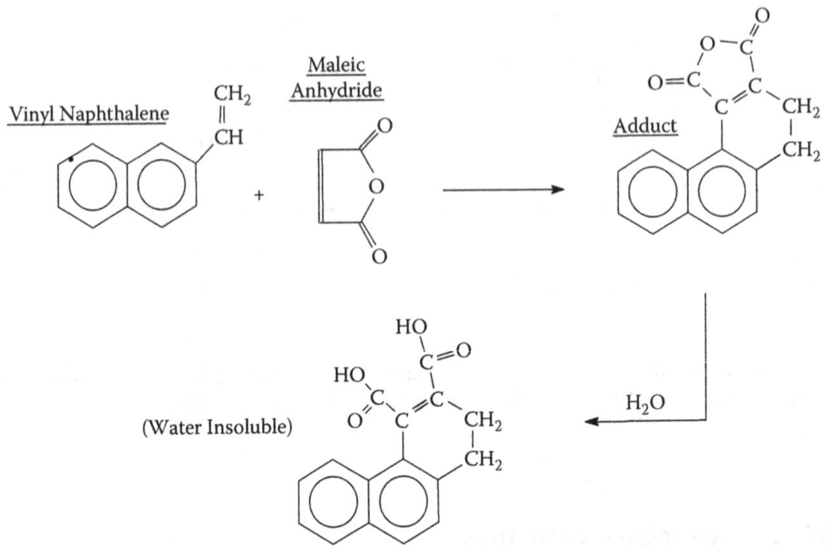

FIGURE 6.2 By UOP Method 326-82, the concentration of conjugated olefins is measured by the formation of a water-insoluble adduct between the conjugated olefin and maleic anhydride. (From IA Wiehe. *Pet Sci Technol* 2: 673–680, 2003. With permission.)

the oil. Thus, when the product is washed with water, the unreacted maleic anhydride (water soluble) reacts to form maleic acid. The amount of resulting maleic acid is determined by titrating with sodium hydroxide. From this, the number of moles of reacted maleic anhydride, equivalent to the moles of conjugated olefins, can be calculated. Because the average molecular weight of the conjugated olefins in an oil is unknown, the molecular weight of iodine is arbitrarily used to convert moles to grams. Thus, the diene value is reported as g I/100 g oil. The measurement of oils from coker fractionator locations that popcorn coke is known to form had diene values higher than 4 g I/100 g oil.

6.3.4 POSSIBLE MITIGATION ACTIONS

The survey of coker fractionators showed that, even when sufficient conjugated olefins were present, popcorn coke only formed in the temperature range of 450°F to 615°F and particularly in 500°F to 580°F. Below this temperature range, the reaction rate is too slow, and above this temperature range the bonds are broken thermally faster than they are formed. Although it is common to lower temperature to reduce the rate of fouling by carbonaceous solids, it is a surprise that raising the temperature can mitigate such fouling. However, once the mechanism was determined to be polymerization, raising temperature to reduce polymerization rate is a logical conclusion. Polymerization reactions are known to have ceiling temperatures.

Another property of polymerization reactions is that they are very concentration-dependent because they require combining a number of reacting species. Thus, another mitigating action is to reduce either the residence time or the concentration of conjugated olefins in the critical temperature range for polymerization. Thus, popcorn coke may be mitigated by increasing the flow rate of the oil in the critical temperature range and redesigning to minimize dead zones where oil can sit for long periods of time. The concentration of conjugated olefins can be reduced by adding an oil that has not been processed in a coker or modifying the coker to reduce the conjugated olefin concentration. Examples are reducing the coking temperature or reducing the vapor phase residence time. All of these possible mitigation actions have been patented.[10]

6.3.5 FLEXICOKER FRACTIONATOR FOULING

Typical refinery fouling is rather slow. With some coker fractionators showing fouling that affected the process in 6 to 12 months, this was considered to be fast. However, not being severe enough to shut down the process, each unit could run for two years before being shut down when there was an opportunity to improve the process by a design change. It would take at least another 6 to 12 months to find if the fouling had been reduced. Although very rapid fouling is very bad for a refinery, it is easier to convince the refinery to try mitigating solutions and to monitor improvement, assuming the cause and possible mitigating solutions have been determined.

In this project the opportunity of rapid fouling occurred just after the cause and possible mitigation solutions were determined. A commercial Flexicoker was suddenly in danger of being shut down because of fractionator fouling. Often, foulant is found on the trays in the bottom pump around (BPA) section of the fractionator. However, in this case, popcorn coke also accumulated in the bottom pool, flowed out with the oil, and plugged the pump suction strainers in the BPA circuit (Figure 6.3). The pump had to be shut down frequently to clean the strainers. Although the typical frequency of cleaning was twice a week, during this time it was less than a day. As a result, the bottom pool temperature was reduced in steps from 575°F to 565°F. The popcorn coke continued to increase, requiring a cleaning frequency of only two hours, even after greatly increasing the size of the strainer. Meanwhile, analysis of the bottom pump around oil with UOP Method 326-82 gave a diene value of 7 to 9 g I/100 g oil, showing a strong potential for polymerization. Analysis of the popcorn coke from the strainer showed little or no vanadium and nickel, and H/C atomic of 0.76. This was similar to the analysis of the coke on the trays of the BPA section when the coker was previously shut down, and different from the coke (H/C 0.20-0.42, Ni and V > 1000 ppm) found in the coker scrubber between the coker and the fractionator. This shows that the fractionator coke was not carried over from the coker, but was formed in the fractionator, most likely by polymerization of conjugated olefins. As there was no spare BPA pump, and the two-hour shutdown frequency endangered the pump, the Flexicoker

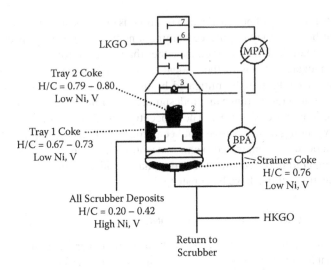

FIGURE 6.3 Schematic of Flexicoker fractionator fouled with popcorn coke. (From IA Wiehe. *Pet Sci Technol* 2: 673–680, 2003. With permission.)

was going to be shut down. Before this was done, the refinery agreed to try the author's idea of increasing the bottom pool temperature to 590°F, the maximum temperature that it could efficiently operate.

It is not easy to convince a refinery to reduce coking by raising temperature, but the team had persuasive evidence, and the only alternative was to shut down the unit. As it turns out, at that time the difference in price between residual fuel oil (vacuum resid) and clean refinery products (transportation fuels) was unusually high. This provided even greater motivation for the refinery not to shut down the Flexicoker that converted vacuum resid to transportation fuels. Fortunately, once the higher bottom pool temperature had been reached, the time between strainer cleanings immediately began to increase to many hours and then once a day. This provided a great method to monitor progress. After several months, the pump strainer cleaning frequency was reduced to the normal rate of twice a week. The coker continued to operate for eight months until being shut down as part of the normally planned maintenance schedule and when the difference in price between residual fuel oil and clean refinery products was back to normal. Therefore, this was one of the few cases of fouling or coking that was mitigated by raising the temperature. Not only was this a commercial success in that it saved the refinery considerable money, but it provided a vivid demonstration that popcorn coke in coker fractionators is caused by the polymerization of olefins conjugated to aromatics. It also points out the strong advantage in determining the cause of fouling. Just about all other causes of coking would have been made much worse by raising the temperature.

6.4 CASE OF HEAT EXCHANGER FOULING AFTER RESID HYDROTREATER[11]

This is an unusual example of fouling caused by insoluble asphaltenes on cooling, but builds on the fundamentals learned in the previous chapters. A fixed-bed hydrotreater in a refinery is used as a pretreater for atmospheric resid before fluid catalytic cracking. The purpose of this hydrotreater is to remove sulfur and metals and to reduce Conradson carbon residue. To hydrotreat asphaltenes in the feed, large-pore catalyst was put on the top of the catalyst bed. Each of the first three runs followed a consistent fouling pattern. Although no fouling was observed during most of the one-year run, near the end of run the heat exchanger that cooled the liquid product severely fouled. Typically, as the catalyst deactivated, the temperature of the hydrotreater was increased during the run to keep the sulfur level below a certain specification.

6.4.1 MICROSCOPIC EXAMINATION OF FEEDS AND PRODUCTS

Although no samples of the heat exchanger deposit were provided, at the end of the third run before the catalyst was removed, a number of commercial tests were done at various temperatures and space velocities. Samples of the feeds and products of these tests were collected. In addition, samples of feed and products were taken at the beginning of the fourth run using fresh catalyst. Examination of the oil samples alone and in mixture with toluene and n-heptane with an optical microscope revealed that all the liquid products at the end of the run contained insoluble asphaltenes (toluene soluble, but n-heptane insoluble) except one operated at low temperatures and at the highest space velocity. This is in contrast with the oil samples of the liquid product of the fresh catalyst and the resid feeds that contained only soluble asphaltenes. Because the oil samples were free of carbonaceous mesophase and other toluene insolubles, it was clear that the asphaltenes were soluble in the reactor, but must have phase-separated on cooling below thermal reaction temperatures. Therefore, the cause was asphaltenes in the oil that only became insoluble after leaving the reactor, and after nearly a year of a run. Recall from chapter 4 that asphaltenes that become insoluble in an oil at thermal reaction temperatures form toluene-insoluble carbonaceous mesophase and that thermal cracking of an oil containing insoluble asphaltenes forms toluene insoluble coke that is isotropic.

6.4.2 PRECURSOR ANALYSES

The precursor analyses for a cause of insoluble asphaltenes on cooling should be the oil compatibility tests and determination of the H/C atomic ratio of the asphaltenes. Although this case occurred before the OCM was developed, the toluene equivalence, the minimum percent toluene required to keep asphaltenes from precipitating when mixed with n-heptane at a concentration of 2 g of oil in 10 mL of toluene–heptane mixture, was measured on each feed and product oil.

TABLE 6.1
Properties of Feed and Product Asphaltenes at the Start and End of Hydrotreater Run

	Asphaltene	Yield, wt%	H/C, Atomic	Sulfur, wt%
Start of run	Feed	5.4	1.25	4.54
	Product	2.8	1.36	2.76
End of run	Feed	6.3	1.23	4.53
	Product	5.7	1.18	3.44

Source: IA Wiehe. *J Dispersion Sci Technol* 25: 333–339, 2004. With permission.

We now know from section 5.3.10 that for self-incompatible product oils (those containing insoluble asphaltenes), the insolubility number and the solubility blending number equals the measured toluene equivalence. As the toluene equivalence of the products with insoluble asphaltenes was much higher (33 to 41) than the toluene equivalence of products without insoluble asphaltenes (26 to 28), we sought the reasons for this difference.

Asphaltenes (*n*-heptane insolubles) were separated from all feeds and products and analyzed for carbon, hydrogen, and sulfur. As shown in Table 6.1, at the start of the run the quantity of asphaltenes decreased significantly from feed to product, the expected conversion by hydrotreating. For the asphaltenes remaining in the product, the H/C atomic ratio actually increased as compared to the asphaltenes in the feed. Also, the sulfur in the asphaltenes significantly decreased. In contrast, the asphaltenes from samples at the end of the run barely decreased concentration, decreased in H/C atomic ratio, and only slightly decreased in sulfur as compared with asphaltenes in the feed. Therefore, asphaltenes were entering the large-pore catalysts at the start of the run and being hydrotreated. At the end of the run, the asphaltenes were mostly bypassing the catalyst and only slightly thermally cracking to reduce sulfur, yield, and H/C atomic ratio (Figure 6.4). As the pore diameter of the spent large-pore catalyst was found to greatly decrease from the deposition of vanadium and nickel, this explanation seemed to be verified. However, a H/C atomic ratio of less than 1.0 is usually required for an asphaltene to be insoluble in an oil. As the end-of-run product still

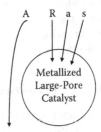

FIGURE 6.4 At end of hydrotreater run, asphaltenes bypass the catalyst. (From IA Wiehe. *J Dispersion Sci Technol* 25: 333–339, 2004. With permission.)

had asphaltenes with a H/C atomic ratio of 1.18, another reason for the greater asphaltene insolubility other than thermal cracking of asphaltenes needed to be determined.

6.4.3 Cause of Asphaltene Insolubility

From chapter 5, insolubility of asphaltenes can also be caused by the increase of saturates, decrease of aromatics, or decrease of resins. Therefore, one possibility is the hydrogenation of aromatics to form saturates. However, this should more likely happen with fresh catalyst at the beginning of a run than with deactivated catalyst at the end of a run. Therefore, by the process of elimination, the possibility of the decrease of resins was examined.

The quantity of resins (polar aromatics) was measured by high performance liquid chromatography (section 2.3) on each of the remaining deasphalted oils. As is shown in Figure 6.5, at end of the run (below a resin-to-asphaltene weight ratio of 1.7), the toluene equivalence (thus, the insolubility number) increases approximately linearly with decreasing resin-to-asphaltene weight ratio. Thus, the asphaltene insolubility was caused by the removal of resins, natural dispersants

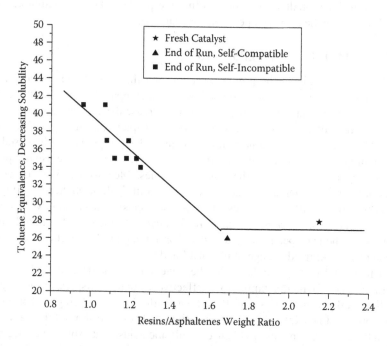

FIGURE 6.5 The asphaltenes become less soluble (higher toluene equivalence) at end of run for resid hydrotreater as the resin-to-asphaltene ratio decreases. (From IA Wiehe. *J Dispersion Sci Technol* 25: 333–339, 2004. With permission.)

for asphaltenes (chapters 2 and 5). This decrease in resins probably happened when the reactor temperature was increased because asphaltenes bypassed the catalyst and increased the sulfur content of the product. This caused the resins, small enough to enter the pores, to be converted even more to compensate for the unconverted asphaltenes.

6.4.4 CASE STUDY CONCLUSION

In conclusion, the heat exchanger was fouled by asphaltenes on cooling without sufficient natural dispersant near the end of run. The decrease in resin-to-asphaltene ratio resulted from the greater conversion of resins than asphaltenes when the large-pore catalysts were partially plugged with deposited vanadium and nickel, and when the hydrotreater temperature was increased for greater desulfurization reactivity. Although it is extremely difficult, if not impossible, to physically separate resins without separating asphaltenes, the resid hydrotreater achieved this by chemical reaction with size-selective catalysts. Nevertheless, once the mechanism of fouling was known, a number of ways to interrupt the mechanism and mitigate the fouling became obvious. These included increasing the proportion of large-pore catalyst, not raising the reactor temperature above a given value, and, once the toluene equivalence of the product significantly increased, injecting either asphaltene dispersant or a fluid catalytic cracking cycle oil (solvent for asphaltenes) before cooling the product. The mitigating actions that best fit into the refinery operation were successfully applied.

6.5 ASPHALTENE DISPERSANTS[12]

With the case study in the previous section and the case studies in chapter 5 involving crude oils that are incompatible, self-incompatible, and nearly incompatible, it is clear that one needs to be able to increase the concentration of resins as a possible mitigation action. As discussed in chapter 2, crude oils contain the resin fraction that acts as a natural dispersant to keep asphaltenes dispersed as approximately 40 Å in diameter particles. However, the resins are usually present in larger concentrations than the asphaltenes. Meanwhile, synthetic additives are sold commercially that are reported to mitigate asphaltene fouling at concentrations of 100 ppm or less. One possible class of such additives is synthetic dispersants, which are much more effective than the resin fraction. In this section, synthetic dispersants are tested and optimized for their ability to disperse asphaltenes, taking advantage of the OCM and tests.

Chang and Fogler[13] studied alkyl-benzene sulfonic acids as asphaltene dispersants. They found this family to be effective asphaltene dispersants as long as the linear alkyl chain is less than 16 carbons, with 12 carbons being the optimum. They provided no explanation for this optimum. They also showed that, at high-enough concentrations of alkyl benzene sulfonic acids, that asphaltenes become soluble in *n*-heptane and, thus, by definition are no longer asphaltenes. This dramatically points out that asphaltene solubility is a phenomenon that is based on both the presence of solvents and dispersants.

6.5.1 MEASUREMENT OF DISPERSANT EFFECTIVENESS

The action of dispersants is to reduce the asphaltene insolubility number. As the insolubility number is the toluene equivalence at infinite dilution in oil, the toluene equivalence at the concentration of 2 g of oil and 10 mL of test liquid is a better measure of the dispersant effect than the heptane dilution test that is at a much higher concentration of oil. Therefore, the ability of dispersants to reduce the toluene equivalence was used as the measure of dispersant effectiveness. To accurately measure the effect, high concentrations of dispersant (1 to 5 wt%) were used to obtain large reductions in toluene equivalence. In applications, often the insolubility number needs only to be reduced by 1 or 2 out of 30 to 40 to make two oils compatible. Thus, much lower concentrations of dispersant are required, but usually this still needs much more than 100 ppm. Nevertheless, this test is much better at comparing a set of asphaltene dispersants for effectiveness than typical thermal fouling tests. The toluene equivalence test is faster, more accurate, and builds on the learnings of the OCM. For much of the research, the toluene equivalence was run on Maya crude oil. This is a high-volume, heavy crude oil from Mexico with a high toluene equivalence of 38. Limited studies were done using a resid hydrotreater product (toluene equivalence of 35) from the case study discussed in section 6.4 and the fractionator bottoms from a fluid catalytic cracker with a toluene equivalence of 87. In section 5.3.11, the OCM was shown to apply to catalytic cracker fractionator bottoms even though they contain low-molecular-weight, very insoluble asphaltenes.

6.5.2 DISPERSANT HEAD GROUP

The general design of an asphaltene dispersant is shown in Figure 6.6 with the important attributes. The head group was briefly investigated for straight-chain

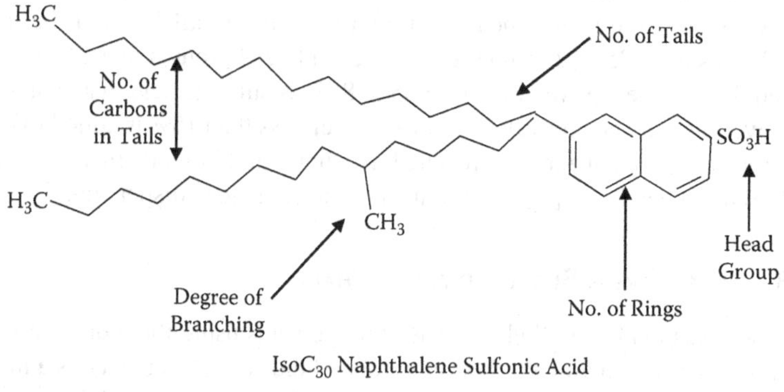

IsoC$_{30}$ Naphthalene Sulfonic Acid

FIGURE 6.6 General design of a synthetic asphaltene dispersant. (From IA Wiehe, TG Jermansen. *Pet Sci Technol* 21: 527–536, 2003. With permission.)

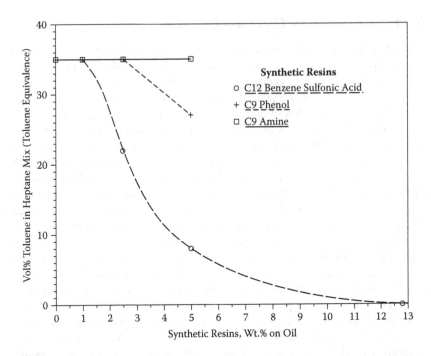

FIGURE 6.7 Toluene equivalence versus wt% dispersant for synthetic dispersants with different head groups for self-incompatible, hydrotreater product.

alkyl-benzenes on a resid hydrotreater product containing 0.25 wt% insoluble asphaltenes as hot filtration sediment (discussed in section 6.4). Therefore, some dispersant was needed to dissolve the insoluble asphaltenes before the toluene equivalence could be lowered. The head groups tried were phenol (–OH), amine (–NH$_3$), and sulfonic acid (–SOOH). The results in Figure 6.7 show with 5% dispersant that the amine head group with a 9 carbon tail did not lower the toluene equivalence, the phenol head group on a 9 carbon tail lowered the toluene equivalences from 35 to 27, and the sulfonic acid head group and a 12 carbon tail lowered the toluene equivalence from 35 to 8. This sulfonic acid dispersant could even lower the toluene equivalence to zero when less than 15% is added to the oil. Thus, the same conclusion was reached as Chang and Fogler[13] that the sulfonic acid (–SOOH) head group gives by far the best asphaltene dispersant.

6.5.3 Linear versus Branched Alkyl Chains

This study was made on alkyl-benzene sulfonic acids using the reduction in toluene equivalence of fractionator bottoms from a fluid catalytic cracker as a measure of dispersant effectiveness. Those with linear tails were purchased from Aldrich, whereas the one with a branched tail is an intermediate in the manufacture of lube oil detergents supplied by Exxon Chemicals. The results are in Table 6.2.

TABLE 6.2

Reduction in Toluene Equivalence of Fractionator Bottoms from Fluid Catalytic Cracking by Alkyl Benzene Sulfonic Acids with Linear and Branched Alkyl Tails

No. of Rings	No. of Tails	Branched Methyls	No. of Carbons in Tails	Dispersant Concentration (wt% on Oil)	Toluene Equivalence Reduction	Asphaltenes Soluble Next Day?
1	1	0	8	1.0	87–60	No
1	1	0	8	5.0	87–55	No
1	1	0	12	1.0	87–60	No
1	1	0	12	5.0	87–55	Yes
1	1	0	18	1.0	Dispersant	Insoluble
1	1	5	24	1.0	87–60	Yes
1	1	5	24	5.0	87–55	Yes

Source: IA Wiehe, TG Jermansen. *Pet Sci Technol* 21: 527–536, 2003. With permission.

Dispersants with long linear alkyl tails (above 16 carbons) were found to be less effective, as reported by Chang and Fogler.[13] However, observation with an optical microscope showed that alkyl-benzene sulfonic acids with long tails are not completely soluble in either catalytic cracker fractionator bottoms or toluene–heptane mixtures at room temperature. This is because the long alkyl chains crystallize like a wax, and these crystals are observed under cross-polarized light with an optical microscope. In addition, even the shorter-chain alkyl-benzene sulfonic acids were found to have reduced dispersant effectiveness over time (24 hours). As is shown in Table 6.2, with 8- and 12-carbon alkyl-benzene sulfonic acids that form borderline soluble mixtures with toluene–heptane–Baytown catalytic cracker fractionator bottoms one day, the asphaltenes are insoluble the following day. To inhibit the long alkyl tails from crystallizing and making the dispersant insoluble, branched alkyl chains were tried. As shown in Table 6.2, a highly branched (5-branched methyls), 24-carbon-tail benzene sulfonic acid was soluble in mixtures of toluene–heptane–Baytown catalytic cracker fractionator bottoms, whereas the straight-chain, 18-carbon-tail benzene sulfonic acid was not soluble. Although in this case, the increased alkyl chain length did not improve the ability of the dispersant to reduce the toluene equivalence, it did improve the stability of the dispersant action over time. In most other cases, as will be shown, the longer alkyl chain also improves the dispersant action as long as the dispersant is soluble.

6.5.4 Synthesis of Alkyl Aromatic Sulfonic Acids

The rest of the asphaltene dispersant studies were done using dispersants synthesized within the author's laboratory by a retired chemist, Ted Jermansen, who was hired on contract for this project. First, an aromatic was reacted with an olefin in the presence of a Friedel Crafts catalyst, such as $AlCl_3$. After purification by washing with water and distillation of the unreacted aromatic and olefin, the alkyl-aromatic was sulfonated by reaction with sulfuric acid. By using an internal olefin, the resulting alkyl-aromatic sulfonic acid had two chains with a distribution of lengths of the two chains, adding up to the total length. This was a result of the internal olefin location moving randomly during the Friedel Crafts alkylation. This distribution of chain lengths is a strong advantage for preventing crystallization and was the idea of Ramish Varadaraj, a scientist with a background in interfacial phenomena and an important member of the team on this project. Branched tails were made by using branched, internal olefins in the synthesis with the numbers of branched methyls determined by ^{13}C NMR.

6.5.4.1 Number of Aromatic Rings

Table 6.3 compares the effect of aromatic ring size of alkyl-aromatic sulfonic acids for two different branched alkyl chain lengths (23 and 33 carbons) on the reduction of toluene equivalence of Maya crude oil. In each case, the dispersant with a two-ring aromatic is a more effective dispersant than either a one- or a three-ring aromatic. As shown, if the aromatic is tetralin with one aromatic ring fused to a saturated ring, the dispersant effectiveness is similar to a one-ring aromatic, but

---

TABLE 6.3

Reduction in Toluene Equivalence of Maya Crude Oil by Alkyl Aromatic Sulfonic Acids with Two Branched Alkyl Tails and Varying Aromatic Ring Sizes

No. of Rings	No. of Tails	Branched Methyls	No. of Carbons in Tails	Dispersant Concentration (wt% on Oil)	Toluene Equivalence Reduction
1	2	0.15	23	1.0	38–31
2	2	0.15	23	1.0	38–23
2[a]	2	0.15	23	1.0	38–29
4[b]	2	0.15	23	1.0	38 to >30
3	2	0.15	23	1.0	38–30
1	2	0.99	33	1.0	38–23.5
2	2	0.99	33	1.0	38–17
3	2	0.99	33	1.0	38–29

[a] Aromatic is tetralin instead of naphthalene.
[b] Aromatic is biphenyl.

Source: IA Wiehe, TG Jermansen. *Pet Sci Technol* 21: 527–536, 2003. With permission.

TABLE 6.4

Longer the Tails, Better the Dispersant with Two Tail Dispersants Tested on Maya Crude Oil

No. of Rings	No. of Tails	Branched Methyls	No. of Carbons in Tails	Dispersant Concentration (wt% on Oil)	Toluene Equivalence Reduction
1	2	0.15	23	1.0	38–31
1	2	0.99	33	1.0	38–23.5
2	2	0.17	18	1.0	38–32
2	2	0.15	23	1.0	38–23
2	2	0.99	33	1.0	38–17
2	2	0.54	37	1.0	38–13
2	2	0.28	47	1.0	38–11

Source: IA Wiehe, TG Jermansen. *Pet Sci Technol* 21: 527–536, 2003. With permission.

worse than a fused two-ring aromatic. However, if the two aromatic rings are not fused (biphenyl), it is no better a dispersant than a single-ring aromatic.

6.5.4.2 Alkyl Chain Length

As shown in Table 6.4, dispersant effectiveness in reducing toluene equivalence increases with the total number of carbons in the alkyl chain for both those containing one and two aromatic rings.

6.5.4.3 Degree of Branching

As is shown in Table 6.5, the greater number of branched methyls at nearly the same alkyl length increases the dispersant effectiveness.

TABLE 6.5

More Branched Methyls, Better the Dispersant on Maya Crude Oil

No. of Rings	No. of Tails	Branched Methyls	No. of Carbons in Tails	Dispersant Concentration (wt% on Oil)	Toluene Equivalence Reduction
2	2	0.15	23	1.0	38 to 23
2	2	1.9	25	1.0	38 to 17

Source: IA Wiehe, TG Jermansen. *Pet Sci Technol* 21: 527–536, 2003. With permission.

TABLE 6.6

Amount of Branched C_{24}-Benzene Sulfonic Acid Required for Zero Toluene Equivalence

Crude	Toluene Equivalence	Wt% Required for Toluene Equivalence = 0
Maya	38	11.5
Arab Light	26	6.5
Arab Heavy	20	5.6
Cold Lake	19	8.5
Kuwait	12	3.0

Source: IA Wiehe, TG Jermansen. *Pet Sci Technol* 21: 527–536, 2003. With permission.

6.5.5 MAKE ASPHALTENES SOLUBLE IN HEPTANE

If sufficient dispersant is added to an asphaltene-containing oil, the asphaltenes become soluble in *n*-heptane.[13] Because asphaltenes are defined to be *n*-heptane soluble, the action of the dispersant is to convert asphaltenes to resins. In Table 6.6 are listed five crude oils, their toluene equivalence, and the amount of highly branched C_{24}-benzene sulfonic acid (see Table 6.2) required to obtain a zero toluene equivalence (soluble in *n*-heptane at 2 g oil and 10 mL *n*-heptane). This points out the power of synthetic dispersants and the colloidal nature of asphaltenes. As the dispersant used in this study is much less effective than those in Table 6.3 to Table 6.5, one could reduce the quantity required for *n*-heptane solubility by an order of magnitude. However, this is still too much to be practical. Instead, the practical use of dispersants is to add the minimum quantity to reduce the toluene equivalence and, thus, the insolubility number, by a few units. This still can make incompatible crude oil mixtures become compatible. Also, dispersants at higher concentrations can be added periodically, rather than continuously, to clean insoluble asphaltenes off walls and still stay within economic constraints.

6.5.6 CONCLUSIONS ON ASPHALTENE DISPERSANTS

It has been shown that asphaltene dispersants can be selected on their ability to reduce the toluene equivalence of an asphaltene-containing oil. By adding sufficient dispersant, the oil can be made to be soluble in *n*-heptane. However, the practical use of dispersants is to add minimum quantity to make incompatible crude oil mixtures become compatible. The limitation on long normal alkane side chains (tails) on dispersants was discovered to be a result of their tendency to crystallize and become insoluble in oils. Therefore, dispersants were used with branched, double tails with a distribution of how the total length was split between the two tails. This successfully suppressed crystallization and overcame any limitation on side-chain length. The optimum asphaltene dispersant of this type was found to have a sulfonic acid head, fused two-ring aromatic, and a branched tail with a total length as long as possible, but at least 30 carbons. An example is the structure in Figure 6.6. This family of asphaltene dispersants has been patented.[14]

6.5.7 FUTURE NEEDS FOR SYNTHETIC DISPERSANTS

Although synthetic dispersants can be a mitigating solution for both when asphaltenes become insoluble on cooling and for oil incompatibility, in many cases, more dispersant would be required than is economical to keep the asphaltenes in solution. Another possibility is to use less dispersant with the objective to keep the insoluble asphaltene particles small. If this is used in combination with maintaining a high velocity of the oil through the fouled unit, reduction, but not elimination, of fouling should be expected. This objective would require a different test than the toluene equivalence test to optimize the dispersant. A method[15] that measures the settling rate of asphaltenes in oils mixed with toluene, *n*-heptane, and dispersant (ASTM D7061-04) may be well suited to this approach.

6.6 OTHER CASES OF INSOLUBLE ASPHALTENES ON COOLING

6.6.1 MECHANISM AND IDENTIFICATION

As discussed in chapters 4 and 5, asphaltenes become more insoluble during thermal conversion, such as in a visbreaker or during hydroconversion, because the cracking off of paraffinic side chains and small-ring aromatics. In addition, resins are converted to more asphaltenes and lighter fractions. This increases the concentration of asphaltenes while reducing the amount of natural dispersant. Also, the volatile products that are formed are usually higher in saturates than the starting resid, decreasing the solvency (solubility-blending number) of the oil for asphaltenes. If the combination of these effects causes asphaltenes to become insoluble at thermal cracking temperatures, asphaltenes phase separate and form liquid crystalline coke, called the carbonaceous mesophase, which can be identified using optical microscopy with cross-polarized light. This was discussed in chapter 4 and will also be discussed later in this chapter. However, if the reaction is stopped just short of when asphaltenes become insoluble at reaction temperatures, it is common for asphaltenes to precipitate on cooling. These asphaltene sediments can be identified in optical microscopy by typical yellow-brown, curved chain agglomerates that are black in cross-polarized light. An example is shown in Figure 6.8 where a drop of the liquid product of thermally reacting Brent atmospheric resid is observed with an optical microscope. The large, round, black particle near the center of the micrograph is partially light (orange, if it were in color) in cross-polarized light and is identified as carbonaceous mesophase. The rest of the small particles are agglomerates, yellow-brown if in color, and

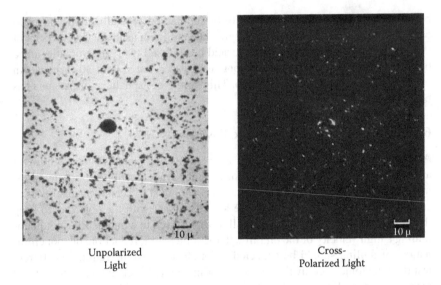

Unpolarized Light	Cross-Polarized Light

FIGURE 6.8 Asphaltene sediments and carbonaceous mesophase seen by an optical microscope in transmitted, unpolarized light, and cross-polarized light of a drop of the total liquid product from reacting Brent atmospheric resid at 400°C in a closed reactor.

black in cross-polarized light, which are identified as asphaltene sediments. This identification can be confirmed by mixing the liquid product with a large excess of toluene, quinoline, or dichlorobenzene, and by optical microscopy, finding that all but the carbonaceous mesophase dissolves. If the asphaltene sediment appears when thermal conversion is not expected, the asphaltenes can be separated by *n*-heptane precipitation or by filtering the sediment from the oil at about 100°C. If the hydrogen-to-carbon atomic ratio of the asphaltenes so separated is less than one, the asphaltenes have been thermally converted.

6.6.2 MITIGATING SOLUTIONS

If asphaltene sediments precipitate on cooling after conversion, some possible mitigating solutions are

1. Only cool the oil enough to slow the thermal reaction.
2. Flash off volatile nonsolvents before cooling.
3. Cool by blending with a solvent oil.
4. Add dispersants before cooling.
5. Decrease conversion.

Of course, after a conversion unit, one needs to cool the oil enough to stop the thermal reaction, but if the cooling is limited, asphaltenes may not precipitate. This is particularly good if the next step is a hot separator or a fractionator that

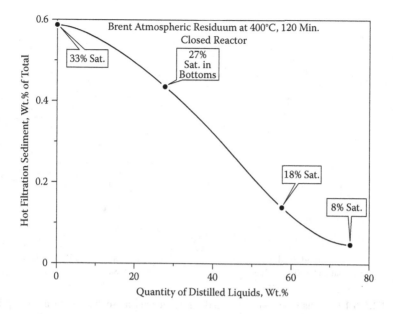

FIGURE 6.9 The amount of hot filtration sediment of the remaining liquid decreases with increasing quantity of liquids distilled out of the total liquid product from reacting Brent atmospheric resid for 120 min at 400°C in a closed reactor.

flashes off volatile liquids. Because the volatile liquids are often nonsolvents for asphaltenes, their removal frequently prevents asphaltenes from precipitating. Figure 6.9 shows that the total liquid product from reacting Brent atmospheric resid for 120 min at 400°C, contained 0.59 wt% hot filtration sediments and the oil contained 33 wt% saturates. Distillation of the liquid product in steps reduced the hot filtration sediment of the distillation bottoms even though the asphaltenes were more concentrated in the bottoms. This is because asphaltene sediments were dissolved by reducing the saturates in the oil and making it a better solvent for asphaltenes. By distilling off 75 wt% of the oil, the amount of hot filtration sediment was reduced to 0.05%, and the rate of decrease of the sediments in the bottoms was low. The micrographs in Figure 6.10 of the initial total liquid product and the bottoms after distilling off 75% show that the only insolubles that remain are the large round particles of carbonaceous mesophase that cannot be redissolved.

An oil that contains insoluble asphaltenes was termed self-incompatible in chapter 5. For such an oil, the insolubility number and the solubility blending number equals the toluene equivalence. Therefore, to select an oil to quench the oil-containing asphaltene sediment, one first needs to measure the toluene equivalence of the oil containing the asphaltene sediment. Then the quench oil needs to have a solubility blending number much higher than this toluene equivalence to limit the quantity of quench oil required. This mitigating solution is only

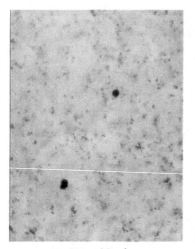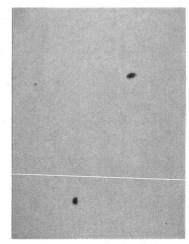

10 μ

Total Liquid Product
33% Saturates, 0.59% HFS

Bottoms after 75% Distilled
8% Saturates, 0.20% HFS

FIGURE 6.10 Asphaltene sediments and carbonaceous mesophase seen by an optical microscope in transmitted light of the total liquid product from reacting Brent atmospheric resid for 120 min at 400°C in a closed reactor, before and after removing 75% by distillation.

economical if the quench oil is one that would be blended anyway downstream of the conversion unit.

The use of synthetic dispersants to prevent asphaltenes from becoming insoluble on cooling is often limited by the thermal stability of the synthetic dispersant. Thus, thermal stability should be used in addition to the ability to reduce toluene equivalence to select the dispersant.

Reducing conversion is usually the mitigating solution when none of the others are successful, and when the fouling is sufficiently severe because this is often a costly solution. However, if conversion is sufficiently reduced, the insoluble asphaltenes will always disappear. Therefore, this solution is often employed on a temporary basis until a better solution is developed. Other solutions involving changing the feed and reactor conditions will be discussed in later chapters on resid conversion processes.

6.7 THERMAL COKE FORMATION

6.7.1 Definition, Mechanism, and Identification

Thermal coke often forms in furnace tubes of delayed cokers and distillation units, visbreakers, high-temperature heat exchangers, and hydroconversion reactors. It can occur in any location where the oil is held above the temperature of about 350°C for sufficient time, or even lower if the oil contains insoluble asphaltenes. Coke can even be formed in one unit, and flake off and foul a downstream unit. This was the case discussed in section 5.3.12 when coke formed in a high-temperature heat exchanger because the feed mixture was incompatible and flaked off and plugged the inlet of a hydrotreater.

The significance of coke formation is that it is irreversible. Once formed, "true" coke cannot be melted or redissolved in any solvent. In chapters 2 and 5, coke that was toluene insoluble but soluble in chlorobenzene, dichlorobenzene, trichlorobenzene, or quinoline was discussed. This is because the operational definition of toluene insolubility was used for coke. Speight[16] uses the name "carbine" for the fraction that is toluene insoluble, but soluble in some other solvent. This differentiation rarely matters in refinery foulant as any black organic solid that adheres to a surface and is kept hot for a long period of time meets the definition of "true" coke. Although foulant by insoluble asphaltenes can be redissolved if kept below thermal cracking temperatures and if a solvent is introduced quickly enough, the insoluble asphaltene foulant will eventually form coke if kept at moderate temperatures (250°C to 350°C) for a long time.

The cause, thermal coke, is meant to be coke that forms by the insolubility of asphaltenes during thermal cracking as discussed in chapter 4. However, there are at least two other mechanisms for forming coke from oil. These are by polymerization of conjugated olefins (section 6.3), and by insoluble asphaltenes in the oil being heated to thermal cracking temperatures. Once it is established that a foulant is greater than 90 wt% carbon, hydrogen, and sulfur (less than 10% inorganic) and is insoluble in a convenient solvent, such as toluene or methylene chloride, then one needs to determine which of these three mechanisms formed the coke. As asphaltenes typically contain nickel and vanadium, coke formed from asphaltenes contains even higher concentrations of nickel and vanadium. Because stainless steels contain nickel that could contaminate the sample, the presence of greater than 50 ppm vanadium in the coke is better to use as an indicator that the mechanism is one of the two from asphaltenes. The polymerization of conjugated olefins can also be eliminated if the hydrogen-to-carbon ratio of the solvent-washed coke is below 0.6. Coke from polymerization of conjugated olefins can also be identified by being brittle and by significant concentrations of conjugated olefins in the oil.

Once it is established that the coke is formed from asphaltenes, one needs to determine whether the asphaltenes were insoluble before thermal cracking or became insoluble during thermal cracking. The presence of carbonaceous mesophase as determined by cross-polarized, optical microscopy definitely means that the asphaltenes phase separated during thermal cracking. In transmitted, cross-polarized light, the carbonaceous mesophase usually appears orange in color with black streaks. As one rotates the microscope stage, the black streaks move and form a Maltese cross. If the coke is only a piece of a larger carbonaceous mesophase, one only observes part of the pattern of the black streaks as one rotates the microscope stage. As the carbonaceous mesophase is a result of a liquid–liquid phase separation (chapter 4), if it stays in the bulk, interfacial tension causes it to form spherical drops before further reaction sets it up as spherical solid

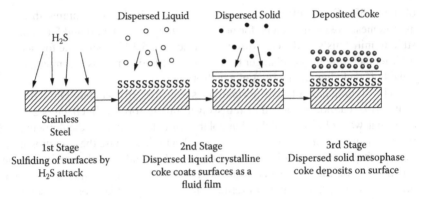

FIGURE 6.11 Adhesion of coke to stainless steel requires sulfiding for the carbonaceous mesophase to wet as a liquid film.

particles that cannot be melted or dissolved. The more interesting case is when a steel surface is involved as shown in Figure 6.11. The carbonaceous mesophase does not wet an oxide metallic surface and, therefore, only weakly adheres. However, thermal cracking of petroleum oils forms hydrogen sulfide, which forms iron sulfide on the surface of iron and steel surfaces, even stainless steels, and the carbonaceous mesophase wets iron sulfide surfaces. Thus, when liquid drops of carbonaceous mesophase contact an iron sulfide surface, they coat the surface as a liquid film and then set up as a very strongly adhering solid. This provides the adhesive to adhere other particles in the fluid, including dispersed coke solids, to the surface. As a result, there is potential for coatings that cannot be sulfided or wetted by carbonaceous mesophase to mitigate fouling by coke, which will be discussed in section 6.7.2.

If the asphaltenes are determined to be insoluble before thermal cracking, the cause is oil incompatibility, which was discussed in chapter 5 and will be discussed in section 6.8. Clearly, this requires observing insoluble asphaltenes in the oil flowing into the unit and use of the OCM for mitigation.

6.7.2 Mitigating Solutions

If coke forms by the phase separation of converted asphaltenes during thermal cracking of the oil, some possible mitigating solutions are

1. Flash off volatile nonsolvents.
2. Add hydrogen-donor solvent oil.
3. Coat steel surfaces.
4. Reduce conversion.

In section 4.4.8, we found that the coke induction period is much shorter for a closed reactor as compared to an open reactor. This is because the volatile liquids are nonsolvents for converted asphaltenes. Thus, a convenient mitigating solution for a thermal conversion reactor is to operate more like an open reactor than a closed reactor. This might be achieved by lowering the pressure or adding a stripping gas, such as steam. However, for tubular reactors operating in turbulent flow, one needs to be careful, as two-phase flow can undergo flow regimes that promote coking. This mitigating solution has been most successful for catalytic hydroconversion with dispersed or slurry catalysts, as special reactors with hydrogen as the stripping gas have been used. As will be discussed in chapter 10, this mitigation solution enabled increasing the conversion from 70 to 80% to 90 to 95%.

As examined in chapter 4, adding an aromatic oil to increase the solvency of converted asphaltenes will not necessarily increase the coke induction period for visbreaking if it decreases the concentration of hydrogen donors. This is not a serious issue for resid hydroconversion units because the aromatics are hydrogenated to form hydrogen donors. If a hydrogen donor solvent is added to a visbreaker, it will increase the coke induction period. However, if it is a more reactive hydrogen donor than the visbreaker feed, it will slow down the thermal reaction rate unless the temperature is increased. On the other hand, if the need is to mitigate coking where the objective is to heat the oil without thermal cracking, such as the furnace tubes of a distillation unit or the heater of a delayed coker, adding a reactive hydrogen donor that is a good solvent for asphaltenes would be quite beneficial.

Another way to mitigate coke fouling is to reduce the adhesion between the coke and the metal surface so that the shear stress of the flowing oil will remove the coke faster than it deposits. One method is to coat the steel surface with a material that does not sulfide and is not wetted by the carbonaceous mesophase. This can be more complex than it initially seems for distillation furnace tubes. The vaporization of the oil can deposit solids in the oil, such as clay, sand, salt, rust, and iron sulfide, on the coated wall. These solids or their sulfided products can be wetted by carbonaceous mesophase and actually increase coke deposition by the higher surface-area deposit. In addition, the practice of decoking by pigging[17] can scrape off the coating, requiring a method to coat the furnace tubes *in situ*. Actually, furnace tubes are often made with high chromium steels that form an oxide layer that is more resistant to sulfiding than other steels. However, pigging can scrape off the chromium oxide layer from the surface of the furnace tube and make it much easier to sulfide. Therefore, the furnace tubes should be conditioned with steam, air, and heat to replace the chromium oxide surface after pigging and before adding oil. Despite these complications, coating steel surfaces for coke mitigation has considerable promise and has not yet been fully explored.

Coking is usually caused by brief excursions away from steady state as steady-state formation of coke at a surface would require rapid shutting down of the unit. However, if the steady state is operated too close to the end of the induction period, the rate of coke deposition can be too high because of the higher frequency of these excursions beyond the coke induction period. As a result, one of the other mitigating solutions needs to be implemented or the conversion in the unit needs to be reduced. If the unit is one where conversion is not desired, such as a high-temperature heat exchanger, a reduction in the temperature or the residence time of the oil may be a very practical solution.

6.8 OIL INCOMPATIBILITY

6.8.1 MECHANISM AND IDENTIFICATION

As discussed in Chapter 5 and shown in Figure 6.12, there are three modes of fouling heated surfaces with asphaltenes: self-incompatible, incompatible on blending, and nearly incompatible. Once at the heated surface of a heat exchanger or furnace tube, depending on the time and temperature, the insoluble asphaltenes can reactively combine to form coke that cannot be redissolved. If a sample of the foulant is fresh enough and one finds that the major fraction is soluble in toluene or methylene chloride, it is clear that the deposit is caused by insoluble asphaltenes. More commonly, this cause can only be determined by applying the OCM to samples of the oil flowing in and out of the fouled unit. If the oil flowing into the unit, by the use of optical microscopy, is found to contain insoluble asphaltenes that dissolve in toluene, the oil is self-incompatible for that unit. Then one needs to trace back to previous units to determine where and why the asphaltenes became insoluble. As discussed in chapter 5, one possibility is that the crude oil was self-incompatible. More likely, the asphaltenes became insoluble because of blending incompatible oils or because of thermal conversion and cooling. Of these possibilities, the correct one becomes clear when the upstream unit

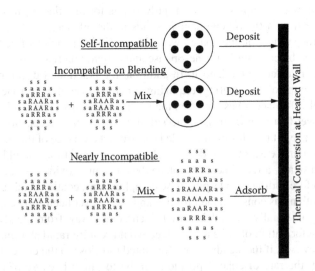

FIGURE 6.12 Three modes of fouling heated surfaces with asphaltenes: self-incompatible, incompatible on blending, and nearly incompatible. (From IA Wiehe, TG Jermansen. *Pet Sci Technol* 21: 527–536, 2003. With permission.)

is located where the oils flowing into the unit contain no insoluble asphaltenes, whereas the oil flowing out of the unit contains insoluble asphaltenes. If the oils are blended in the unit, application of the OCM will confirm that the oils are incompatible. If thermal conversion in the unit is suspected, it can be confirmed by asphaltenes leaving the unit having a much lower hydrogen-to-carbon atomic ratio than those entering the unit. If none of these mechanisms are confirmed, look for increases in the saturates or decreases in the aromatics or resins from the oil at the outlet as compared with the oil at the inlet of the unit. Recall that in section 6.4 the asphaltenes became insoluble because of the decrease in the concentration of resins.

If no insoluble asphaltenes are found to enter or leave a fouled unit, it is still possible that asphaltenes deposited by adsorption form a nearly incompatible oil. This can only be identified by running the oil compatibility tests and applying the OCM and determining that the solubility blending number of the oil is less than 1.4 times (especially less than 1.3 times) the insolubility number. This type of fouling usually occurs only over a long period of time (six months to several years) before noticing.

6.8.2 MITIGATING SOLUTIONS

If oil incompatibility is identified as a cause of fouling, possible mitigating actions are

1. Solve cause of insoluble asphaltenes of self-incompatible oils at source.
2. Do not blend oils incompatible in any proportion.
3. Blend in the right order; descending solubility blending number.
4. Blend in the right proportions; solubility blending number less than 1.4 times insolubility number.
5. Add solvent oils.
6. Use asphaltene dispersants.

If the oil flowing into the fouled unit is self-incompatible, section 6.8.1 describes tracing the insoluble asphaltenes to the source. There are three possibilities: (1) the crude oil is self-incompatible, (2) incompatible oils were blended upstream of the fouled unit, and (3) asphaltenes became insoluble because of conversion and cooling. The simplest solution to self-incompatible crude oils is to reject purchase. There are other possibilities, but the incentives need to be unusually high to make it worthwhile. These possibilities include blending with a hot solvent oil to dissolve the asphaltenes without a heat exchanger, processing at high velocities to minimize the deposition, and solving the insoluble asphaltene cause at the production site. Solutions to asphaltenes on cooling after a conversion unit has already been discussed, and the blending of incompatible oils within a unit are covered next.

If the insoluble asphaltenes are fouling a unit because of incompatibility or near incompatibility within the unit, the best solution and the one outside Exxon patents,[18,19] is to change the process not to blend any oils that are incompatible in any proportion. The oil compatibility tests and model may be used to determine the oil with the maximum insolubility number in the blend, and which of the oils have solubility blending numbers below this maximum insolubility number. Thus, one choice is to schedule the processing of the oils so that they are not processed together. This is usually easier to do for crude oils than process oils. For process oils, determine why the process oil with the maximum insolubility number has such a high value and if the process that created it can be changed. Thus, use the insolubility number as the precursor, trace it to the source, and follow the general mitigation strategy to devise methods to reduce it. Otherwise, solutions may be used that fall within the Exxon patents[18,19] to prevent incompatibility or near incompatibility by blending the oils in the right proportions and right order using the OCM as described in chapter 5, including adding a high solubility blending number oil. Finally, dispersants may be used to lower the maximum insolubility number in the blend as discussed in section 6.5.

6.9 ENGINEERING METHODS FOR FOULING MITIGATION

Although we have concentrated on chemical methods to mitigate fouling, one should recognize that there are engineering methods that often reduce fouling even if the cause is not known, particularly for heat exchangers. Wayne Ebert of Chevron and C. B. Panchal[2] of Argonne National Labs have determined that the range of fouling depends on the film temperature (average of bulk and wall temperatures) and the Reynolds and Prandel numbers. Thus, fouling can be mitigated by increasing the velocity through the heat exchanger tubes. This increases the shear stress at the wall and removes the foulant faster than it deposits. For this reason, smaller heat exchangers are often better than large heat exchangers.[20] The addition of tube inserts, such as vibrating springs[21] or using twisted tubes[22] that increase turbulence at the wall, often reduce fouling. Also, the Ebert and Panchal correlation teaches us that selection of a lower-temperature liquid to heat exchange with a process oil can reduce fouling.

6.10 LIMITATIONS AND FUTURE DEVELOPMENTS

A systematic procedure has been given for mitigating fouling by determining the cause of refinery fouling and tracing its precursors to the source. These are based on understanding the chemical and physical mechanisms of the most common causes of refinery fouling and the use of key indicators that enable quick diagnosis and focusing on the correct precursors. The understanding of organic fouling mitigation is in good shape because of the new understanding in oil compatibility, the phase-separation mechanism of coke formation, and the mechanism of the polymerization of conjugated olefins to form popcorn coke. Previous to these advances, organic fouling was very mysterious, but now most organic fouling can be mitigated by the systematic procedures outlined. Of course, the understanding of organic fouling can still be improved. For example, ways to improve the OCM were discussed in chapter 5. Another need is to combine the chemical and engineering methods into one strategy. How does the ratio of the solubility blending number to the insolubility number affect the Ebert and Panchal correlation? Which tube insert is best depending on the fouling cause? However, at present the mitigation of inorganic fouling is lagging behind. Therefore, inorganic fouling mitigation needs further emphasis to improve the removal of water-soluble and oil-insoluble inorganics in desalting and prevent corrosion by naphthenic acids and hydrogen sulfide. Desalting and problems of stable oil–water emulsions are discussed in the next chapter. It has already been examined how the sulfiding of steel surfaces promotes the adhesion of coke and that the presence of inorganic fouling accelerates organic fouling. Therefore, it is recommended that those who study petroleum corrosion and petroleum fouling work closely together to make the next significant advance in the mitigation of petroleum fouling.

REFERENCES

1. G Brons, IA Wiehe. Mechanism of reboiler fouling during reforming. *Energy and Fuels* 14: 2–10, 2000.
2. CB Panchal. Refinery fouling research. Proceedings of Refinery Fouling Mitigation, Argonne National Laboratory, Aogonne, IL, 1998, pp. 1–8.
3. HK Lemke. Fouling in refinery equipment: An overview. Proceedings of the 1st International Conference on Petroleum Phase Behavior and Fouling, AIChE, New York, 1999, pp. 375–382.
4. IA Wiehe. The chemistry of petroleum fouling. Proceedings of the 4th International Conference on Refinery Processing, AIChE, New York, 2001, pp. 204–210.
5. IA Wiehe. Mechanisms of coke and sediment resulting from resid conversion. Spring ACS Meeting, 1997, San Francisco.
6. G Brons, TM Rudy. Logical approach to troubleshooting crude preheat exchanger fouling. Proceedings of the 2nd International Conference on Petroleum Phase Behavior and Fouling, Copenhagen, 2000, paper 50.
7. WK Robbins. Challenges in the characterization of naphthenic acids in petroleum. Preprints, ACS, *Div Pet Chem*, 43: 137–140, 1998.
8. WK Robbins. Challenges in the analytical characterization of non-thiophenic and thiophenic sulfur compounds in petroleum. Preprints, ACS, *Div Pet Chem* 45: 68–71, 2000.
9. IA Wiehe. Mitigation of fouling by popcorn coke. *Pet Sci Technol* 21: 673–680, 2003.
10. IA Wiehe, G Brons, LS Cronin. Mitigation of Fouling by Thermally Cracked Oils. U.S. Patent 6,210,560, 2001.
11. IA Wiehe, Self-incompatible crude oils and converted petroleum resids. *J Dispersion Sci Technol* 25: 333–339, 2004.
12. IA Wiehe, TG Jermansen. Design of synthetic dispersants for asphaltenes. *Pet Sci Technol* 21: 527–536, 2003.
13. C-L Chang, HS Fogler. Stabilization of asphaltenes in aliphatic solvents using alkylbenzene-derived amphiphiles. *Langmuir* 10: 1749–1766, 1994.
14. IA Wiehe, R Varadaraj, TG Jermansen, RJ Kennedy, CH Brons. Branched Alkyl-Aromatic Sulfonic Acid Dispersant for Solubilizing Asphaltenes in Petroleum Oils. U.S. Patent 6,048,904, 2000.

15. H Buron, I Cohen, J-A Ostlund. New ASTM method for heavy fuel oil stability, Preprints, ACS, *Div Fuel Chem* 50(1): 230–231, 2005.
16. JG Speight. *The Chemistry and Technology of Petroleum*, 3rd ed., New York: Marcel Dekker, 1999, p. 885.
17. RJ Parker, RA McFarlane. Mitigation of fouling in bitumen furnaces by pigging. *Energy and Fuels* 14: 11–13, 2000.
18. IA Wiehe, RJ Kennedy. Process for Blending Potentially Incompatible Petroleum Oils. U.S. Patent 5,871,634. Assigned to Exxon, 1999.
19. IA Wiehe, RJ Kennedy. Process for Blending Petroleum Oils to Avoid Being Nearly Incompatible. U.S. Patent 5,997,723. Assigned to Exxon, 1999.
20. H Joshi, G Brons. Effect of fouling on design of oil refinery heat exchangers. Proceedings of the 4th International Conference on Petroleum Phase Behavior and Fouling, Trondheim, Norway, 2003.
21. F Jardin, AW Krueger. Examples of successful fouling mitigation in crude refining operations by utilizing mechanical online cleaning devices. Proceedings of the 3rd International Conference on Petroleum Phase Behavior and Fouling, AIChE, New York, 2002, pp. 466–472.
22. B Ljubicic. Heat exchanger design: a manufacturer's perspective. Proceedings of the 4th International Conference on Refinery Processing, AIChE, New York, 2001, pp. 211–223.

7 Separation of Petroleum

In this chapter, the commercial separation processes (desalting and solvent deasphalting) will be discussed in some detail and the constraints on separation processes as primary upgrading processes will be determined. Desalting is covered first because of its relationship to fouling mitigation in chapter 6. If desalting is not done well, it can result in refinery fouling by inorganics. Distillation is not discussed because it was covered in sufficient detail in sections 1.5.1 and 1.5.2, considering that distillation provides few new opportunities for improved processing of petroleum macromolecules. On the other hand, solvent deasphalting has begun to be used for fuels and is showing promise as an upgrading process, even though in the past it was primarily used for production of heavy lubricating oils.

In this chapter, the stage for much of the rest of the book will be set. Section 1.6.2 contained a discussion on how to upgrade the heavy oil, extra heavy oil (bitumen), or petroleum resid in the primary upgrading process so that it becomes suitable for secondary upgrading with relatively expensive catalysts: fixed bed hydrotreating, fluid catalytic cracking, and hydrocracking. None of the primary upgrading processes will be able to yield 100 wt% of conversion to feed for catalytic processing because of the presence of polynuclear aromatic cores. Therefore, we want to determine the maximum feed to catalytic processing that is possible and the maximum that one can achieve by ideal laboratory experiments. This will determine if there is potential for improving commercial upgrading processes and, if so, how one might accomplish this improvement. In this chapter, separation processes will be considered as the upgrading process.

7.1 DESALTING[1,2]

In the petroleum industry, desalting is often the forgotten separation process. However, with the desalter at the front of the refinery, poor desalting causes problems throughout the rest of the refinery. The fouling of heat exchangers and furnace tubes by sea salts increases energy costs. Extreme fouling of these units with sea salts can reduce the feed rate by plugging. Hydrolysis of calcium and magnesium chloride during crude oil distillation and resid conversion forms hydrochloric acid, which not only causes corrosion, but also combines with ammonia formed in the same process to form solid ammonium chloride on overhead surfaces. By leaving sodium

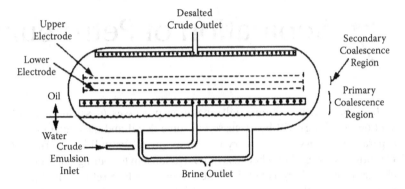

FIGURE 7.1 Schematic of an electrostatic desalter. (Reproduced courtesy of Cameron Petreco®.)

chloride in resid, the sodium attacks zeolite catalysts in resid cat cracking and forms a low-melting eutectic with vanadium in gasifiers, such as in Flexicoking.

In desalters, as shown in Figure 7.1, freshwater is mixed with the oil that naturally contains saltwater to form water in oil emulsion. An electrostatic field induces the brine droplets to be charged positive on one end and negative on the other. This causes the droplets to become attracted that promotes coalescence and rapid settling into a bottom brine layer. With two stages, the brine from the second stage being the wash water in the first stage (Figure 7.2), desalting is greatly improved without any increase in the freshwater requirement. Desalters are commonly operated under pressure and above the boiling point of water for low oil viscosity and rapid settling rates. Surface-active agents (3 to 10 ppm) are added to the oil as demulsifiers and to make suspended solids water wet.[2] By regulating the

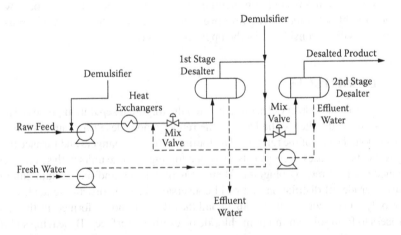

FIGURE 7.2 Flow diagram of a typical two-stage electrostatic desalter system. (Reproduced courtesy of Cameron Petreco®.)

pressure drop across the mixing valve, the compromise between having a good but unstable emulsion can usually be obtained. However, a stable water-in-oil emulsion can greatly disrupt the separation and form a rag layer (emulsion band) between the oil and the water layers. This is encouraged by high oil viscosity, suspended solids (clays, rust, iron sulfide, etc.), natural surface-active agents, such as naphthenic acids and salts of naphthenic acids, and insoluble and nearly insoluble asphaltenes. Another common problem is too high oil conductivity caused by suspended solids, naphthenic acids, and salts of naphthenic acids. This limits the effect of any electrical field that can be applied. Finally, variations in crude oil and wash water can greatly upset desalter performance.[3]

Methods to eliminate the rag layer depend on determining the cause. With the exception of incompatible and nearly incompatible oils, the cause may be difficult to diagnose.

7.1.1 RAG LAYER CAUSED BY INCOMPATIBLE OR NEARLY INCOMPATIBLE OILS

McLean and Kilpatrick[4] have determined that insoluble or soluble, but nearly insoluble asphaltenes, are drawn to the oil–water interface and form a strong shell around water drops. These strong shells prevent emulsion water drops from coalescing because colliding water drops cannot break through the strong shells. Nearly incompatible oils actually form more stable emulsions than insoluble asphaltenes because submicroscopic associated asphaltenes pack tighter around water droplets than do the much larger insoluble asphaltene particles.

The diagnosis and elimination of rag layers caused by incompatible or nearly incompatible oils are the same as described in section 5.3 on the oil compatibility model, although even greater emphasis should be placed on eliminating nearly incompatible crude oils. Thus, if a rag layer is obtained when running a blend of crude oils through a desalter, measure the insolubility number and solubility blending number of each crude oil in the blend. If the solubility blending number of the blend is less than 1.3 times the largest insolubility number in the blend, near incompatibility should be a suspected cause. Of course, if the solubility blending number of the blend is less than the largest insolubility number in the blend, incompatibility is the likely cause. In either case, the preferred solution is to eliminate either the crude oil with the highest insolubility number or the crude oil with the lowest solubility blending number from the blend.

7.1.2 RAG LAYER CAUSED BY INORGANIC PARTICLES

Inorganic particles that are wetted by both oil and water tend to accumulate at the oil–water interface and to stabilize emulsions by steric stabilization. Clays and iron sulfide are the most common inorganic particles that stabilize oil–water emulsions and cause rag layers. Crude oils may be analyzed for inorganics, such as by inductively coupled plasma spectroscopy (ICPS). High clay content can be suspected if greater than 100 ppm aluminum and silicon are observed. High iron sulfide content can be suspected if greater than 100 ppm iron is observed. Often, if these are the cause, a sample of the rag layer under the microscope can be observed to have inorganic particles at the oil–water interface. The solutions are to reduce or eliminate the crude oil containing the inorganic particles or to add chemical agents to make the particles water-wet, but not oil-wet. This causes the particles to go to the water phase and flow out of the desalter with the water.

7.1.3 RAG LAYER CAUSED BY NAPHTHENIC ACIDS AND NAPHTHENIC ACID SALTS

Naphthenic acids were discussed in section 2.6.2 and can be suspected to cause a rag layer if the total acid number (TAN) is greater than 2. Salts of naphthenic acids, such as calcium naphthenate, are soaps and result in even more stable oil–water emulsions than naphthenic acids. In both cases, they stabilize oil–water emulsion by lowering the interfacial tension between oil and water. One solution is to dilute the crude oil high in naphthenic acids and naphthenic acid salts with crude oils low in both. Another solution is to add napthenate inhibitors that go to the oil–water interface and prevent cations in the water from reacting with naphthenic acids in the oil.[5]

7.2 MAXIMUM POTENTIAL YIELD BY SEPARATION[6,7]

It is always worthwhile determining the very best that a process can achieve. If this does not meet one's objective, one does not waste time and effort trying to do the impossible. Sometimes, by understanding the cause of a barrier to achieve the maximum, one can find ways of going around the barrier.

7.2.1 PRIMARY AND SECONDARY UPGRADING

In Figure 1.3 and section 1.6.2, solvent deasphalting was included as a possible primary upgrading process for producing a higher-quality feed for secondary upgrading by catalytic processing. Extra heavy oils and bitumen are often similar to a low-quality atmospheric resid and as a result, also fit into this scheme. The advantage of secondary upgrading by catalysts is that they give much better selectivity to high-quality transportation fuels than the processes considered for primary upgrading. However, these catalysts are relatively expensive and are poisoned by poor-quality feeds containing Conradson carbon (polynuclear aromatics), basic nitrogen, vanadium, nickel, sodium, etc. More generally, any separation process more selective than distillation is to be considered as a primary option if development of a new separation process can be justified. Because the only commercial separation process meeting this description is solvent deasphalting, it is listed in Figure 1.3.

Given that the feeds and upgrading requirements vary considerably with different locations and different refineries, one can only work with examples. Therefore, in this section, we consider separating from an example vacuum resid, Arabian Heavy or from an example heavy oil, Cold Lake bitumen, the fraction that can meet fluid catalytic cracking specifications with the remainder going to coking. In the innovation process a good first step is to determine the potential under ideal conditions. This will tell us if the prize is worth going after; also, any limitations that exist will become known. Therefore, without committing to any particular separation process, a small laboratory procedure is used to determine the maximum separation possible for any separation process.

Although refinery separation of heavy oil is often restricted to distillation, fluid catalytic cracking (FCC) feed is not required to be volatile, but should contain acceptable levels of catalyst poisons: Conradson carbon, vanadium, nickel, and basic nitrogen. Therefore, the laboratory separation of Cold Lake bitumen and Arabian Heavy vacuum resid was done to determine the potential for a more molecularly selective separation. The combination of distillation, deasphalting, and adsorption was used to determine the ultimate separation that is possible to approximate the molecular limits, or the molecular separability. Finally, thermal conversion before separation was investigated to determine how conversion changed the separability.

The feed specifications for fluid catalytic cracking of resid-containing oils depend greatly on the design and practice of the FCC unit, on the catalyst, and on other feeds available for blending. Nevertheless, the typical guidelines of Barnes[8] shown in Table 7.1 will be used. Of these, Conradson carbon and vanadium content are often the most restrictive and will be emphasized.

7.2.2 Experimental

The general laboratory procedure used for separating heavy oils is shown in Figure 7.3. The first possible step is mild thermal conversion, which was done in tubing bombs by immersing them in a sand bath preheated to 400°C. The second step was batch distillation at 1.4 mmHg, which was done directly out of tubing bombs immersed in a sand bath at 315°C. The third and fourth steps were repeated using several solvents, but the same solvent for each step for a given trial. The third step removed the solvent insolubles by mixing 25 parts solvent to one part oil, waiting eight hours, and filtering. In the fourth step Attapulgus clay was mixed with the oil dissolved in the solvent and allowed to sit for eight hours. This mixture was filtered with a fine glass frit and washed with additional solvent

TABLE 7.1
Feed Specifications for Resid FCC[8]

Feed Property	Limit
Conradson carbon residue	3–8 wt%
Nickel content	20–25 ppm
Vanadium content	10–15 ppm
Sodium content	5–10 ppm
Basic nitrogen	800 ppm

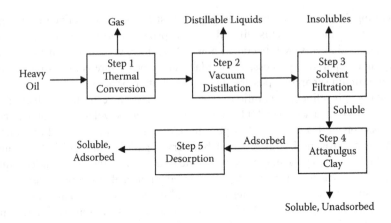

FIGURE 7.3 General procedure for laboratory selective separation.

until the solvent passed through clear of any color. The heavy oil dissolved in the solvent was recovered by rotary evaporation and vacuum drying. The fraction remaining on the clay was recovered by washing on the glass frit with a mixture containing 50% acetone and 50% toluene, followed by 10% methanol and 90% toluene. Finally, the adsorbed fraction of heavy oil was recovered from the solvents by rotary evaporation and vacuum drying. The solvents used in the separation trials included *n*-heptane, cyclohexane, toluene, and carbon disulfide. Once collected, analytical data were measured on each of these fractions.

7.2.3 Separation of Unconverted Heavy Oils

For the separation of each heavy oil, each quality measurement — Conradson carbon, vanadium, nickel, etc. — in the higher-quality fraction was plotted against the yield of the higher-quality fraction. Because vacuum distillation was the highest-quality fraction, it was the point at the lowest yield for Cold Lake bitumen. The soluble, but unadsorbed, fraction was the next highest quality fraction for each trial solvent. Thus, the quality measurement and yield were calculated as if this fraction and the distillable liquids were mixed to form the second point for each trial solvent. Likewise, the soluble adsorbed fraction and the insolubles were added. This type of data for Conradson carbon is shown in Figure 7.4 for Cold Lake bitumen and in Figure 7.5 for Arabian Heavy vacuum resid. In each case, a curve is drawn through those of highest yield at a given quality measurement as this determines the best way the heavy oil could be split into two fractions, the separability.

Of the solvents tried, cyclohexane is the best because it gave the greatest yield of high-quality oil (soluble, unadsorbed fraction) in step 4 without any insolubles in step 3. The yields and analytical data for the cyclohexane separations are shown in Table 7.2 and compared with the starting feed data. Thus, for Arabian Heavy vacuum resid that does not require distillation, the separation is achieved in two steps, adsorption and desorption; and about half of the resid

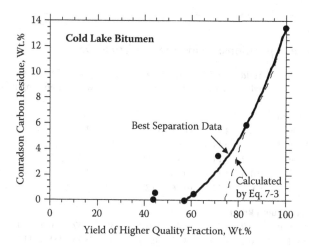

FIGURE 7.4 Selective separation of Conradson carbon for Cold Lake bitumen.

would meet the specifications for resid FCC feed as shown in Table 7.1. On the other hand, combining vacuum distillation with cyclohexane adsorption separates 70% of Cold Lake bitumen into a fraction that meets the example specifications for resid FCC feed.

7.2.4 Conradson Carbon Separability Limit

In this section, we will focus on the problem of maximizing the yield of a low Conradson carbon fraction. One hundred parts of heavy oil of C_F Conradson

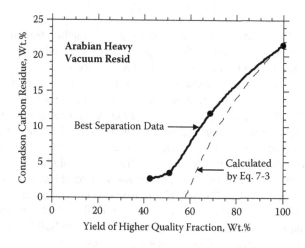

FIGURE 7.5 Selective separation of Conradson carbon for Arabian Heavy vacuum resid.

TABLE 7.2

Cyclohexane Separations: Yields and Analytical Data

Fraction	Yield wt%	C wt%	H wt%	N wt%	O wt%	S wt%	V ppm	Ni ppm	Conradson Carbon wt%
Cold Lake Bitumen									
Distillable liquids	44.5	85.14	11.67	0.092	0.00	2.63	0.6	0.0	0.57
Soluble, unadsorbed	26.7	84.08	10.96	0.17	0.60	4.11	29	10	8.4
Soluble, adsorbed	30.2	81.07	8.73	0.99	2.39	6.55	359	138	43.8
Total	101.5	83.57	10.60	0.38	0.87	4.19	115	41	15.5
Full feed	100	83.82	10.46	0.38	0.68	4.57	152	62	13.5
Arabian Heavy Vacuum Resid									
Soluble, unadsorbed	50.9	85.07	11.47	0.030	0.50	3.57	0.0	0.0	3.4
Soluble, adsorbed	50.1	82.63	8.53	0.80	1.63	6.78	451	112	36.9
Total	101.0	83.86	10.01	0.41	1.06	5.16	224	56	20.2
Full feed	100	83.51	9.93	0.45	0.57	5.80	165	40	22.3

carbon is separated into β parts of a higher-quality fraction containing a Conradson carbon of C_H and $100 - \beta$ parts of a lower-quality fraction containing a Conradson carbon of C_L. As Conradson carbon is conserved for separations (chapter 3)

$$100\,C_F = \beta\,C_H + (100 - \beta)\,C_L \qquad (7.1)$$

Solving for β

$$\beta = 100\,[1 - (C_F - C_H)/(C_L - C_H)] \qquad (7.2)$$

This shows that to maximize the yield of a high-quality fraction meeting a Conradson carbon specification from a given heavy oil feed (C_H), one needs to concentrate the Conradson carbon in the low-quality fraction (C_L). Therefore, for Cold Lake bitumen (Conradson carbon = 13.5 wt%) to separate 85 wt% of a fraction meeting a specification of 5 wt% Conradson carbon, one would need to isolate 15 wt% containing 61.7 wt% Conradson carbon. However, the highest Conradson carbon fraction that was isolated from Cold Lake bitumen is the 14.1 wt% yield of n-heptane insoluble asphaltenes with a Conradson carbon of 52.5%. Thus, it is unlikely that 85 wt% of a fraction of Cold Lake bitumen with a Conradson carbon of 5 wt% or less exists because significant fractions of Conradson carbon much greater than 50 wt% Conradson carbon are not present in heavy oils. This is because the coke precursors are chemically linked to noncoke precursors in the same molecule, as shown by the Pendant–Core Building Block Model (chapter 3). For instance, Arabian Heavy vacuum resid with a Conradson carbon of 21.5 wt% and an asphaltene Conradson carbon of 51.6 wt% at a specification of 5 wt% Conradson carbon:

$$\beta = 100\ [1 - (21.5 - 5)/(51.6 - 5)]$$

$$\beta = 64\ \text{wt\%}$$

Thus, the separation of 64 wt% of Arabian Heavy vacuum resid with a Conradson carbon of 5 wt% or less is not possible because there are only 20 wt% asphaltenes and not the 36 wt% required. As a result, equation (7.2) with the asphaltene Conradson carbon substituted for C_L is an upper limit on the yield of a fraction of a given Conradson carbon value. Equation (7.1) can be rearranged and the asphaltene Conradson carbon, C_A, substituted for C_L to give

$$C_H = \{C_F - (1 - \beta/100)\ C_A\}/\{\beta/100\} \tag{7.3}$$

This upper limit, equation (7.3), is plotted on Figure 7.4 and Figure 7.5 as dashed curves. The curve through the best separation data approaches the upper limit at high yields of the higher-quality fraction for Cold Lake bitumen. On the other hand, a gap remains between the two curves for Arabian Heavy vacuum resid, indicating that further improvements in separability may be possible.

This concept of molecular limited separability of Conradson carbon is shown more graphically in Figure 7.6 on the basis of the Pendant–Core Building Block Model. Assume we can design a selective separation process that could separate on the basis pendant and cores. First, we separate all the molecules containing cores into the low-quality stream and those only containing pendants into the high-quality stream. The high-quality steam is of much better quality than required, but it is a small fraction of the starting heavy oil or resid. We decide to lower our quality standard to get a greater fraction into the high-quality stream. Therefore, we separate, as shown in Figure 7.6, so that all molecules containing one core or less are put into the high-quality stream. If we continue to relax our standard and

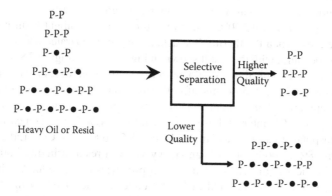

FIGURE 7.6 Hypothetical separation of pendant–core molecules of one core or less.

continue to increase the yield of the higher-quality stream, we will obtain a plot of quality versus yield, such as Figure 7.4 and Figure 7.5. Also, from chapter 3, the fraction of the molecule that comprises cores is equal to the Conradson carbon residue. Thus, the generated curve would be the maximum yield possible of the higher-quality fraction at any given average Conradson carbon of that fraction. We cannot do any better because of the covalent bonds between pendants and cores. The lower-quality stream has to contain pendants because all molecules that contain cores also contain pendants. Although there is a fraction that contains only pendants (zero Conradson carbon residue), to obtain any greater yield of the higher-quality fraction, the higher-quality fraction will contain some cores and, thus, some Conradson carbon residue. On the other hand, if the pendant and cores existed in separate molecules, one could separate a large fraction containing only pendants (zero Conradson carbon residue) as the higher-quality fraction and the remaining lower-quality fraction would only have cores (100% Conradson carbon residue). Only in this case would there not be a molecular limit to the separation. Even if the asphaltenes have an appreciable fraction of molecules greater than 50% cores, one could put more pendants in the higher-quality fraction. Thus, for a given average quality of feed at each quality of the higher-quality fraction, the yield of that fraction could increase if the cores were more concentrated in the lower-quality fraction (fewer pendants).

7.2.5 SEPARATION OF THERMALLY CONVERTED HEAVY OILS

The barrier on the yield of the higher-quality fraction of heavy oils now has been identified as the chemical links between the Conradson carbon precursors and the high-quality parts of the same molecules. It makes no sense to study a wide range of separation processes to obtain a better separation of Conradson carbon when the barrier is caused by the heavy oils themselves. To overcome this barrier, the chemical links need to be broken. In terms of the Pendant–Core Building Block Model, the bonds between the pendants and core building blocks need to be broken. The easiest method to do this is by thermal conversion. Thus, Cold Lake bitumen and Arabian Heavy vacuum resid were thermally treated for 45 and 60 min, respectively, at 400°C, the maximum for each without forming toluene insoluble coke. Then the thermally cracked oils were separated according to the general procedure, with the result shown in Figure 7.7 and Figure 7.8 for Conradson carbon. As is typical, the total Conradson carbon increases for thermal conversion (section 3.7). Despite this, after thermal conversion, the yield is higher for the best separation than without thermal conversion for Conradson carbon values in the range of FCC interest. For instance, at the 5 wt% Conradson carbon level, this corresponds to an increase in yield from 77% to 85% for Cold Lake bitumen and from 58% to 66% for Arabian Heavy vacuum resid, bringing both well in excess of typical liquid yields from coking processes on these feeds (chapter 8). However, this increase in yield of 8% absolute for each feed by mild thermal conversion is less than the increase of distillable liquids of 13% for Cold Lake bitumen and 28% for Arabian Heavy vacuum resid.

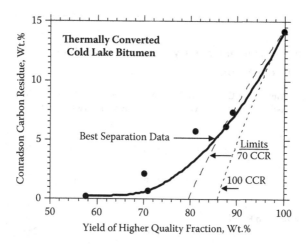

FIGURE 7.7 Selective separation of Conradson carbon on thermally converted Cold Lake bitumen.

The increase in yield of the higher-quality fraction at the same Conradson carbon limit after thermal conversion also causes an increase in Conradson carbon of the asphaltene fraction as the pendants are cracked off the asphaltene core. As shown in Table 7.3, thermal conversion forms significant cyclohexane insolubles of 70% Conradson carbon that were not present in the feeds. Therefore, the separation limits for 70% Conradson carbon were calculated with equation (7.3) and added to Figure 7.7 and Figure 7.8 and shown to provide a good approximation to

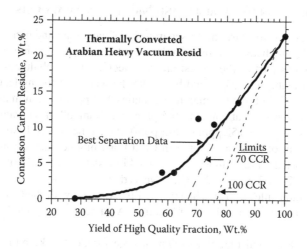

FIGURE 7.8 Selective separation of Conradson carbon for thermally converted Arabian Heavy vacuum resid.

TABLE 7.3

Comparison of *n*-Heptane and Cyclohexane Insolubles after Thermal Conversion

Fraction	Yield (wt%)	C (wt%)	H (wt%)	N (wt%)	S (wt%)	V (wt%)	Ni (wt%)	Conradson Carbon (wt%)
Cold Lake Bitumen								
n-Heptane asphaltenes	13.1	82.84	6.42	1.40	7.24	949	439	62.1
Cyclohexane insolubles	11.1	82.27	6.35	1.46	7.62	1120	451	68.3
Arabian Heavy Vacuum Resid								
n-Heptane asphaltenes	23.9	82.36	6.24	1.05	8.12	711	252	65.8
Cyclohexane insolubles	15.9	82.86	6.05	1.27	8.51	672	250	70.1

the data at high yields. The absolute upper limit for the combination of thermal conversion and selective separation is the 100% Conradson carbon limit shown in Figure 7.7 and Figure 7.8 (no bonds between pendants and cores). Therefore, a coker that separates product selectively, instead of by distillation, would show significant further yield advantages at the same product Conradson carbon. How this insight can be taken advantage of will be described in chapter 8.

7.2.6 SEPARATION OF VANADIUM AND NICKEL FROM HEAVY OILS

Unlike for Conradson carbon, the separability of vanadium and nickel is not limited much by the heavy oil macromolecules. Vanadium with an atomic weight of 50.9, even as part of the largest macromolecules, the asphaltenes of average molecular weight of 3000, would be 1/60th the weight of the molecule bonded to it. Thus, in Cold Lake bitumen that contains 152 ppm vanadium and Arabian Heavy vacuum resid that contains 165 ppm vanadium, all of the vanadium is in less than 1 wt% (60 × 152 ppm or 165 ppm) of each heavy oil. Likewise, nickel (atomic weight of 58.7) at 62 ppm in Cold Lake bitumen and 56 ppm in Arabian Heavy vacuum resid must be contained in less than 0.5 wt% of each heavy oil (3000/58.7 × 62 ppm or 56 ppm). Nevertheless, the vanadium- and nickel-containing molecules (porphyrins) physically associate with the Conradson carbon precursors, the polynuclear aromatics. As a result, they tend to separate together.

As Figure 7.9 and Figure 7.10 demonstrate for Cold Lake bitumen, a little more zero vanadium fraction or zero nickel fraction can be separated than zero Conradson carbon fraction (see Figure 7.4). If we assume limits a little better than

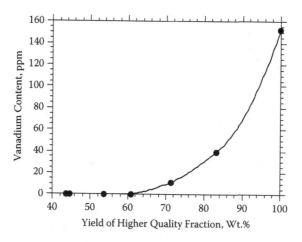

FIGURE 7.9 Selective separation of vanadium in Cold Lake bitumen.

the midpoint of limits in Table 7.1, we arrive at 5% Conradson carbon residue, 12 ppm of vanadium, and 22 ppm of nickel. Based on these for Cold Lake bitumen, one can obtain a yield of the higher-quality fraction of 77% at the limit for Conradson carbon residue, 72% at the limit for vanadium, and 82% at the limit for nickel. Thus, vanadium restricts the yield of fluid catalytic cracking feed for Cold Lake bitumen. Similarly, using Figure 7.5, Figure 7.11, and Figure 7.12 for Arabian Heavy vacuum resid, the yield of the higher-quality fraction is 58% at the limit for Conradson carbon residue, 64% at the limit for vanadium, and 91% at the limit for nickel. Thus, Conradson carbon residue restricts the yield of fluid catalytic cracking feed for Arabian Heavy vacuum resid.

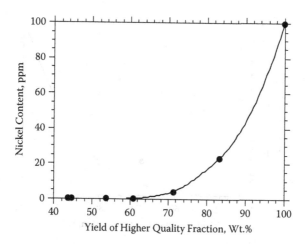

FIGURE 7.10 Selective separation of nickel in Cold Lake bitumen.

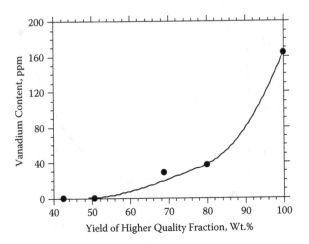

FIGURE 7.11 Selective separation of vanadium in Arabian Heavy vacuum resid.

As shown in Figure 7.13, thermal cracking increases the yield of fluid catalytic cracking feed from Cold Lake bitumen at the limit of 22 ppm vanadium from 72% to 79% as compared with the increase from 77% to 85% for Conradson carbon residue. Thus, vanadium is also the limit in the yield of fluid catalytic cracking feed from thermally cracked Cold Lake bitumen. On the basis of nickel, a higher yield would be obtained even without thermal cracking. By thermally converting Arabian Heavy vacuum resid at the vanadium limit of 12 ppm, the yield of fluid catalytic cracking feed is increased from 64% to 74% (Figure 7.14).

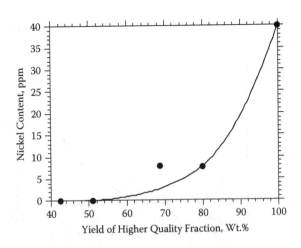

FIGURE 7.12 Selective separation of nickel in Arabian Heavy vacuum resid.

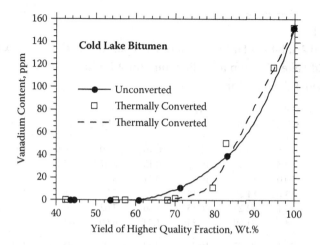

FIGURE 7.13 Selective separation of vanadium in unconverted and thermally converted Cold Lake bitumen.

However, this is beyond the yield of 66% limited by 5% Conradson carbon residue on thermally converted Arabian Heavy vacuum resid. Again, nickel imposes no limitation, as without thermal conversion the yield of the high-quality fraction could be 91% and still meet the nickel specification of 22 ppm.

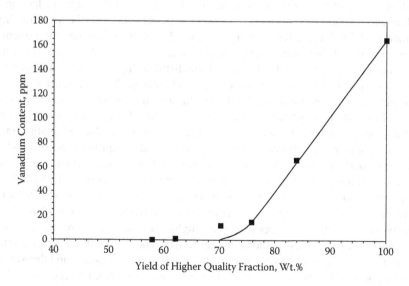

FIGURE 7.14 Selective separation of vanadium in thermally converted Arabian Heavy vacuum resid.

TABLE 7.4

Yields and Analytical Data on Fractions Separated with Cyclohexane from Cold Lake Bitumen after Reacting for 45 min at 400°C in a Closed-Tube Reactor

Fraction	Yield (wt%)	C (wt%)	H (wt%)	N (wt%)	S (wt%)	V (wt%)	Ni (wt%)	CCR (wt%)
Dist. Liq.	42.5	85.51	11.17	0.10	2.60	0.0	0.0	0.29
Unadsorbed	27.5	84.38	10.57	0.13	3.87	3.4	0.0	5.2
Nonpolars	70.0	85.07	10.93	0.11	3.10	1.3	0.0	2.2
Adsorbed	13.0	81.75	8.18	1.22	5.73	306	121	41.3
Insoluble	11.1	82.27	6.35	1.46	7.62	1120	821	68.3
Polars	24.1	81.99	7.34	1.33	6.60	681	443	53.7
Total	94.1	84.42	10.01	0.42	4.00	175	113	15.4
Full feed	100	83.82	10.46	0.38	4.57	152	62	13.5

7.2.7 DO EVEN BETTER: THERMAL CONVERSION WITH RECYCLE

In Figure 7.7 and Figure 7.8, the curves for the best separation after thermal conversion deviates the most from the 70% Conradson carbon residue limits at low values of Conradson carbon residue for the higher-quality fraction. How can we move the separability curves toward the 70% Conradson carbon residue limits? There is a fraction of intermediate quality that is not good enough for feed to a fluid catalytic cracker, but not bad enough to be rejected with the lower-quality fraction. Table 7.4 and Table 7.5 show the yields and analytical data for the separated fractions using the general procedure for laboratory selective separation with cyclohexane as the solvent. The sum of the distillate liquids and the unadsorbed fraction, which is called nonpolars, makes a higher-quality fraction that is well within specifications for feed to fluid catalytic cracking for both thermally cracked Cold Lake bitumen and Arabian Heavy vacuum resid. The cyclohexane insolubles with Conradson carbon residues of 68% and 70% make good lower-quality fractions. However, the adsorbed fractions with Conradson carbon residues of 41% are not good enough to add to the better-quality fraction nor bad enough to add to the lower-quality fraction as done to make the polar fraction with Conradson carbon residues of 54%. What makes more sense is to recycle this moderate Conradson carbon fraction to be further thermally converted into a low Conradson carbon residue product (nonpolars) and a high Conradson carbon residue product (cyclohexane insolubles) as shown in Figure 7.15. The result will be two products, with the better-quality product having the properties of nonpolars and the lower-quality product having the properties of cyclohexane insolubles. Thus, equation (7.2) can be used to calculate the yields of fluid catalytic cracker feed at a Conradson carbon residue of 5% from Cold Lake bitumen using the value of Conradson

TABLE 7.5

Yields and Analytical Data on Fractions Separated with Cyclohexane from Arabian Heavy Vacuum Resid after Reacting for 60 min at 400°C in a Closed-Tube Reactor

Fraction	Yield (wt%)	C (wt%)	H (wt%)	N (wt%)	S (wt%)	V (wt%)	Ni (wt%)	CCR (wt%)
Dist. Liq.	30.2	84.42	11.61	0.12	3.28	2.0	0.0	1.3
Unadsorbed	31.9	84.78	11.01	0.022	4.24	0.0	0.0	6.1
Nonpolars	62.1	84.60	11.30	0.07	3.77	1.0	0.0	3.8
Adsorbed	21.8	82.79	8.06	0.83	6.65	210	53	41.6
Insolubles	15.9	82.86	6.50	1.27	8.51	672	250	70.1
Polars	37.7	82.82	7.40	1.02	7.43	404	136	53.6
Total	99.8	83.93	9.83	0.43	5.15	153	51	22.6
Full feed	100	83.51	9.93	0.45	5.80	165	40	22.3

carbon residue of 15.4% for thermally converted Cold Lake bitumen and of 68.3% for cyclohexane insolubles (Table 7.4):

$$\beta = 100\,[1 - (15.4 - 5)/(68.3 - 5)] = 83.8\%$$

Likewise, the Conradson carbon of 23.2% for thermally converted Arabian Heavy vacuum resid (Table 3.4) and 70.1% for cyclohexane insolubles (Table 7.5):

$$\beta = 100\,[1 - (23.2 - 5)/(70.1 - 5)] = 72.0\%$$

These yields are higher than will ever be attained, even with ideal conditions, because the thermal conversion of the recycled adsorbed fraction will produce more hydrocarbon gases and further raise the total Conradson carbon residue.

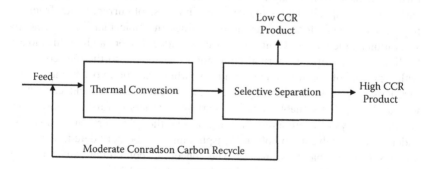

FIGURE 7.15 Recycling of the fraction with moderate Conradson carbon residue for further conversion can greatly increase the yield of product with low Conradson carbon residue.

However, these yields are far in excess of current yields from current commercial visbreaking (28%) and even coking processes (61% for Fluid Coking) of Arabian Heavy vacuum resid. In chapter 8, we will find that coking processes also have large potential improvement. Nevertheless, this section shows that by combining visbreaking for the thermal conversion step with a selective separation process, such as deasphalting, and recycling a moderate Conradson carbon residue stream, one can potentially greatly increase the yield over current processing of extra heavy oils and vacuum resids. It will be shown later in this chapter that solvent deasphalting indeed can produce deasphalted oil (DAO), resin, and asphalt products. Thus, the resin stream would be the moderate Conradson carbon that could be recycled to the visbreaker. This also promises to improve the solubility of asphaltenes and thereby increase the coke induction period. Although this process is promising, the author is not aware it being attempted, even at the pilot-plant scale. In chapter 10, the recycling of resins from deasphalting the product of hydroconversion will also be suggested as a way to improve the hydroconversion yield. The bottom line is that resid conversion processes have plenty of room for improvement over the processes that are currently practiced, and even the current level of understanding gives an indication as to how most of the full potential can be obtained by determining and overcoming the barriers to improvement.

7.2.8 CONCLUSIONS ON MAXIMUM POTENTIAL YIELD BY SEPARATION

Selective separation alone has the potential to provide substantial fluid catalytic cracking feed from heavy oils, but could only compete with resid conversion processes by achieving the highest yields obtained in the laboratory. Further separation of Conradson carbon is limited by the restricted concentration of Conradson carbon in the lower-quality fraction. Thermal conversion before selective separation breaks some of the chemical bonds between high- and low-quality parts of the same molecule to increase the yield of the low Conradson carbon fraction. Meanwhile, the lower-quality fraction increases in Conradson carbon. By recycling a moderate Conradson carbon fraction back to thermal cracking, further yields of fluid catalytic cracker feed can be obtained within Conradson carbon residue specifications that are well in excess of current yields from coking processes. Therefore, this research identified the molecular limit to selective separation on the basis of Conradson carbon residue. Understanding this molecular limit pointed first to thermal conversion and second to thermal conversion with recycling of the moderate Conradson carbon fraction to overcome this barrier. This is the value of first determining, by research, the best one can achieve with a process. This enables one to determine barriers to further improvement and devise ways of overcoming or going around these barriers. In section 7.3, a wider range of solvents in solvent deasphalting will be considered, to improve it as a selective separation process. The concept of combining conversion with selective separation will be utilized in chapters 8, 9, and 10.

On the other hand, there is no molecular limit to the separation of all the molecules in extra heavy oils and resids containing vanadium and nickel in less than 1.5 wt% of the feed. Commercially, these metals are removed either by coking or by hydrotreating and, thus, are not as great a problem as separating the polynuclear aromatics (Conradson carbon residue). Still, this lack of a molecular barrier to separation of the porphyrins that contain vanadium and nickel shows that their separation is possible. One needs to either overcome the physical attraction between porphyrins and other polynuclear aromatics in the extra heavy oil or resid or to use a chemistry that is specific for metal-containing porphyrins to alter them enough for separation. Oxidation[9] and electrochemical reduction[10] are possibilities.

7.3 LABORATORY SOLVENT DEASPHALTING WITH A WIDER RANGE OF LIQUIDS

In this section, the range of liquids for separation is greatly increased with the hope that Cold Lake bitumen might be more selectively separated than with the noncomplexing and low-complexing liquids tried previously. The research discussed in this section was a side project of David Jennings in 1995, who worked with the author as a postdoctoral fellow. He sought to determine if a wider range of deasphalting liquids and liquid mixtures as well as varying temperature could improve the selectivity in separating catalyst poisons from Cold Lake bitumen.

7.3.1 EXPERIMENTAL

The principal part of the apparatus was two 100 mL jacked Buchner filter funnels (Figure 7.16) equipped with a medium glass frit filter (10 to 15 μm). The filter funnels were stoppered, inserted into stoppered Erlenmeyer filter flasks, and

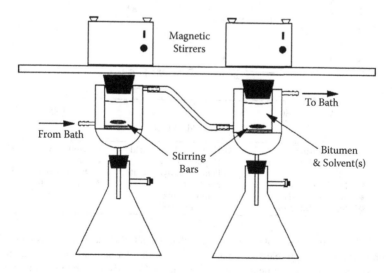

FIGURE 7.16 Apparatus for performing laboratory solvent deasphalting.

connected in series to a circulating bath to maintain constant temperature. Aluminum foil covered the rubber stoppers used for the filter funnels to protect them from solvent vapors. For each experiment, a weighed amount of bitumen (about 3 g) and liquid solvent were charged directly on the glass frit of each of the filter flasks. Magnetic stirring bars in the liquid–bitumen mixture enabled mixing—first by an external hand-held magnet until the liquid lowered the viscosity of the bitumen and, second, by inverted magnetic stirrers mounted above the filter funnels. During this part of the procedure, not a drop flowed through the filters because of the small pore size and because the stoppers did not let air flow in above the filters. The mixing continued for seven to nine hours, and then a vacuum was pulled on the back side of the filters to initiate filtering. The remaining solids on the filters were washed with 10 to 15 g of additional liquid. These insolubles (asphaltenes) were dried in a vacuum oven at 90°C for 2 h. The bitumen liquid was removed from the bitumen solubles (DAO) by rotovaping. Typically, four filters were used for each experiment to obtain enough sample for analyses.

7.3.2 Results for Separation of Conradson Carbon Residue

The result for the separation of Conradson carbon residue (same as microcarbon) is shown in Figure 7.17. Those from this study were done with eight parts by

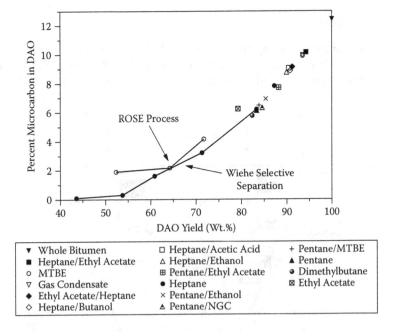

FIGURE 7.17 Comparison of laboratory deasphalting of Cold Lake bitumen using various solvents with laboratory selective separations and with pilot plant ROSE (Residuum Oil Supercritical Extraction™) deasphalting in terms of Conradson carbon residue (microcarbon) of the deasphalted oil (DAO) versus yield of DAO.

weight liquid and one part Cold Lake bitumen. The binary liquid mixtures were seven parts of the alkane by weight and one part of complexing liquid. The points marked "Wiehe Selective Separation" are from the selective separation procedure discussed in section 7.2. The points marked "ROSE (Residuum Oil Supercritical Extraction™) Process" are pilot plant solvent deasphalting that will be discussed in section 7.4.3. The yield of soluble DAO was higher for the liquids containing complexing liquids than for selective separation that included adsorption and also higher for deasphalting with light alkanes under pressure. Nevertheless, the best separation results from the three data sources follow the same curve of the Conradson carbon residue of the DAO versus yield of DAO. This is because of the molecular limit on Conradson carbon residue separation, as discussed in section 7.2. The use of complexing liquids provides no improvement in separation of Conradson carbon residue over use of noncomplexing and low-complexing liquids.

Figure 7.18 shows the Conradson carbon residue (same as microcarbon) of the asphaltenes (insolubles) as a function of yield corresponding to the DAO in Figure 7.17. All the data follow a band showing that at a given yield the asphaltenes have nearly the same Conradson carbon residue, no matter what the liquid used to obtain the precipitation was. This also verifies the conclusion in section 7.2 that there is a ceiling on the Conradson carbon residue of asphaltenes. These data indicate that this ceiling is below a Conradson carbon residue of 55% for Cold Lake bitumen.

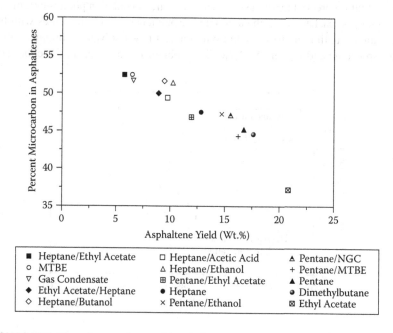

■ Heptane/Ethyl Acetate	□ Heptane/Acetic Acid	▲ Pentane/NGC
o MTBE	△ Heptane/Ethanol	+ Pentane/MTBE
▽ Gas Condensate	⊞ Pentane/Ethyl Acetate	▲ Pentane
◆ Ethyl Acetate/Heptane	● Heptane	● Dimethylbutane
◇ Heptane/Butanol	× Pentane/Ethanol	⊠ Ethyl Acetate

FIGURE 7.18 Conradson carbon residue (microcarbon) of the insolubles (asphaltenes) from laboratory deasphalting of Cold Lake bitumen with various solvents as a function of asphaltene yield.

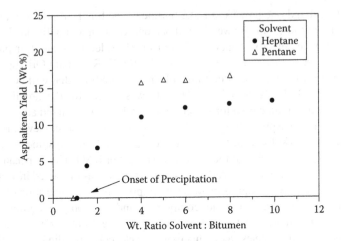

FIGURE 7.19 The yield of asphaltenes as a function of solvent to Cold Lake bitumen ratio using *n*-heptane and *n*-pentane as deasphalting solvents.

Figure 7.19 to Figure 7.21 show deasphalting results with *n*-pentane and *n*-heptane by changing the solvent-to-bitumen ratio. The quantity of precipitated asphaltenes (Figure 7.19) follows the typical case described in chapter 5, in which initial dilution precipitates no asphaltenes until the flocculation point (onset of precipitation) is reached. Then the amount of asphaltenes quickly increases with further dilution until it approaches an asymptote. At lower solvent-to-bitumen ratios, less asphaltene and higher DAO yields are obtained. However, in the end, about

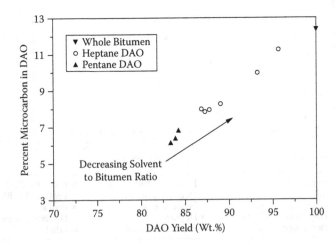

FIGURE 7.20 The Conradson carbon residue (microcarbon) of the deasphalted oil (DAO) versus yield of DAO obtained by deasphalting Cold Lake bitumen with varying solvent-to-bitumen weight ratios.

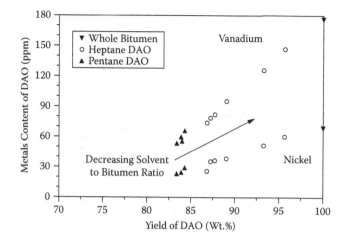

FIGURE 7.21 The vanadium and nickel content of the deasphalted oil (DAO) versus yield of DAO obtained by deasphalting Cold Lake bitumen with varying solvent-to-bitumen weight ratios.

the same relationship between Conradson carbon residue in the DAO and the yield of DAO (Figure 7.20) is obtained as changing the deasphalting solvent at constant solvent-to-bitumen ratio. Thus, it is clear that in solvent deasphalting the best separation of Conradson carbon residue depends only on yield and not on how that yield was obtained. Of course, by not following good experimental procedures one can obtain poorer, but not better separations. Again, the best separation runs up against the molecular limit. Meanwhile, Figure 7.21 shows how the vanadium and nickel in the DAO varies with DAO yield by varying the solvent-to-bitumen ratios.

7.3.3 Effect of Temperature

Figure 7.22 shows the asphaltene yield as a function of temperature for *n*-pentane, *n*-heptane, and *n*-octane solvents at an 8:1 solvent-to-bitumen volume ratio. In this limited temperature range, it is clear that asphaltenes are more soluble (lower asphaltene yield) with increasing temperature. Again, Figure 7.23 shows that nearly all the data collected with varying temperature fall on the same curve of Conradson carbon residue of the DAO versus yield of DAO as the data in Figure 7.17 and Figure 7.20. The exception is one data point for *n*-octane at the maximum temperature of 94°C that is at slightly lower Conradson carbon residue for the yield of DAO. However, this is still at a much higher Conradson carbon residue than the specification of 5% for fluid catalytic cracker feed.

7.3.4 Results for Separation of Metals

Figure 7.24 shows that even though there is no molecular limit for the separation of vanadium and nickel, all the data of the better separations fall on one curve for

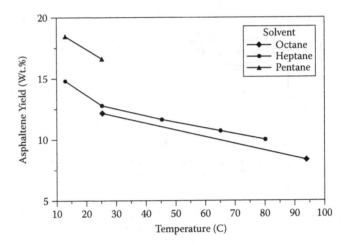

FIGURE 7.22 The yield of insolubles (asphaltenes) decreases with increasing temperature for *n*-pentane, *n*-heptane, and n-octane for the laboratory deasphalting of Cold Lake bitumen at an 8:1 solvent-to-bitumen weight ratio.

each metal of metal content of DAO versus yield of DAO. No reduction in nickel or vanadium concentration in the high quality fraction is obtained at a given yield of that fraction by using a gaseous alkane under pressure, *n*-heptane with Attapulgus clay adsorption, or a mixture containing complexing liquids. Likewise, in Figure 7.25, all the data of each metal fall on the same curve of the metal content

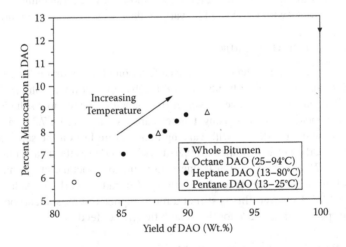

FIGURE 7.23 The Conradson carbon residue (microcarbon) of the deasphalted oil (DAO) versus yield of DAO obtained by deasphalting Cold Lake bitumen with varying temperatures at an 8:1 solvent-to-bitumen weight ratio.

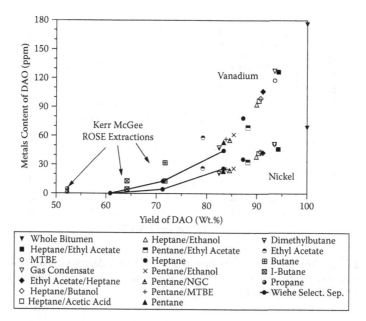

FIGURE 7.24 The content of vanadium and nickel in the deasphalted oil (DAO) from laboratory deasphalting of Cold Lake bitumen with various solvents and by the pilot plant ROSE process give about the same results as a function of DAO yield.

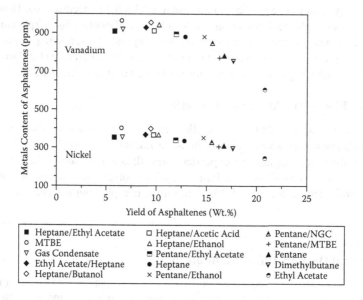

FIGURE 7.25 The content of vanadium and nickel in the insolubles (asphaltenes) from laboratory deasphalting of Cold Lake bitumen with various solvents give about the same results as a function of asphaltene yield.

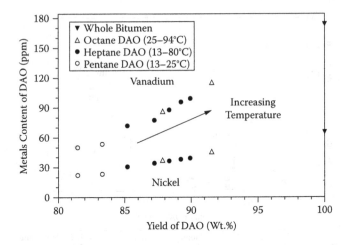

FIGURE 7.26 The content of vanadium and nickel in the deasphalted oil (DAO) from laboratory deasphalting of Cold Lake bitumen with varying temperatures at an 8:1 solvent-to-bitumen weight ratio.

of the insoluble asphaltenes versus yield of asphaltenes. In each case, the metal content of the asphaltenes does not vary from an asphaltene yield of 5 wt% to 10 wt%, seemingly to indicate that these are the highest concentrations present (918 ppm for vanadium and 357 ppm for nickel). On the contrary, as discussed, chemically it should be possible to achieve much higher concentrations. However, there is such a physical attraction between the metal-containing porphyrins and the polynuclear aromatics that they cannot be separated, even using complexing liquids. Figure 7.26 shows that modest changes in the temperature (13°C to 94°C) for the deasphalting also does not improve the separation of the metals.

7.3.5 RESULTS ON ATHABASCA BITUMEN

David Jennings also obtained a smaller amount of deasphalting data on Athabasca bitumen with analytical data in Table 7.6. As this is such a vast resource with production being actively expanded, these data are included. The conditions were the same as for Cold Lake bitumen, with a solvent-to-bitumen weight ratio of 8:1 and a temperature of 25.2°C. Only three deasphalting solvents were used:

TABLE 7.6
Analytical Data on Athabasca Bitumen

C (wt%)	H (wt%)	N (wt%)	S (wt%)	V (ppm)	Ni (ppm)	Fe (ppm)	CCR (wt%)
82.89	10.31	0.52	5.05	220	87	822	14.8

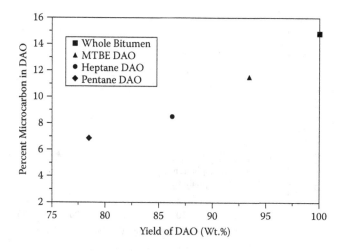

FIGURE 7.27 The Conradson carbon residue (microcarbon) of the deasphalted oil (DAO) versus yield of DAO from laboratory deasphalting of Athabasca bitumen with a solvent-to-bitumen weight ratio of 8:1 and a temperature of 25.2°C.

n-pentane, n-heptane, and methyl tertiary butyl ether (MTBE). The Conradson carbon residue (percent microcarbon) in the DAO and metals in the DAO versus yield of DAO are given in Figure 7.27 and Figure 7.28, and the Conradson carbon residue and metals in the asphaltenes versus yield of asphaltenes are given in Figure 7.29 and Figure 7.30. An unusual behavior is observed in Figure 7.29 indicating that the

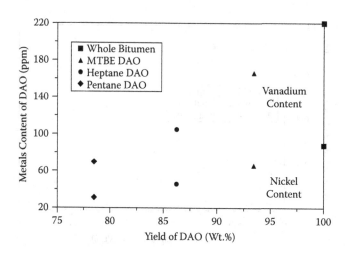

FIGURE 7.28 Vanadium and nickel content versus the yield of deasphalted oil (DAO) from laboratory deasphalting of Athabasca bitumen with a solvent-to-bitumen weight ratio of 8:1 and a temperature of 25.2°C.

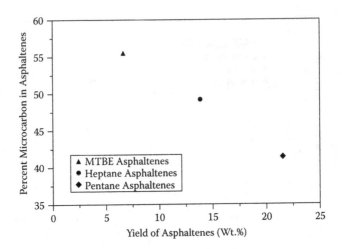

FIGURE 7.29 The Conradson carbon residue (microcarbon) of the insolubles (asphaltenes) versus yield of DAO from laboratory deasphalting of Athabasca bitumen with a solvent-to-bitumen weight ratio of 8:1 and a temperature of 25.2°C.

Conradson carbon residue for Athabasca bitumen has not reached a limit at a 5% yield and may actually contain greater than 60% Conradson carbon residue at asphaltene yields below 5%. The other unusual feature is the high iron content in Athabasca bitumen (822 ppm), nearly all precipitating with the first 6.7 wt% to

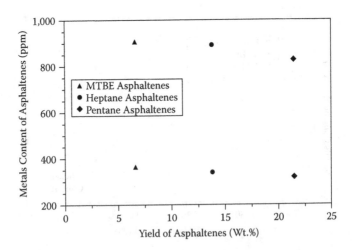

FIGURE 7.30 Vanadium and nickel content versus the yield of asphaltenes from laboratory deasphalting of Athabasca bitumen with a solvent-to-bitumen weight ratio of 8:1 and a temperature of 25.2°C.

give asphaltenes containing 1.12 wt% iron. This indicates that the iron is in the form of mineral matter, not porphyrins.

7.4 SOLVENT DEASPHALTING AS A PROCESS

7.4.1 OBJECTIVES

The traditional use of solvent deasphalting is to remove asphalt from qualified vacuum resid as part of the process of producing bright stock, a heavy lube base stock. The asphalt is either blended with vacuum resid to make asphalt products or is blended into heavy fuel oil. Solvent deasphalting is also used to separate the higher-quality oil from vacuum resids for making a resid feed for fluid catalytic cracking. The DAO is typically blended with the vacuum gas oil and hydrotreated before catalytic cracking. This combination of deasphalting and hydrotreating can enable the production of a good fluid catalytic cracker feed from a low-cost, poor-quality atmospheric resid. The volumes of fuels deasphalting is typically much greater than those of lubes deasphalting. As a result, deposition of the asphalt is a greater problem because asphalt products and heavy fuel oil usually cannot absorb such high volumes. The most common choice is to add the fuels asphalt to coker feed. However, as the resulting coke is becoming difficult to sell, gasification of asphalt is probably the better long-term choice in many locations.

Solvent deasphalting has been considered for upgrading bitumens and other heavy oils, both above ground near the production site and within the reservoir (VAPEX[11]). Above ground, a significant issue is how to dispose of the asphalt and both above ground and in the reservoir, an issue is the degree of upgrade for the amount of product lost as asphalt. Another issue in the reservoir is whether the precipitated asphalt will plug up the formation.

7.4.2 COMMERCIAL PROCESSES

Deasphalting is done in a countercurrent treater (or separator) using n-propane, n-butane, i-butane, or n-pentane as the precipitating liquid (Figure 7.31). Lubes deasphalter treater towers are typically high (on the order of 80 feet high with a 13-foot diameter) with trays as contacting devices (Figure 7.32). They are operated at high solvent (usually n-propane)-to-oil ratios (8 to 13:1) under reflux. The resid can be fed at multiple locations so as not to have low local solvent-to-oil ratios. The temperature at the bottom of the tower is about 100°F (38°C), but is heated at the top with steam coils to as high as 180°F (82°C) while maintaining a pressure of 300 to 400 psig to maintain a liquid regime. In these near-critical fluids, solubility decreases with increasing temperature. Thus, heating near the top of the tower drops out heavier, more aromatic oil and further purifies the DAO, quite important for lube base stock.

For fuels deasphalting, the uniform quality is not as important as for lubes. What is more important for fuels is processing large volumes at low cost, both capital and expense. Hydrotreating further cleans up the DAO, and fluid catalytic cracking can tolerate some impurities. Therefore, fuels deasphalter treating

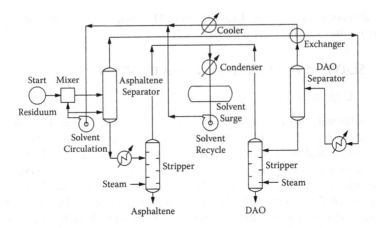

FIGURE 7.31 Schematic of a commercial deasphalting process with supercritical solvent recovery. (From JY Low, RL Hood, KZ Lynch. Preprints *Div Pet Chem*, ACS, 40:780–784, 1995. With permission from Kellogg Brown & Root LLC.)

towers are shorter (on the order of 40 feet high with a 13-foot diameter) and contain packing instead of trays or a combination of packing and trays (Figure 7.32). Fuels deasphalters typically use butane or pentane at a low solvent-to-oil ratio (4:8) and do not heat the top of the tower.

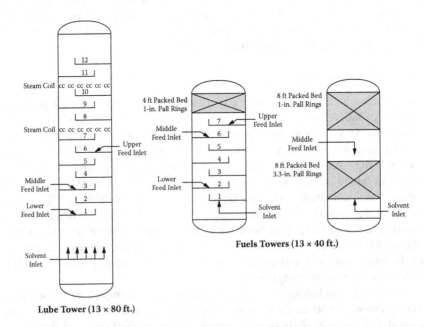

FIGURE 7.32 Comparison of commercial deasphalting towers for lubes with that for fuels.

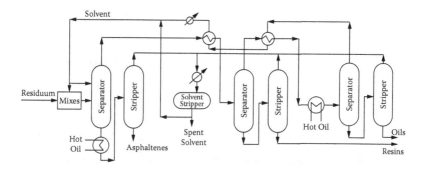

FIGURE 7.33 Schematic of a commercial deasphalting process that separates a resid into deasphalted oil (DAO), resins, and asphaltenes. (From JY Low, RL Hood, KZ Lynch. Preprints *Div Pet Chem*, ACS, 40:780–784, 1995. With permission from Kellogg Brown & Root LLC.)

Both lubes and fuels deasphalters typically use supercritical solvent recovery from the DAO (Figure 7.31). The mixture of DAO and solvent is heated above the critical temperature of the solvent, causing the DAO to precipitate in the DAO separator. This results in significant saving in the cost of energy for the heat of vaporization. The remaining solvent in the DAO and asphalt are separately stripped out with steam. Meanwhile, the recovered solvent is cooled and recycled.

If the mixture of DAO and solvent is heated in two stages (Figure 7.33), the more insoluble DAO can be separated in the first stage at the lower temperature, producing resins. Without the resins, the remaining DAO separated in the second stage at higher temperature is of higher quality. However, this type of solvent deasphalter has been unpopular because not many uses for the resins have been found. Nevertheless, as discussed in section 7.2.7, recycling the resins to visbreaking or to hydroconversion has potential for greatly improving these resid conversion processes.

7.4.3 Pilot Plant Solvent Deasphalting Results

Residuum Oil Supercritical Extraction (ROSE) pilot plant solvent deasphalting results are shown in Figure 7.34 to Figure 7.36 for separating a typical vacuum resid using propane, *i*-butane, *n*-butane, and *n*-pentane on the basis of Conradson carbon residue, vanadium, and nickel, respectively.[12] In each case, propane gives the best-quality DAO at a given yield of DAO, but the most DAO that can be obtained with propane is limited by the solubility in propane to 50 wt%. Above 50 wt% yield of DAO, the best-quality DAO is obtained by *i*-butane up to the solubility limit of 66 wt%. Above 66 wt% yield of DAO, the best-quality DAO is obtained by *n*-butane up to the solubility limit of 74 wt%. Above 74 wt% yield of DAO, the best-quality DAO is obtained by *n*-pentane up to the solubility limit of 84 wt%. If one connected all the best-quality DAO curves up to the feedstock values at 100%, the result is the molecular limit of the vacuum resid. Again, this is a true molecular limit for Conradson carbon residue, but only an apparent

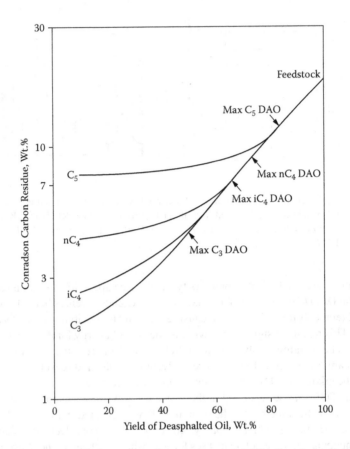

FIGURE 7.34 Residuum Oil Supercritical Extraction (ROSE) pilot plant solvent deasphalting results in terms of Conradson carbon residue of DAO versus DAO yield. (From SB Sprague. How Solvent Selection Affects Extraction Performance. NPRA Annual Meeting. Paper # AM-86-36, 1986. With permission from Kellogg Brown & Root LLC.)

molecular limit for vanadium and nickel. The presence of a molecular limit is similar to what was observed in laboratory separations, which, however, showed no significant difference in DAO quality of different solvents at the same yield of DAO. This is probably the result of the pilot plant being run at a faster rate, but with many stages of separation as compared with the slow rate but single stage in the laboratory.

Fortunately, former colleagues of the author, Glen Brons and Jimmy Yu,[13] obtained ROSE pilot plant solvent deasphalting results on Cold Lake bitumen, enabling direct comparison with laboratory separations on the same feedstock. These were already compared with laboratory results in Figure 7.17 for Conradson carbon residue and in Figure 7.24 for vanadium and nickel. Vanadium and Nickel of DAO and asphaltenes (insolubles) from the pilot plant are also shown

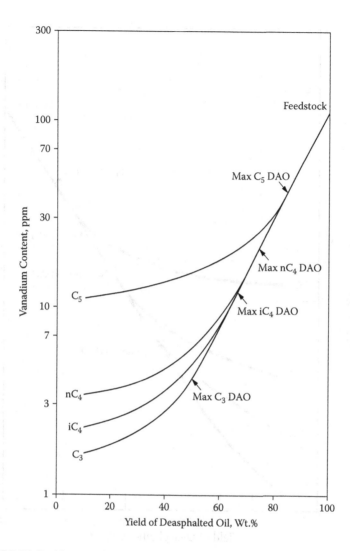

FIGURE 7.35 Residuum Oil Supercritical Extraction (ROSE) pilot plant solvent deasphalting results in terms of vanadium content of DAO versus DAO yield. (From SB Sprague. How Solvent Selection Affects Extraction Performance. NPRA Annual Meeting. Paper # AM-86-36, 1986. With permission from Kellogg Brown & Root LLC.)

separately in Figure 7.37. Whereas the pilot plant separations were obtained with one multistage separation, the laboratory separations in the same range of DAO yield required precipitation followed by adsorption and desorption with Attapulgus clay in one stage (marked Wiehe selective separation). Figure 7.17 shows that only one of the pilot plant separations (*i*-butane) is as good as the laboratory separation for Conradson carbon residue at the same yield of DAO. Figure 7.24 shows that none of the pilot plant separations are quite as good as the laboratory

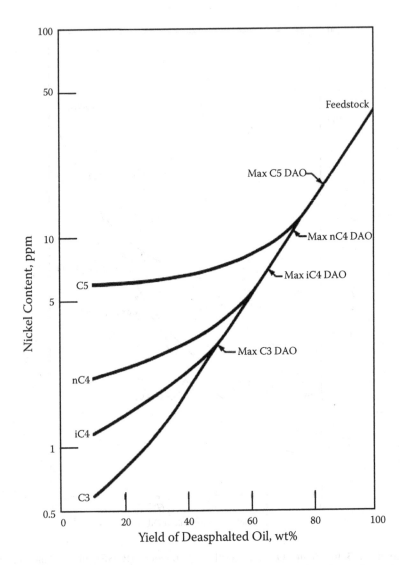

FIGURE 7.36 Residuum Oil Supercritical Extraction (ROSE) pilot plant solvent deasphalting results in terms of nickel content of DAO versus DAO yield. (From SB Sprague. How Solvent Selection Affects Extraction Performance. NPRA Annual Meeting. Paper # AM-86-36, 1986. With permission from Kellogg Brown & Root LLC.)

separations for vanadium and nickel at the same yield of DAO. Thus, the laboratory separation results appear to be an upper limit of the separations that can be obtained by commercial solvent deasphalting, as limited by the macromolecules in vacuum resid for Conradson carbon residue and effectively limited by physical attraction for vanadium and nickel. Nevertheless, solvent deasphalting is a viable,

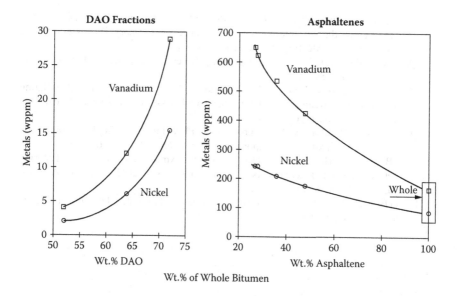

FIGURE 7.37 Vanadium and nickel contents of DAO and of asphaltenes as a function of yield for Cold Lake bitumen in the Residuum Oil Supercritical Extraction (ROSE) pilot plant deasphalting process. (From G Brons, JM Yu. *Energy and Fuels* 9: 641–647, 1995. Reprinted with permission. Copyright 1995. American Chemical Society.)

commercial-scale process that shows promise for selectively separating petroleum macromolecules and combining with resid conversion to improve greatly the yields of feed for secondary processing within a given quality constraint.

REFERENCES

1. SJ Rossetti. Tutorial: Electrostatic desalting fundamentals. Proceedings of the 1st International Conference on Refinery Processing, AIChE, 61–69, 1998.
2. LN Kremer. Challenges to desalting heavy crude oil. Proceedings of the 3rd International Conference on Refinery Processing, AIChE, 3–8, 2000.
3. LN Kremer. Controlling quality variations in feed to desalters. AIChE Spring National Meeting, Conference Proceedings, Vol. 1, Session 21c, 2006.
4. JD McLean, P Kilpatrick. Effects of asphaltene aggregation in model heptane-toluene mixtures on stability of water-in-oil emulsions, *J Colloid Interface Sci* 196: 23–34, 1997.
5. SJ Uppels, M Turner. Diagnosing and preventing naphthenate stabilized emulsions and naphthenate deposits during crude oil processing. Proceedings of the 6th International Conference on Petroleum Phase Behavior and Fouling, Amsterdam, Paper #27, 2005.
6. IA Wiehe. Separability of Cold Lake bitumen and Arabian Heavy vacuum resid. Preprints, ACS, *Div Fuel Chem* 44: 796–804, 1999.
7. IA Wiehe. Separability of petroleum macromolecules. Proceedings of the Conference on Energy and the Environment, AIChE Annual Meeting, 2000, pp. 34–37.

8. PH Barnes. Tutorial: Basic Process Principles of "Residue Cat-Cracking." Proceedings of the 1st International Conference on Refinery Processes, AIChE Spring National Meeting, New Orleans, 1998, pp. 227–236.

9. KA Gould. Oxidative demetalization of petroleum asphaltenes and residua. *Fuel* 59: 733–736, 1980.

10. MA Greaney, MC Kerby, WN Olmstead, IA Wiehe. Method for Demetallating Refinery Feedstreams. U.S. Patent 5,529,684. Assigned to Exxon Research & Engineering Co., 1996.

11. RM Butler, IJ Mokrys. A new process (VAPEX) for recovering heavy oils using hot water and hydrocarbon vapor. *J Can Pet Technol* 30: 97–106, 1991.

12. SB Sprague. How solvent selection affects extraction performance. National Petrochemical and Refiners Association (NPRA) Annual Meeting, Los Angeles. Paper # AM-86-36, 1986.

13. G Brons, JM Yu. Solvent deasphalting effects on whole cold lake bitumen. *Energy and Fuels* 9: 641–647, 1995.

8 Coking

In this chapter the coking process is discussed in general, the types of commercial cokers are reviewed, and some other interesting coking process concepts are described. Finally, the innovative procedure, discussed initially in chapter 7 for generating fluid catalytic cracker feed from resids and extra-heavy oils, is applied to coking and is carried farther than for separations. This will include determining the maximum potential and barriers to achieving it with this approach, devising and demonstrating ways to overcome these barriers, and trying to develop and to commercialize the resulting process concept.

8.1 PROCESS OBJECTIVES

In coking processes, petroleum resids are thermally converted at high temperatures (450 to 525°C or 842 to 977°F) with high extent of conversion. The gases and volatile liquids are allowed to escape the reactor, leaving solid coke behind. Because coke is an expected by-product, procedures are designed to remove it from the reactor. Usually the distillation cut-point in the coking reactor is higher than the lowest boiling point of the resid feed. Therefore, a heavy fraction is evaporated out of the coker with little or no reaction, condensed in the coker fractionator, and recycled back to the coker until it is cracked to a mixture of lighter liquids, gas, and coke. The volatile liquid products, after fractionation into different boiling ranges, are typically hydrotreated to remove heteroatoms, sulfur, nitrogen, and metals, and to saturate olefins and aromatic rings. Also, since these volatile liquids are converted from resids, they typically are more aromatic and have higher heteroatom content than unprocessed oils distilled from the same feed in the same boiling range.

The coker vacuum gas oil needs to be further converted to lower boiling range liquids that can be sold by a refinery as by-products (e.g., transportation fuels). After hydrotreating, coker vacuum gas oil may be converted by catalytic processing, such as fluid catalytic cracking (FCC) or hydrocracking. Therefore, the only real restriction on the recycle cut-point for cokers should be the minimum quality that is acceptable for feed after hydrotreating for the fluid catalytic cracker or the hydrocracker. The fluid catalytic cracker can accept some resid in its feed as long as it is not too high in catalyst poisons: basic nitrogen, metals, and coke precursors (polynuclear aromatics measured by Conradson carbon residue or CCR). The optimum coker recycle cut-point is specific to a particular refinery. It depends on the coker feed, the type of and operating conditions of the coker, the type of and operating conditions of the fluid catalytic cracker, the catalyst used in the fluid catalytic cracker, the degree of hydrotreating for the coker gas oil, and the other feeds going into the fluid catalytic cracker.

8.1.1 CATALYST POISON REJECTION PROCESS

Coking is the most popular choice for the conversion of resids high in aromatic cores (low hydrogen, high heteroatom, and high Conradson carbon residue) because it reactively separates the aromatic cores and the catalyst poisons into a low value by-product (coke), leaving the pendants for catalytic upgrading. Therefore, coking should be called a "catalyst poison rejection" process because the principal reason for coking is to improve the quality of liquids for further upgrading by catalytic processing.[1,2] Unfortunately, for a very long time coking has been mistakenly called a "carbon rejection" process, a term that continues to be used without critical examination. For example, Table 8.1 shows that the coke product in a microcarbon tester has a much higher concentration of sulfur, nitrogen, and metals than the vacuum resid feed and the volatile liquid products. However, the concentrations of carbon in the vacuum resid in the coke and in the liquid product are nearly the same. The rejection of heteroatoms is counterbalanced by an increase of hydrogen content in the liquid product. This highly desirable result is because most of the polynuclear aromatic building blocks in the petroleum macromolecules contain heteroatoms that are reactively separated with the polynuclear aromatics into the coke.

Calling coking a "carbon rejection process" is not only wrong, it greatly undervalues the coking process. Catalytic processes, hydrotreating, hydrocracking, and fluid catalytic cracking, are much more selective to the desired products than is coking. However, the catalysts used in these processes are relatively expensive, and one cannot afford to process feeds high in catalyst poisons including polynuclear aromatics. Thus, cokers by reactively separating catalyst poisons

---

TABLE 8.1
Elemental Balance around the Microcarbon Test Showing that Coking Is a Catalyst Poison Rejection Process

Element	Arabian Heavy Vacuum Resid	22.3% MCR Coke	77.7% Liquid + Gas Balance	% Change
Carbon	83.5%	85.8%	82.9%	−0.8
Hydrogen	9.9%	3.7%	11.7%	+18
Sulfur	5.8%	8.6%	5.0%	−14
Nitrogen	0.45%	1.3%	0.21%	−53
Vanadium	165 ppm	808 ppm	0 ppm	−100
Nickel	40 ppm	165 ppm	4 ppm	−90

(including polynuclear aromatics) as coke, produce liquids that can be economically processed by catalysis. As a result, coking is the most popular choice for conversion of almost all vacuum resids. The aim is to produce only enough coke to achieve the desired reactive separation. Unfortunately, an excess of coke is usually the result. Therefore, we usually want to reduce the yield of coke in a coker, but it is certainly not desirable to eliminate the yield of coke.

8.1.2 METALLURGICAL QUALITY COKE

The coke from delayed cokers may have higher value if it meets the specifications for anode-grade coke (used in production of electrodes for aluminum manufacture) or needle-grade coke (used in production of electrodes for steel manufacture). However, anode-grade coke needs to be low in sulfur and metals and, thus, requires high quality resid feeds. Needle coke is made from low sulfur fluid catalytic cracker bottoms, and not from resid. Although about one pound of carbon anode is required for every pound of aluminum produced, the production rate of anode coke is still small, compared to the total coke production.

8.2 COMMERCIAL COKERS

There are commercially three types of cokers: delayed cokers, Fluid Cokers, and Flexicokers®. The liquid products from coking tend to be aromatic, especially coker gas oil, and contain sulfur and nitrogen. They also may contain olefins conjugated to olefins (diolefins) and to aromatics that tend to polymerize and cause fouling (see section 6.3). Therefore, coker liquids usually require hydrotreating before being processed in fluid catalytic cracking or hydrocracking.

8.2.1 DELAYED COKERS

The first delayed coker was built by Standard Oil of Indiana at Whiting, Indiana in 1929.[3] As shown in Figure 8.1, fresh resid feed is pumped into the bottom of the coker fractionator where it is initially heated, and any light oil is flashed off. This resid and the bottoms of the coker vapor product are further heated to about 935°F (502°C) by passing through a coil in a furnace (heater). Steam is often injected into the coil to increase the velocity and vaporization, and to minimize coking in the coil. The hot mixture is fed to the bottom of a tall, insulated coker drum (about 28 ft in diameter and 80 ft high) that is under a pressure of about 25 psig. The heavy liquid thermally reacts in the coke drum to produce volatile liquids and gases that bubble up and pass out of the top of the drum. Meanwhile, the solid coke that forms accumulates in the drum.

After about 16 hours the drum is filled with coke, and the feed is switched to a second drum. The hot coke is cooled by injecting steam and water. Heads at the top and bottom of the drum are removed. A hole is drilled down the center

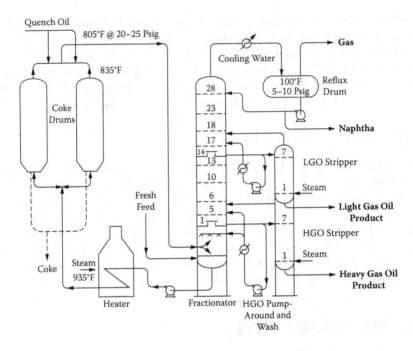

FIGURE 8.1 Schematic of a delayed coker. (From F Self, E Ekholm, K Bowers. *Refining Overview—Petroleum Processes and Products*. CD-ROM. AIChE, South Texas Section, 2000, pp. 22-1–22-11. With permission.)

with a drilling rig, and then the coke is cut out with high-pressure water. The coke drops out of the bottom into one of four collection systems. One collection system is just hopper rail cars that allow the water to drain off. Another is an inclined slide that directs the coke to a pile on the side of the drum. Another is a pit under the drum that uses a crane to load the coke onto the rail cars. The most sophisticated collection system drops the coke into a crusher where it is ground, mixed with water, and pumped as slurry into dewatering bins. During the drum filling period, the overhead product is quenched with coker gas oil and pumped to the fractionator. The vapors are washed with spray nozzles between the feed inlet and the heavy gas oil draw to remove coke fines and entrained resid. Light gas oil and naphtha are also drawn off, leaving gas that is removed from the top of the fractionator.

The common problems in delayed coking are foaming in the drum, shot coke, coking of the heater tubes, and coking/fouling of the fractionator (see section 6.3). Foaming is commonly controlled by injecting silicone oils. Shot coke is in the form of 2 to 5 mm round particles containing hot resid that results in mesophase coke on the outside and isotopic coke on the inside. It can be a safety hazard if, during quenching of the coke with water, the shot coke breaks open and is propelled by steam when the water hits the hot center. The mechanism for formation of shot coke is not yet known, but probably involves a liquid–liquid phase separation. Shot coke separation is usually observed with resid feeds that have high asphaltenes to Conradson carbon residue (thus, low resin to asphaltene ratio). The best solution is to blend the resid with a different feed that has high resin to asphaltene ratio or one with very high concentrations of small ring aromatics, such as fluid catalytic cracker bottoms.

Coking of heater coils is not a big problem for four-pass, double-fired heaters[4] that allow online decoking by pigging[3] in two passes while continuing to run with the remaining two passes. Injecting steam and increasing recycle maintains high velocity/shear rate in the coils and minimizes coking or the accumulation of coke. Resids high in aliphatic sulfur thermally react faster than other feeds. When running these feeds, the coil outlet temperature should be decreased to prevent coking in the heater tubes. In addition, feeds with little reserve solvency will be prone to coking in the heater tubes as in visbreaking. Coking in the fractionator can be caused by entrained resid/coke particles or by polymerization of olefins conjugated to aromatics (section 6.3).

Recent improvements in delayed cokers are automatic deheaders that shorten cycle time by 30 min,[6] operating at lower pressure (15 psig) and higher temperature for higher liquid yield and lower quality,[4] and recycling distillate product rather than heavy product for much higher liquid yield and much lower quality.[4] An example for improvement with this last advancement (Conoco-delayed coking advantage) is shown in Table 8.2 for an Arabian Heavy vacuum resid. As can be seen, the coke yield has reduced, and the liquid and gas yields have increased. The coke yield (1.2 CCR) decreased to match the amount that was initially formed in Fluid Coking before 20% of this was burned. Of course, also operating Fluid Coking in a once-through operation (not recycling the heavy product) reduces the coke yield to 1.1 CCR. It is difficult to compare these yields without a measure of liquid quality, such as hydrogen content.

TABLE 8.2
Conoco-Delayed Coking on Arab Heavy 1050°F⁺

Product, wt%	Conventional	Conoco
Butane and lighter	9.2	10.4
C5-335 °F	11.3	11.6
335–650 °F	21.9	16.3
650 °F+	25.4	32.9
Coke	32.2	28.8

Source: Data from GC Hughes, BJ Doerksen, AM Englehorn, S Romero. Tutorial: Delayed Coking Commercial Technology. Proceedings of 3rd International Conference on Refinery Processing, AIChE, 66–81, 2000.

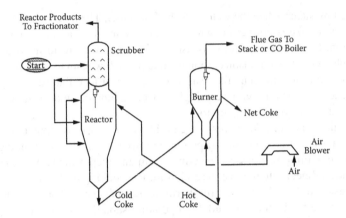

FIGURE 8.2 Schematic of a fluid coker. (From CJ Mart, EG Ellis, DS McCaffrey, Jr. ExxonMobil Resid Upgrading Technologies. PETROTECH 2005, 6th International Petroleum Conference and Exhibition, New Delhi, India, 2005. Reprinted with permission of ExxonMobil Research & Engineering Co.)

8.2.2 Fluid Coking and Flexicoking

Fluid Coking and Flexicoking[7,8] are processes licensed by ExxonMobil. The first Fluid Coker was started up in Billings, Montana, in 1954, and the first commercial Flexicoker was started up in Japan in 1976.[8] As of this writing, there are nine operating Fluid Cokers and five operating Flexicokers. In Fluid Coking, shown in Figure 8.2, the resid is sprayed on a hot, fluidized bed of coke particles in a vessel, the reactor. The volatile products go overhead to a fractionator, whereas the coke particles are removed out of the bottom and transferred to another vessel, the burner, where the coke is partially burned with air to provide the heat for the process. This coke is then recirculated back to the reactor. As this process produces much more coke than is required for heat, coke is withdrawn at the bottom of the reactor. This is high sulfur, fuels-grade coke of low value. A heavy liquid product is recycled back to the reactor from the fractionator. This is used to scrub coke fines from the reactor vapors and to improve the quality of the liquid product.

In Flexicoking, shown in Figure 8.3, a third vessel, the gasifier, is added to the fluid coker and the second vessel is called the "heater." In the gasifier, coke is gasified with steam and air in net reducing conditions to produce a low BTU gas containing hydrogen, carbon monoxide, nitrogen, and hydrogen sulfide. After adsorbing out the hydrogen sulfide, this low BTU gas is burned as a clean fuel within the refinery and/or in a nearby power plant. Only a small amount of coke (about 3%) needs to be withdrawn in Flexicoking to keep vanadium and nickel from accumulating in the gasifier. With the trend of decreasing high sulfur coke prices, gasification of coke to form fuel and syngas to be used within the refinery may be the only economical and environmental option for refineries in the future.

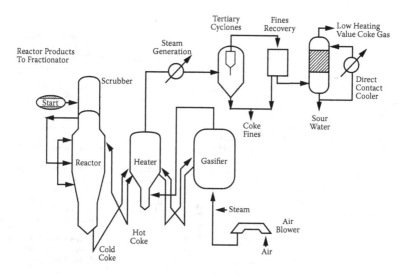

FIGURE 8.3 Schematic of a Flexicoker. (From CJ Mart, EG Ellis, DS McCaffrey, Jr. ExxonMobil Resid Upgrading Technologies. PETROTECH 2005, 6th International Petroleum Conference and Exhibition, New Delhi, India, 2005. Reprinted with permission of ExxonMobil Research & Engineering Co.)

If this happens, Flexicoking may become a more popular choice, but for syngas, oxygen will need to be used for gasification rather than air.

The advantage of Fluid Coking over delayed coking is that it is a continuous process of high capacity (65,000 B/D of 20% CCR in a single reactor) with long run lengths (~24 months). Fluid Coking produces less coke in the reactor (~1.2 CCR) than delayed coking (~1.4–1.6 CCR), and about 20% is burned as fuel. The burner flue gas goes through heat recovery and cleanup before being completely burned in the refinery as fuel. It is not easy to compare liquid yields and qualities from delayed coking and Fluid Coking because they rarely are done at the same recycle cut-point. At the same recycle cut-point, Fluid Coking obtains higher liquid yield of lower quality than delayed coking. However, it is easier to operate delayed cokers than Fluid Cokers with little or no heavy oil recycle. In this mode, delayed cokers can obtain greater liquid yield of much poorer quality than Fluid Cokers. Typical yields for the Fluid Coking of an Arabian Light vacuum resid of 22 Conradson carbon residue is shown in Figure 8.4. Nevertheless, the author prefers to compare resid processing alternatives on a plot of hydrogen content of the liquid product versus the yield of liquid product, as is shown later in this chapter. Delayed coke is of higher quality than fluid coke. Fluid coke is only of fuel grade, lower volatility, and much more difficult to grind than delayed coke. Fluid coke is most useful in the 100 micron size in fluid bed boilers or as fuel in the cement industry.

Flexicoking obtains similar yields as Fluid Coking except a low BTU gas is produced rather than coke. Thus, it is critical to be able to take advantage of this low BTU gas in or near the refinery. The small amount of coke produced can be

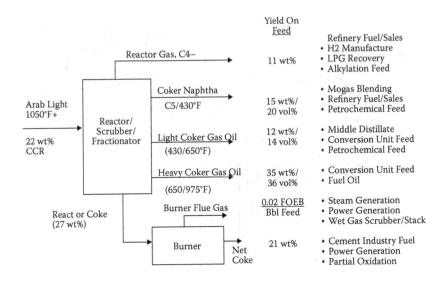

FIGURE 8.4 Example yields from a fluid coker. (From DG Hammond et al. Review of Fluid Coking and Flexicoking Technologies, AIChE, Paper 44c, Spring National Meeting, 2003. Reprinted with permission of ExxonMobil Research & Engineering Co.)

very high in vanadium and nickel, thus it can be more valuable as a source for these metals than as a fuel.

Common problems in Fluid Cokers and Flexicokers are the build-up of coke on reactor walls and on reactor internals, secondary cracking of liquid products, and fractionator fouling (see section 6.3). No operator wants to reach a condition known as "bogging" (a career-ending move), where all the coke particles stick together and the bed loses fluidity. To avoid bogging, one operates the reactor at high temperatures, so that the resid quickly forms dry coke and there is no tendency to stick. However, at high temperatures volatile liquid products recrack in the vapor phase before they can exit the reactor to form lower-valued hydrocarbon gases and less of the higher-valued volatile liquids. In addition, vapor-phase cracking forms conjugated olefins that cause fractionator fouling. This is an incentive to operate the reactor at as low a temperature as possible so as not to bog the bed. On the other hand, these reactor temperatures are prone to form reactor wall coke. Thus, these problems are all interrelated, and Fluid Cokers and Flexicokers have to be operated with some amount of compromise in mind. Flexicokers used to have slag problems in the gasifier caused by vanadium and nickel alloys that have been greatly reduced by withdrawing purge coke to reduce the build-up of metal concentrations in the coke. Nevertheless, it is important to have a good desalter operation so as to keep sodium out of the resid feed because sodium forms a low melting eutectic with vanadium in the gasifier. The only recent improvement in Fluid Coking and Flexicoking has been to reduce fouling and increase run lengths to an average of 24 months for Fluid Cokers and 30 months for Flexicokers.

8.3 OTHER COKING/PYROLYSIS TECHNOLOGIES

Over the past 30 years, since Flexicoking was commercialized, there have been other coking/pyrolysis technologies that have been proposed for thermally converting resids and extraheavy oils. Pyrolysis usually implies that heavy liquids volatilizing out of the reactor are not recycled back into it. As a result, the liquid product contains entrained coke particles and resid as well as a high boiling, more aromatic fraction. In addition, much higher liquid yields and much lower coke yields than commercial cokers should be expected because heavy liquid recycle is the greatest reason why commercial cokers produce more coke than Conradson carbon residue. Nevertheless, many of these other techniques have interesting features that may be adopted more widely for commercial use in the future.

What is required of a coker is, first, a method to heat the resid or heavy oil to thermal reaction temperatures. This is usually done either by heating in a coil in a furnace or by spraying the oil on hot solids. Second, there needs to be a way to collect and transport the coke once it forms. Although delayed coking is an exception, it is an advantage to transport the coke continuously. Examples of solid transporting methods are fluidized or moving beds, screws, kilns, and conveyor belts. All of these could be incorporated into cokers. Examples for the use of fluidized beds are Fluid Coking, Flexicoking, the asphalt residue treatment (ART) process, the discriminatory destructive distillation (3D) process, and the fluid thermal cracking (FTC) process. An example that uses a screw is the LR-flash coker. An interesting exception to rule 2 is not to form solid coke, but a very aromatic liquid ("liquid coke") that can flow out of the coker. The Eureka® process is an example of this. The third requirement is to have a method to separate the liquid and gaseous products from the coke. Nearly all methods use vaporization to accomplish this separation.

8.3.1 ART PROCESS

The ART process, shown in Figure 8.5, was invented by David Bartholic of Engelhard Corp.[9,10] The principle behind ART is to upgrade an atmospheric resid or heavy oil that does not quite meet FCC feed specifications to FCC feed by removing metals, nitrogen, and Conradson carbon. The process looks much like a fluid catalytic cracker, however, instead of using an expensive catalyst, inexpensive microporous solids (spray-dried kaolin clay) are circulated to collect coke containing metals and nitrogen and burn it off in a burner. Thus, catalytic poisons are removed before exposing the feeds to fluid catalytic cracking. Coke yield is less than Conradson carbon residue, and most of the liquid product remains in the atmospheric resid boiling range needed for FCC with the CCR of total liquids reduced to 3.7%. This process was commercially operated at an Ashland Petroleum Refinery for a number of years using a revamped FCC unit. However, it was plagued by fouling problems and was eventually shut down. The most likely cause for such fouling is incompletely converted resids (sticky solids) and polymerization of conjugated olefins.

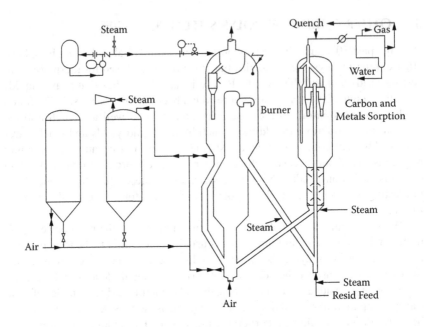

FIGURE 8.5 Schematic of the ART process. (From DB Bartholic. Preparation of FCC Charge from Residual Fractions. U.S. Patent 4,243,514, assigned to Engelhard Corp., filed May 14, 1979, and issued January 6, 1981. With permission.)

8.3.2 3D PROCESS

The 3D process,[11,12] shown in Figure 8.6, was also invented by David Bartholic, but after he left Engelhard Corp. It is similar to ART except the feed is sprayed across a falling curtain of solid particles (coke) to give very short vapor residence times (less than 1 sec) in order to minimize secondary cracking. Resid that is not fully converted falls along with the particles into a fluidized bed that has a vapor residence time similar to a Fluid Coker. This is fine as long as the particles with incompletely converted resid are not sticky. The process became a joint venture of Bartholic and Coastal (Bar-Co) and was demonstrated over a six-month period at a Coastal's El Dorado refinery. There were operational problems with the longest run being six weeks.

8.3.3 FTC PROCESS

The FTC process,[13] shown in Figure 8.7, was invented by Fuji Standard Research, Inc., of Japan. This is similar to Flexicoking except it has only two vessels (reactor and gasifier, but no heater), and microporous particles are circulated rather than coke. As the resid flows into the pores of the particles, the bogging temperature is greatly lowered so that the reactor is operated at 878°F (470°C) instead of 964°F

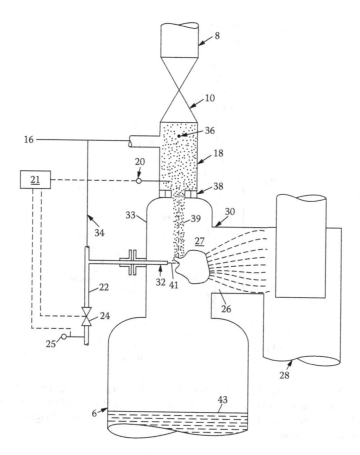

FIGURE 8.6 Schematic of the 3D (discriminatory destructive distillation) process. (From DB Bartholic. Liquid–Solid Separation Process and Apparatus. U.S. Patent 4,859,315, filed November 5, 1987, and issued August 22, 1989.)

(518°C). The lower temperature decreases secondary cracking and increases the liquid yield by about 6 wt% absolute (61 to 67%). However, the reactor needed to be operated under 100 psig hydrogen to suppress dehydrogenation reactions caused by nickel and vanadium accumulating on the microporous solids. This was operated in a three barrel per day pilot plant.

8.3.4 Eureka Process

The Eureka process,[14] shown in Figure 8.8, was invented by Kureha Chemical and Chiyoda of Japan. It is similar to delayed coking, but the heavy by-product is a flowable, aromatic pitch rather than coke. Steam efficiently strips out the volatile liquid products to extend the coke induction period to very high conversions.

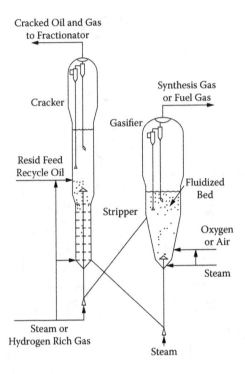

FIGURE 8.7 Schematic of the FTC (fluid thermal cracking) process. (From T Miyauchi et al. A New Fluid Thermal Cracking of Residual Oil with High-Quality Fluidization and Metal Effect Elimination. Proceedings of the 12th World Petroleum Congress, 335–341, 1987. With permission.)

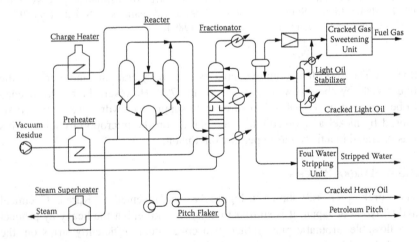

FIGURE 8.8 Schematic of the Eureka process. (From T. Aiba et al. *Chem Eng Progress*, 37–44, February 1981. With permission.)

The final product does have fine particles of mesophase dispersed in the pitch. The process alternates between reactors. When one reactor is filled with pitch, feed is sent to the second reactor as pitch from the first reactor flows to a holding vessel and then to a conveyor belt where it cools, solidifies, and is flaked. When compared with a delayed coker, less gas (4 to 5%) and more liquids (66 wt% on CCR = 20%) are produced in the Eureka process and, remarkably, less pitch (1.5 CCR) is formed than coke in a delayed coker (1.6 CCR). The flaked pitch is used as a binder to make electrodes for the iron and steel industry. This has been commercialized at two refineries in Japan.

8.3.5 LR-Flash Coker

This process,[15-17] shown in Figure 8.9, was developed by Lurgi of Germany. The reactor is a mechanically-driven twin screw that provides good mixing, plug flow, and intermeshing self-cleaning. Refiners do not like mechanical devices that tend to break down, develop leaks around seals, and wear out. This reactor has been used with sand or coke as the circulating solid. It can achieve 0.5 to 1.0 sec vapor and 5 to 20 sec solid residence time. Actually, the solid residence time is short, but the length is limited by mechanical considerations. Thus, the solids are dropped in a bin where the reaction is completed, and coke is burned to provide the heat, then recirculated back to the screw reactor. The power requirements are much greater to transport solids without resid. The reacting resid volatizes liquid products and forms gases that fluidize the solids, greatly reducing the power requirements. This has been operated on the lab, pilot plant, and commercial scale, but volumes are low for refinery standards. However, the LR coker holds promise for operations near heavy oil resources if there is a way to dispose or use the coke that is produced.

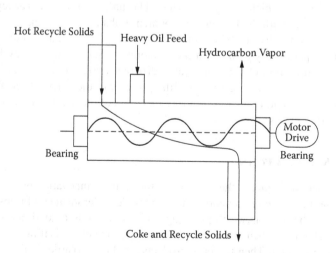

FIGURE 8.9 Schematic of the LR-Flash Coker.

8.4 THE INNOVATION PROCEDURE APPLIED TO COKING

This section is used to further demonstrate an approach to develop an innovation[18,19] that was discussed in chapter 7. There is a mistaken belief that invention and innovation are a result of either an accident or some out-of-body creative flash of genius. As a result, many scientists and engineers wait for invention or innovation just to happen rather than to work toward it as a goal. Edison, one of the greatest inventors of all time said, "Invention is one percent inspiration and ninety-nine percent perspiration." The author discovered late in his career that invention can be achieved by setting it as a goal and working hard toward that goal, using the full set of analytical tools and knowledge at one's disposal. Of course, there are many paths to achieve the goal, but the author will give an example path that has been successful for him. However, this example will also demonstrate some of the pitfalls in going from idea to commercialization.

In research, projects fall in and out of favor. The author was alone in Corporate Research when he first started research on petroleum resids; the project then grew to include nearly half of the scientists in the division. However, eventually, this research topic, as well as refining research in general, fell out of favor. The author then worked part-time on research to understand and mitigate engine deposits. He accepted a temporary assignment with Exxon Engineering to develop a new process model for coking, but the refineries changed their minds and dropped it. However, he remained convinced that improvement in resid conversion processes had to be important in the future for a petroleum company. Exxon management decided to form an interdivisional team of five scientists and engineers to investigate new leads to keep the knowledge in resid conversion (particularly coking) alive, and to conceive new paths to innovation. In this way, when economic justification returned, resid conversion research could be quickly restaffed, which is what happened. The author was asked to accept a two-year assignment with the Exxon Development Labs in Baton Rouge, Louisiana, where three of the five members of the team were to be located. This gave the author the opportunity to work in the Exploratory Section with Roby Bearden, one of two inventors of the Microcat hydroconversion process described later in this book, and in the highest scientific position in the corporation. With this background, the author, as well as other members of the team, sought new paths to innovation in coking.

8.4.1 MAXIMUM POTENTIAL

As discussed in chapter 7, the first step in the author's innovation procedure is to determine the maximum potential that is possible under ideal conditions in order to assess if the prize is worth pursuing. This is certainly a good question for a mature technology, such as coking. The ideal case in coking is if no hydrocarbon gases are produced. Then the only products would be volatile liquids and coke.

The Pendent-Core Building Block Model (chapter 3) tells us that the hydrogen content of the coke will be about 3.8 wt%. A process model of a Fluid Coker predicts that Arab Heavy vacuum resid (hydrogen content = 10.18 wt%, CCR = 22 wt%) will yield 61.4 wt% volatile liquids with a hydrogen content of 11.28 wt%. Calculating a hydrogen balance for the ideal no-gas case:

$$10.18\,(100) = 11.28\,(L) + 3.8\,(100 - L)$$

$$L = 85.3 \text{ wt\%}$$

The highest possible yield of liquids is 85.3% of 11.28 wt% hydrogen. In some cases the potential is not large enough to warrant further effort, but in this case the increase in yield of 23.9 wt% absolute is certainly a huge prize worth pursuing.

8.4.2 Reasons for Not Obtaining Maximum Potential

The next step in the author's innovation procedure is to determine the reasons why the maximum potential is not currently achieved. The three limitations to coking chemistry are shown in Figure 8.10 and listed below:

1. Secondary cracking of volatile liquid products.
2. Formation of five-plus ring polynuclear aromatics by the combination of reaction fragments containing smaller ring aromatics.
3. Aromization of hydroaromatics to form five-plus ring polynuclear aromatics.

Most of the hydrocarbon gases are formed by the first limitation as discussed in section 4.4.11.

The second limitation was demonstrated by the Phase-Separation Kinetic Model, which showed that a substantial amount of asphaltenes are formed from heptane solubles (chapter 4). In addition, pendant-core model compounds showed that highly substituted, small ring aromatics form coke in the tests for measuring Conradson carbon residue by combining aromatics (section 3.9). Forming coke from PNAs (polynuclear aromatics) of five rings or more is desirable (intrinsic coke), but forming coke from aromatics of four rings or less is undesirable (extrinsic coke) and should be avoided. High-performance liquid chromatography (HPLC) (chapter 2) enables tracking of the polynuclear aromatics as well as measuring the aromatic carbon (Table 8.3) for a feed to a commercial coker. In this case the entire resid was injected in the HPLC instead of just the deasphalted oil. As a result, the heavy polars (mostly asphaltenes) irreversibly adsorbed on the column. Therefore, the yield of this fraction was calculated by mass balance, and the aromatic carbon of the fraction was calculated to give the 29.9% aromatic carbon of the feed as measured by ^{13}C NMR. All the aromatics of five rings and larger are in the polar fractions. Therefore, in

1. Secondary Cracking of Liquid Product

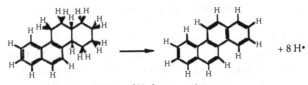

2. Combination of Aromatic Fragments

3. Aromization of Hydroaromatics

FIGURE 8.10 The three limitations to coking.

TABLE 8.3
HPLC Data on Commercial Coker Feed

Fraction	HPLC Yield (wt% of Feed)	HPLC Aromatic Carbon (wt% of Feed)
Saturates	7.2	0.06
1-Ring arom.	3.3	0.5
2-Ring arom.	5.7	1.2
3-Ring arom.	5.7	1.4
4-Ring arom.	17.7	4.4
Light polars	32.9	10.1
Heavy polars	27.5	12.2
TOTAL	100.0	29.9 = NMR

terms of HPLC, the intrinsic coke would contain all of the polar aromatic cores that were in the feed, as calculated below:

$$\text{Intrinsic coke} = \frac{\text{Aromatic carbon in polar cores}}{\text{Weight fraction of carbon in coke}}$$
$$\text{Intrinsic coke} = [10.1 + 12.2]/0.8914$$
$$\text{Intrinsic coke} = 25 \text{ wt\%}$$

This may be compared to the actual coke yield of 35 wt% and the Conradson carbon residue of the feed of 26 wt%. Thus, the extrinsic coke in this case is 10 wt% and is the amount that we would like to prevent from forming. The similarity between the intrinsic coke and the Conradson carbon residue in this case is a coincidence. Later, it will be shown that the resid is degraded in the microcarbon tester used to measure Conradson carbon residue, and that polar aromatics are found in the liquid product from the microcarbon tester.

The third limitation is demonstrated by the aromatic carbon balance around the same commercial coker as described earlier, where the aromatic carbon content increased from 29.9% in the feed to 46.5% in the products, as shown in Table 8.4. Aromization is a consequence of resids having high concentrations of natural hydrogen donors, described in section 4.7. The example in Table 8.4 also shows that 10% of the coking product was hydrogen and hydrocarbon gases that averaged hydrogen content of 22% or 21.8% of the hydrogen in the starting resid. As not all the hydrogen in the resid feed was accounted for by the products, it is quite likely that the actual hydrocarbon gas yield was even higher. If eliminated, these hydrogen-rich fragments that form gas could remain in the liquid product to

TABLE 8.4
Commercial Coker: Increases in Aromatic Carbon

Product	Yield (wt%)	H (wt%)	Aromatic and Olefinic C (wt% Fraction)	(wt.% Feed)
$CO + CO_2 + N_2$	0.84	0.00	0.0	0.0
H_2S	0.93	5.92	0.0	0.0
H_2	0.20	100.0	0.0	0.0
$C_2= + C_3= + C_4=$	2.3	14.37	0.0	0.0
$CH_4 + C_2 + C_3 + C_4$	7.5	22.25	0.0	0.0
Naphtha	14.5	13.22	25.3	3.7
Light coker gas oil	11.4	11.33	26.6	3.0
Heavy coker gas oil	26.6	10.37	32.1	8.5
Coke	35.1	3.73	89.1	31.3
Total	99.4	9.59		46.5
Feed	100.0	10.11	29.9	29.9

TABLE 8.5
Commercial Coker: Degradation of Saturated Carbon

Fraction	Yield (wt%)	Carbon (wt%)	Saturated C (% of carbon)	Saturated C (wt% of Feed)
C4- Gases	14.1	74.04	92.2	9.8
Naphtha	18.4	84.87	70.8	11.2
Light coker gas oil	21.4	84.75	65.1	11.9
Heavy coker gas oil	16.3	84.99	60.7	8.6
Coke	28.5	87.62	5.0	1.3
Total	98.5	83.28		42.8
Feed	100	84.54	68.4	57.8
Liquid product	56.1	85.18	65.7	31.3

increase the yield and quality of the liquid product or to reduce the aromization of hydroaromatics that form extrinsic coke.

Table 8.5 provides an alternate view of similar data from a different commercial coker. Instead of balancing aromatic carbon, the balance on saturated carbon is shown, where the 58% saturated carbon in the feed was reduced to 43% in the products and only 31% in the liquid product. Also, the concentration of saturated carbon is lower in the liquid product than in the vacuum resid feed. The saturated carbon in the feed becomes distributed among products with only 55% going to liquid product, 22% converted to aromatics, 17% going to hydrocarbon gases, 4% converted to olefins, and 2% going to coke. Clearly, the resid feed was degraded by the coker.

8.4.3 WAYS TO ELIMINATE/REDUCE BARRIERS AND DEMONSTRATE ON SMALL LAB SCALE

8.4.3.1 Reduce Secondary Cracking

Any significant improvement in the yield of high quality liquids depends on one devising ways to overcome these limitations. Secondary cracking degrades the quality of the most desirable product while forming hydrocarbon gases, an undesirable by-product. Secondary cracking of volatile liquid products can be minimized by using a coking reactor with a very short vapor residence time, but with a long resid residence time. This latter feature is necessary to go through complete conversion to coke of 100% CCR.

Figure 8.11 shows the result for the use of four different ideal laboratory coking reactors with these two features and a feed of Arabian Heavy vacuum resid. Two of these laboratory reactors were based on tubing bombs, developed and run by Ray Kennedy who worked with the author in Corporate Research. Vacuum coking was a tubing bomb reactor, attached by a tube to a condenser in ice water,

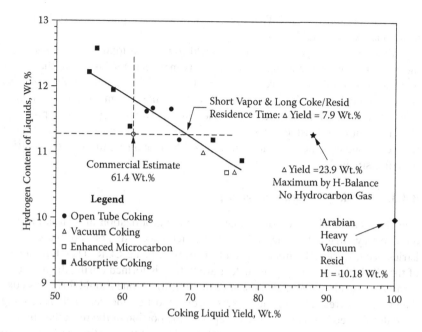

FIGURE 8.11 The yield–quality plot for small laboratory coking with short vapor and long resid residence time of Arabian Heavy vacuum resid.

that pulled a vacuum on the other side of the condenser. Another laboratory reactor, open tubing bomb, was also a tubing bomb attached by a tube to a condenser in ice water, but it was connected by a peristaltic pump back to the tubing bomb. Thus, it recirculated vapors and inert gas through the condenser and back to the tubing bomb. The remaining two laboratory reactors used an enhanced microcarbon tester, developed and run by Cyntell Robertson in the development labs, and described next in section 8.4.3.2. In one case, the enhanced microcarbon tester was used directly and in a second case, adsorptive coking, the resid was coated on porous solids, as will be described in section 8.4.3.3.

In Figure 8.11, the hydrogen content of the volatile liquids is used as the measure of quality, and the resulting yield-quality from the different reactors is compared with a coking model prediction for Fluid Coking. Either the quality can be improved at the same yield or the yield can be increased at the same quality. The attempt to improve quality will be discussed in section 8.4.3.3. In this section, the incentive is a greater yield of liquids of the same quality as Fluid Coking, the method preferred by refineries. Data were fit with a least square line to determine it, at a hydrogen content of 11.28, the liquid yield could be increased by 7.9 wt%. The open tubing bomb with recirculating vapors was the closest at matching this combination because its effective cut-point was close to 975°F (524°C). Both vacuum coking and the enhanced microcarbon tester produced liquids with substantial liquid boiling above 975°F. Therefore, they produced greater yield of liquids of lower quality than Fluid Coking.

Although the 7.9 wt% increase in yield (69.3 wt% total yield) provides an incentive to improve existing processes, it is marginal for justifying the development of a totally new coking process. Of course, one must realize that no commercial coker is going to match the yields obtained from a small, ideal laboratory reactor. Therefore, to justify developing a new coking process, a potential improvement of at least 10 wt% was being sought. As a result, methods both to reduce secondary cracking and to minimize the remaining limiting chemistries were investigated.

8.4.3.2 Coking–Separation Synergy

The chemical combination of aromatics and the aromization of hydroaromatics are two ways that coking degrades the resid feed. Actually, both of these chemistries occur during the microcarbon test for measuring CCR. Therefore, a part of the CCR (five-plus ring polynuclear aromatics) is formed during the measurement. As a result, if these chemistries are minimized, one should form less coke than CCR, whereas commercial cokers form 1.2 to 1.6 times as much. As coking degrades the feed, it makes sense to separate the portion of the resid that already meets the quality requirement before coking. At least this portion will be spared from being degraded.

In the laboratory it was decided to use high cut-point vacuum (Distact) distillation as the only separation process. Deasphalting could also have been used on the feed. The overhead of high cut-point distillation to an atmospheric equivalent boiling point of 1300°F (704°C) is combined with the overhead product of coking the 1300°F+ resid in an enhanced microcarbon tester (Figure 8.12) that is capable

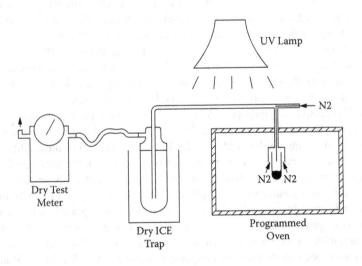

FIGURE 8.12 Schematic of the enhanced microcarbon tester.

of short vapor and long resid residence times while collecting the volatile liquid product. The amount of resid was scaled-up from the usual 0.1 g to 1.5 g in the glass sample vial of a commercial microcarbon tester. A glass chimney was put over the sample vial to collect the product vapors. However, the flow of hot nitrogen had to be greatly decreased to prevent aerosol formation. Room temperature nitrogen with very controlled flow was introduced above the surface of the resid to quench the vapors without condensing. An ultraviolet (UV) lamp heated and evaporated any liquid that condensed in the transfer tube. This took advantage of the fact that the aromatic volatile liquid products are strong UV absorbers. The volatile liquids were condensed and collected in the dry ice trap. Gas chromatography (GC)-simulated distillation showed that 10 to 20 wt% of the collected liquids had boiling points above 975°F (524°C).

Nevertheless, Figure 8.13 shows that the Conradson carbon residue (coke yield) of the enhanced microcarbon tester with 1.5 g of resid agreed with that measured with the normal microcarbon tester on 0.1 g of resid. Tracer studies showed that the vapor residence time of the vapor–liquid product is about 0.1 sec before being quenched, and the residence time of the resid is the usual 15 min at 500°C of the microcarbon test.

Because the microcarbon tester measures Conradson carbon residue, the yield of coke is the Conradson carbon residue of the 1300°F⁺ resid, or on a resid basis it is this value times the fraction of the resid that is 1300°F⁺. A schematic of the process and the calculation of the coke yield for three different vacuum resids

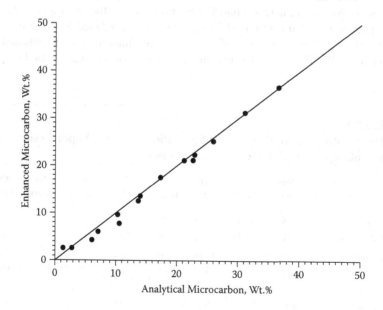

FIGURE 8.13 The coke yield in the enhanced microcarbon tester versus the Conradson carbon residue as measured by an analytical microcarbon tester.

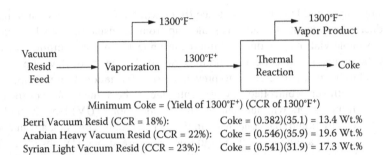

Minimum Coke = (Yield of 1300°F⁺) (CCR of 1300°F⁺)

Berri Vacuum Resid (CCR = 18%):   Coke = (0.382)(35.1) = 13.4 Wt.%
Arabian Heavy Vacuum Resid (CCR = 22%):   Coke = (0.546)(35.9) = 19.6 Wt.%
Syrian Light Vacuum Resid (CCR = 23%):   Coke = (0.541)(31.9) = 17.3 Wt.%

FIGURE 8.14 The coke yield of three different vacuum resids from Distact distillation of each resid and only coking the 1300°F⁺ fraction in the enhanced microcarbon tester.

are shown in Figure 8.14. For each resid the coke yield is less than the Conradson carbon residue of the resid feed (average of 0.8 times Conradson carbon residue). However, it also depends on the fraction of 1300°F⁺ resid, so the coke yield is not proportional to resid Conradson carbon residue. Therefore, Syrian Light resid that has a higher Conradson carbon residue than Arabian Heavy resid, has a lower coke yield.

In order to measure the yield and quality of the liquid product, the 1300°F⁻ fraction is added to the liquids collected in the enhanced microcarbon tester as shown in Table 8.6. The result, shown in Figure 8.15, is the liquid yield has increased by 14.8 wt% over a commercial Fluid Coker at only a slightly lower quality (11.20 wt% hydrogen as compared to 11.28 wt% hydrogen predicted for a commercial coker). The 1300°F⁻ fraction is of a lower quality than the liquids obtained by coking the 1300°F⁺ fraction because the separated liquids contain some five-plus

TABLE 8.6
Yields and Analyses of Products of Separation Plus Low Vapor Residence Time Coking on Arabian Heavy Vacuum Resid

	Yield (wt%)	C (wt%)	H (wt%)	S (wt%)	N (wt%)	V (ppm)	Ni (ppm)	CCR (wt%)
1300 °F- feed	45.4	84.33	11.11	4.12	0.23	45	7	7.4
EMCR liquids	31.4	84.33	11.32	3.99	0.24	0	0	0
Total liquids	76.8	84.33	11.20	4.07	0.24	27	4	4.4
Coke	19.6	84.99	3.85	8.85	1.5	835	260	100
Gas (by diff.)	3.6							
Arab Heavy resid	100	83.67	10.18	5.13	0.42	190	55	22.4
Percentage reduction			20.7	43.8		85.9	92.6	80.5

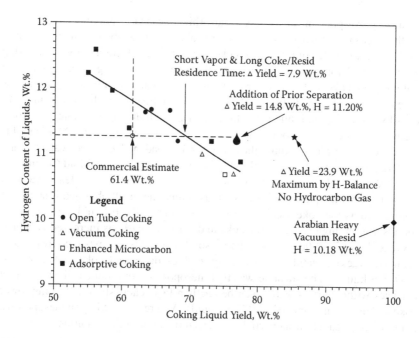

FIGURE 8.15 Combining separation of 1300°F⁻ fraction with coking the 1300°F⁺ fraction increases the liquid yield by 14.8 wt% over commercial Fluid Coking at almost the same quality.

ring polynuclear aromatics that would have been reactively separated by coking. Nevertheless, by combining these liquid products, we obtain almost the desired quality and avoid degrading the better 45% of the resid. Even when separating part of the resid without coking, the process removes 21% of the sulfur, 44% of the nitrogen, 86% of the vanadium, 93% of the nickel, and 80% of the Conradson carbon residue. As the gas yield is 3.6 wt%, there is still room for further improvement toward the upper bound of a liquid yield of 85.3 wt% for a no-gas case using hydrogen balance. Therefore, this result shows that the combination of separation of the higher quality fraction of resid without coking and coking of the lower quality fraction of the resid at long residence time, but with a short vapor residence time, can achieve a substantially higher liquid yield over commercial cokers at nearly the same liquid quality. Thus, despite the current perception, the coking of petroleum resids has the potential for significant improvement.

8.4.3.3 Coking on Different Porous Solids

As there is still room for further improvement, we ask ourselves how may we better improve coking. Instead of the coking being done on coke, we could do the coking on microporous particles as used in the FTC process.[13] Initially, the incentive was to use the microporous particles as adsorbents in order to improve the quality of coker gas oil. Perhaps, one could operate a coker once-through without volatizing five-plus ring aromatics by adsorbing these large polynuclear aromatics on the microporous particles. The author is also skeptical about four-ring aromatics being a valuable component of coker gas oil. The selection between five-ring aromatics and four-ring aromatics is only based on five-ring aromatics having boiling points above 975°F, whereas four-ring aromatics can have boiling points below 975°F. The prediction tools were not accurate enough to determine if, after hydrotreating and fluid catalytic cracking, the four-ring aromatics are worth including in the liquid product of coking instead of forming coke. Although this was called "adsorptive coking," the concept was broadened to include catalysis and the target for increasing the yield of liquids of similar quality as Fluid Coking.

Using the enhanced microcarbon tester and reacting Arab Heavy vacuum resid mixed with equal weights of various solids, the result obtained is shown in Figure 8.16 as the coke yield versus the yield of C_5-975°F liquids. As expected for adsorption, high surface area coke and activated carbons increased coke and decreased C_5-975°F liquids over the no-solids case. Also as expected, resid fluid catalytic cracking catalysts containing zeolites decreased coke and increased C_5-975°F liquids. The surprise was that mesoporous, amorphous silica–alumina particles without zeolites did even better than FCC catalysts in obtaining lower coke and higher amounts of C_5-975°F liquids. As a result, coking on mesoporous, amorphous silica–alumina particles received the bulk of the attention.

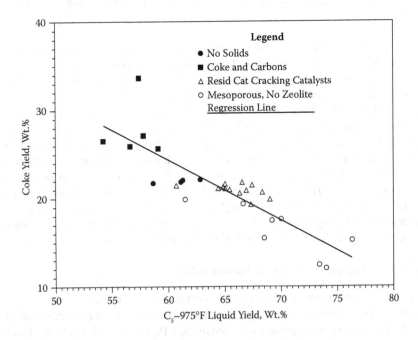

FIGURE 8.16 Coke yield versus yield of distillable liquids from coking Arabian Heavy vacuum resid on various porous particles using the enhanced microcarbon tester.

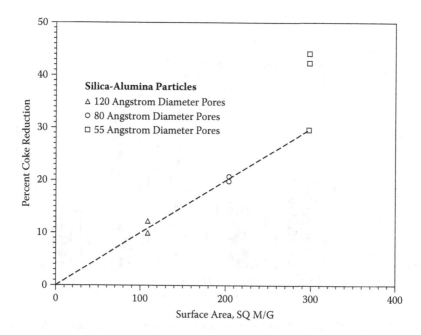

FIGURE 8.17 Coke reduction by silica–alumina particles: Particles with smallest meso-pores and highest surface area give the greatest coke reduction.

Amorphous silica–alumina particles are known to have moderate hydrogen transfer activity, but low cracking activity.[20] Fluid catalytic cracking obtains lower additive coke than Conradson carbon residue because of hydrogen transfer catalysis. Hydrogen transfer catalysis would certainly be helpful in reducing the combining of aromatics to make large PNAs. The pores should be large enough to accept resid molecules but small enough to exclude the asphaltenes and larger resins that have the PNAs of five rings and more. Therefore, when the percent coke reduction is plotted versus surface area, as in Figure 8.17, we find that the larger surface area and smaller mesopores (55 Å) are better than the smaller surface area and larger pores (80 Å and 120 Å) at coke reduction. Silica–alumina particles in the 25 to 50 Å pore range were not available to find the optimum pore size.

Figure 8.18 and Figure 8.19 show that the increase in liquids for 55 Å silica–alumina is for all boiling ranges except for 975°F⁺ from GC-simulated distillation, but that the liquids contain more three-ring aromatics, more four-ring aromatics, and more polar cores from HPLC. These are what the hydrogen transfer is preventing from forming coke. In each case the total concentration of three- and four-ring aromatics and polar (includes five-plus ring aromatics) cores is about constant. Those that do not go to coke become part of the liquid product. Thus, on Flexicoker gasifier coke (surface area = 177 sq m/g of which 166 sq m/g is greater than 20 Å pores), greater coke is formed than with no solids, but the liquids

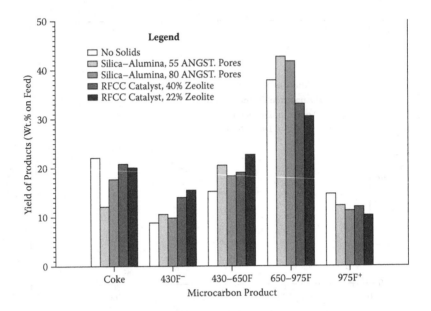

FIGURE 8.18 The yield of coke and various boiling point fractions from coking Arabian Heavy vacuum resid on two silica–alumina particles and on two resid fluid catalytic cracking catalysts as compared with no particles using the enhanced microcarbon tester.

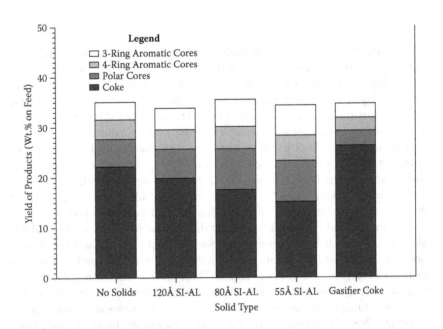

FIGURE 8.19 The three-ring, four-ring, and polar aromatic cores that do not form coke end up in the liquid product.

have much lower concentrations of three-plus ring aromatic cores. Fluid coke and Flexicoker reactor coke have very low surface area. Consequently, they give the same results as with no solids.

Spreading the resid over large surface areas to form thinner films would also improve mass transfer so that molecular free radical fragments formed by cracking in the liquid phase would have less time for addition and recombination reactions. This is how Gray, Le, and McCaffrey[21] interpret their data on coking of thin films of Athabasca vacuum resid and obtaining much lower coke yields. Although mass transfer may play a role, it does not appear to be the dominant phenomenon in this case. Mass transfer does not explain the different coke and liquid yields (higher and lower) of the different porous solids nor the balancing of three-plus aromatic ring cores between liquids and coke. Also, if fast mass transfer prevented free radicals from interacting with other molecules, one would expect a large increase in olefin formation as in gas phase cracking. Because this was not observed in this case, hydrogen transfer must be occurring.

When the 1200 to 1300°F Distact fraction of Arabian Heavy vacuum resid is coked with and without 50 wt% 55 Å pore silica–alumina, an even greater reduction in coke (81%) and in 975°F+ liquids (37%) are obtained, as shown below:

$$\begin{array}{ccccc} & 11.7\% & 75.7\% & 12.1\% & 0.6\% \end{array}$$

$$\text{HAVR 1200 to 1300°F} \xrightarrow{\text{MCR}} \quad \text{COKE} + \text{C5-975°F} + \text{975°F}^+ + \text{Gas}$$

$$\begin{array}{ccccc} \text{HAVR 1200 to 1300°F} & 2.2\% & 89.9\% & 7.6\% & 0.3\% \end{array}$$

$$\text{+50 wt\% 55 A Si-Alumina} \xrightarrow{\text{MCR}} \quad \text{COKE} + \text{C5-975°F} + \text{975°F}^+ + \text{Gas}$$

This is consistent with hydrogen transfer catalysis being most effective for the compounds containing aromatics smaller than five aromatic rings. In Figure 8.20, a repeat of the plot of hydrogen content versus liquid yield, the combination of short vapor residence time and hydrogen transfer catalysis increased the yield of coking liquids from 61.4 to 78.2 wt% (an absolute difference of 16.8 wt%) for a run that gave liquids containing 11.30 wt% hydrogen. However, this is only 2 wt% absolute better than the combination of separation and coking with short vapor residence time and, thus, there still remains further room for improvement.

8.4.4 Demonstrate on Pilot Plant Scale

There are two coking concepts that show enough promise on the lab scale to move to the pilot scale. One needs to look ahead and evaluate which would fit better in the refinery or upgrading world. Both require a reactor that can achieve a short vapor and long resid/coke residence time. The LR-Flash Coker was selected as the one that most closely achieves these needs. In addition, it has the potential to flash off the lightest part of the vacuum resid into the vapor phase with little chemical reaction, similar to Distact distillation. If possible, both the separation and

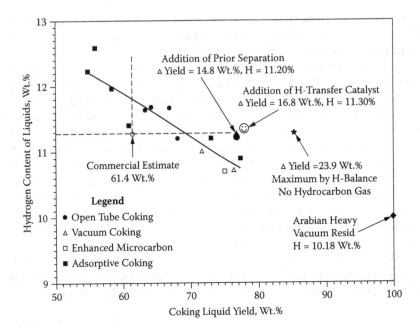

FIGURE 8.20 Coking Arabian Heavy vacuum resid with the low vapor residence time of the enhanced microcarbon tester and with a hydrogen transfer catalyst (silica–alumina) increases the liquid yield by 16.8 wt% at equal quality over commercial Fluid Coking.

the coking with short vapor residence time could be accomplished in the same reactor. Meanwhile, closer examination of hydrogen transfer catalysis indicated that it was more difficult to implement. It did not fit directly in Fluid Coking or Flexicoking because all the coke could not be burned or gasified off to expose fresh surface on each cycle, and the vapor phase residence time is too long. However, it might enable lower reactor temperatures with porous particles suppressing bogging as in the FTC process. Nevertheless, it needed to be practiced more like the ART or the 3D process using the fluid catalytic cracking method to burn off all the coke on each cycle. The FTC process experience pointed out the problem with nickel and vanadium accumulating on the particle surface and causing dehydrogenation reactions. Thus, the examination pointed to a better commercial fit for the LR-Flash Coker than hydrogen transfer catalysis, and so it was moved to the pilot plant.

The champion refinery was more interested in direct coupling such a coker to a fluid catalytic cracker without hydrotreating than having it be a typical coker, much like the objective of the ART and 3D processes. The need was to remove the metals and most of the Conradson carbon residue from a vacuum resid to convert a moderate quality vacuum resid into fluid catalytic cracking feed. Therefore, it was called SATCON[17] for satellite converter or, alternatively, a *thermal deasphalting process*.[22,23] The process completed a short pilot plant stage, but was unsuccessful in the commercial demonstration stage because design constraints prevented it from operating as a short vapor, long resid residence time coker at the desired production rates. This points out the difficulty in commercializing innovation in the refining industry with its many obstacles.

In order for a new process to be a success, it needs to be demonstrated on a large scale. With limited funds for development and commercialization, the choice often is to spend the funds on the pilot scale and hope a refinery will pay for a refinery scale process without demonstration on the large scale, or to do limited pilot scale runs and spend most of the funds on a demonstration scale in a refinery. In this case, the latter choice was made without success. For the Microcat hydroconversion process that will be discussed in chapter 10, the former choice was made and no refinery or upgrader selected the process primarily because of the lack of a large scale demonstration. Hopefully, with the recent large profits of petroleum companies, they will again fully invest in both development and demonstration scale for the same new process.

Although this innovation procedure on coking has yet to produce a commercial success, this section demonstrates some of the steps to create an invention that may lead to an innovation. The most important point is that invention may be achieved by deductive logic and hard work. However, it is written from the perspective of a researcher. It downplays the critical need for a champion refinery and a strong engineering effort that were the real drivers in this particular case. In addition, this case demonstrates adapting technology already developed, the LR-Flash Coker. One should not be so focused on developing completely new technology that he/she ignores existing technology. Adapting existing technology is considerably more efficient in time and cost than developing a completely new technology from scratch. However, the adapted technology needs to be engineered with the involvement of the researchers all the way through commercialization to prevent it's being engineered away from the original concept, as demonstrated in this case.

8.4.5 CONSIDER ALTERNATIVES

If a concept based on sound research is not successfully developed and commercialized because of faulty technology, consider alternative technology. The LR-Flash coker has the mechanical disadvantages previously discussed plus it is limited in length because of its weight and alignment requirements. This constrains the throughput at low temperatures because the resid needs a relatively long residence time to convert to dry particles at the discharge end. High temperatures cause secondary cracking of liquid product that reduces its yield and quality while forming conjugated olefins that promote fouling.

ETX Systems Inc.[24,25] has devised an alternative reactor design, shown in Figure 8.21, that overcomes these disadvantages. Their reactor (ETX Upgrader) is a shallow, cross flow, fluidized bed that is commonly used to dry solids, but not previously used in the petroleum industry. Thus, the solids flow perpendicular to the flow of fluidizing gas. The reactor is not limited in length, has no mechanical

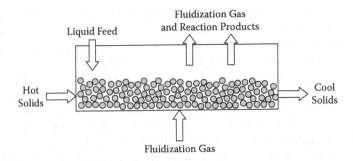

FIGURE 8.21 Schematic of the shallow, cross flow, fluidized bed coker design of the ETX Upgrader. (Used with permission of ETX Systems, Inc., Calgary, Alberta.)

internals, and yet has short vapor residence time and plug flow characteristics like the LR-Flash coker. In addition, they expect that a higher flow of solids will enable operating the coker at much lower temperatures than fluid cokers without bogging. Not only will this obtain still lower vapor phase cracking, but they expect to obtain better advantage of mass transfer in reducing liquid phase recombination reactions.[21] It remains to be seen if this new coking reactor concept is successful, but it shows that there are always alternatives when the potential prize is large. If the alternatives are not devised by the original researcher, they are open to others. When anyone is successful in producing a step-out improvement, the petroleum industry and, hopefully, everyone benefits.

8.5 CONCLUSIONS AND FUTURE DEVELOPMENTS

In this chapter, it was shown that coking has considerable room for improvement. The current limitations were determined and ways to overcome these limitations were devised. To obtain greater yields of liquids of equal quality, it is clear that one needs to reduce or eliminate hydrocarbon gas formation by minimizing vapor phase residence time and allowing for long resid residence time. Forming coke from five-plus ring aromatics is a must, but minimizing the formation of five-plus ring aromatics from smaller aromatics is to be avoided. This may be done by separating the higher quality fraction of the resid and only coking the lower quality fraction or by taking advantage of hydrogen transfer catalysis with pore size selectivity for the smaller macromolecules. Deasphalting is a strong candidate for the separation of the high quality fraction, even for existing cokers. While hydrogen transfer catalysis did not fit the need for coking, it is a strong candidate for resid catalytic cracking. If a way could be devised to segregate mesoporous silica–alumina catalyst from zeolite catalyst in the same fluid catalytic cracking unit, then resid could be sprayed on the silica–alumina catalyst and the volatile products and vacuum gas oil could be contacted with zeolite catalysts. In addition, although adsorptive coking is not a good fit at present, as the petroleum industry improves techniques of molecular engineering in refining, the need for improving coker product quality will come and adsorptive coking may be a strong candidate.

The market for fuel-grade coke will soon come to an end; however, refineries and upgraders will not be permitted to stockpile coke. The only alternative will be gasification to produce hydrogen, electric power, and syngas. Perhaps most cokers will become attached to gasification similar to Flexicoking. More likely, regional gasifiers could be built to accept coke from many refineries and/or upgraders while supplying their hydrogen requirements.

REFERENCES

1. IA Wiehe. Tutorial on resid conversion and coking. Proceedings of the Second International Conference on Refinery Processing, AIChE, 499–505, 1999.
2. IA Wiehe. Petroleum resid conversion and coking. *Rev Process Chem Eng* 2: 263–276, 1999.
3. PJ Ellis, CA Paul. Tutorial: Delayed coking fundamentals. Proceedings of 1st International Conference on Refinery Processes, AIChE, 151–169, 1998.
4. GC Hughes, BJ Doerksen, AM Englehorn, S Romero. Tutorial: Delayed coking commercial technology. Proceedings of 3rd International Conference on Refinery Processing, AIChE, 66–81, 2000.
5. F Self, E Ekholm, K Bowers. *Refining Overview—Petroleum Processes and Products.* New York: AIChE, 2000, pp. 22-1–22-11.
6. LP Antalffy, MB Knowles, SA Martin, WR Sines. A new improved automated drum deheading system, Proceedings of 1st International Conference on Refinery Processing, AIChE, 206–213, 1998.
7. DG Hammond. Tutorial: Fluid coking/flexicoking fundamentals. Proceedings of 1st International Conference on Refinery Processing, AIChE, 170–181, 1998.
8. DG Hammond, LF Lampert, CJ Mart, SF Massenzio, GE Phillips, DJ Sellards, AC Woerner. Review of fluid coking and flexicoking technologies, AIChE, Paper 44c, Spring National Meeting, 2003.
9. DB Bartholic. Preparation of FCC Charge from Residual Fractions. U.S. Patent 4,243,514. Assigned to Engelhard Corp., filed May 14, 1979, and issued January 6, 1981.
10. DB Bartholic, RP Haseltine. New crude/resid treating process offers savings. *Oil Gas J* 79: 242–252, 1981.
11. DB Bartholic. Liquid-Solid Separation Process and Apparatus. U.S. Patent 4,859,315, filed November 5, 1987, and issued August 22, 1989.
12. DB Bartholic, MR Kleim, M Soudek. The simplified approach to residual oil upgrading. NPRA Meeting, AM-91-44, 1991.
13. T Miyauchi et al. A new fluid thermal cracking of residual oil with high-quality fluidization and metal effect elimination. Proceedings of the 12th World Petroleum Congress, 335–341, 1987.
14. T Aiba, H Kaji, T Suzuki, T Wakamatsu. Residue thermal cracking by eureka process. 88th National AIChE Meeting, Preprint #51e, 1980.
15. H Weiss et al. Coking of oil sands, asphaltenes and residual oils in the LR-Process. 5th UNITAR Conference, 1988.

16. H Weiss et al. Coking of residue oils by the LR-Process. *Sonderdruck aus Erdol & Kohle-Erdgas-Petrochemie/Hydrocarbon Technology*, 42: 235–237, 1989.

17. H Weiss, J Schmalfeld. Low cost process for refinery residue conversion. *PTQ*, Summer: 79–83, 1999, and presented at the European Refinery Technology Conference, Berlin, 1998.

18. IA Wiehe. Resid conversion–separation synergy. Proceedings of 4th International Conference on Refinery Processing, AIChE, 245–252, 2001.

19. IA Wiehe. Resid conversion: Limitations and maximum potential. Proceedings of the 6th International Conference on Refinery Processing, AIChE, 281–290, 2003.

20. BR Cook, SG Colgrove. Anthracene isomerization over amorphous silica–alumina: a novel hydrogen transfer reaction. ACS Preprints, *Div Pet Chem* 39: 372–378, 1994.

21. MR Gray, T Le, WC McCaffrey. Coupling of mass transfer and reaction in coking of thin films of athabasca vacuum residue. *Ind Eng Chem Res* 40: 3317–3324, 2001.

22. DG Hammond et al. Process for Deasphalting Residua. U.S. Patent 5,714,056. Assigned to Exxon, 1998.

23. W Serrand et al. Integrated Residua Upgrading and Fluid Catalytic Cracking. U.S. Patent 5,919,352. Assigned to Exxon, 1999.

24. R Pinchuk, W Brown, G Monaghan, W McCaffrey, O Asprino, P Gonzalez. Tailoring RTDs in fluidized beds for improved coker yields: The Envision Technologies Corp. cross-flow coker. Proceedings of 8th International Conference on Refinery Processing, AIChE, 66a, 2005.

25. WA Brown, RJ Pinchuk, GV Monaghan. A Process for Converting a Liquid Feed Material into a Vapor Phase Product. Canadian Patent 2505632, Assigned to Envision Technologies Corp., 2006.

9 Visbreaking

9.1 TECHNOLOGY

Visbreaking is a low conversion thermal process used originally to reduce the resid viscosity for heavy fuel oil applications. However, more often today it uses a resid that exceeds minimum heavy fuel oil specifications. It converts just enough to obtain 20 to 30% transportation boiling range liquids and still have the heavy product (visbreaker tar) to meet heavy fuel oil specifications. Continuous flow processes, such as visbreaking and hydroconversion, cannot tolerate significant coke formation, or it quickly insulates heat transfer surfaces and plugs up the flow. Thus, visbreaking is required to operate within the coke induction period that may limit conversion. In addition, the conversion might be limited if the visbreaker tar is unsuitable for blending into heavy fuel oil.

Visbreaking is a rather minor resid conversion process because of its low-resid conversion. As a result, it is not a popular choice in North America even though it is still extensively practiced in Europe and Asia. However, it is still worthwhile including in this book because it provides a steppingstone to understand other resid conversion processes. For instance, the heater of a delayed coker is similar to a coil visbreaker, and the same methods can be applied to suppress coke formation. In addition, hydroconversion is similar to visbreaking, but with hydrogen pressure and catalysts to maintain hydrogen donor concentration. Although it has a much longer coke induction period than visbreaking and higher resid conversion, these two processes have similar cracking kinetics, methods to detect coke formation, and ways to mitigate coke and sediment formation.

A visbreaker reactor may be similar to a delayed coker with a furnace tube followed by a soaker drum[1,2] as shown in Figure 9.1. However, the drum is much smaller in volume to limit the residence time with the entire liquid product flowing overhead. Alternatively, the entire visbreaker may be a long tube coiled within a furnace as shown in Figure 9.2. Coil visbreakers are operated at higher temperatures (885 to 930°F [473 to 500°C]), but lower residence times (1 to 3 min) compared with soaker visbreakers at 800 to 830°F (427 to 433°C) for 10 to 20 min. In either case, the last portion of the reactor is most prone to coke formation. Thus, the oil is rapidly heated in the beginning and kept nearly isothermal at the end for the coil visbreaker.[3] Although the soaker is insulated, the temperature decreases as the oil flows through the soaker because of the heat of reaction. In addition, recirculation currents (backmixing) in soaker visbreakers

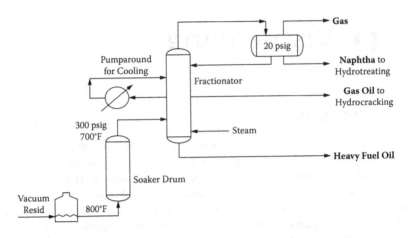

FIGURE 9.1 Schematic of a soaker visbreaker. (From F Self, E Ekholm, K Bowers. *Refining Overview—Petroleum, Processes and Products*, CDROM. AIChE-South Texas section. With permission.)

can cause an undesirable residence time distribution that can be minimized with baffles or perforated plates.

As turbulent flow in coil visbreakers approximates the desired plug flow except near the wall, care needs to be exercised to avoid undesirable two-phase flow regimes, such as slug flow. Nevertheless, even with these precautions, process upsets still cause coke to form and accumulate on visbreaker walls so that they need to be decoked periodically. The soaker visbreaker needs to be decoked less frequently, but the coil visbreaker is much easier to decoke with pigging. If a spare distillation furnace is available, a coil visbreaker may be built quite cheaply. However, visbreakers are scarce in North America because of the low price for residual fuel oil. The conversion is not great enough. On the other hand, it still is

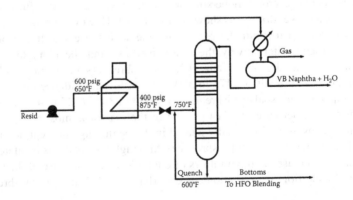

FIGURE 9.2 Schematic of an all coil visbreaker. (Used with permission from ExxonMobil Research & Engineering Co.)

a relatively low-cost method to reduce the viscosity of a heavy crude oil without rejecting a low-valued by-product.

9.2 PROCESS CHEMISTRY OF VISBREAKING

Most of the process chemistry of visbreaking has been covered in previous chapters. The kinetics of the thermal cracking of resids was covered in detail in chapter 4. We learned that the coke is triggered by the phase separation of converted asphaltenes, and the presence of the carbonaceous mesophase is an indicator of this mechanism. The distillable liquid product of thermal conversion is a nonsolvent for converted asphaltenes so that the coke induction period is shorter for a closed reactor than an open reactor. Even if the conversion in a visbreaker is within the coke induction period, converted asphaltenes in the resid product can become insoluble on cooling to form asphaltene sediments (soluble in toluene). We learned that even though asphaltenes have high concentrations of natural hydrogen donors, they are so reactive that their concentration is low near the end of the induction period. As a result, converted asphaltene free radicals are terminated by the higher concentration of hydrogen donors when in the oil phase, but allow for molecular weight growth reactions to form coke when in the asphaltene phase.

9.3 LIMITATIONS TO VISBREAKER CONVERSION

The resid conversion in visbreaking is limited by the coke formation, asphaltene sediment in the visbreaker tar (such as <0.07 wt% hot filtration sediment), or incompatibility of visbreaker tar with other heavy fuel oils. These limitations are very dependent on the visbreaker feed. It was already learned in chapter 4 that different feeds have varying reaction rates, stoichiometric coefficients, and solubility limits that result in different coke induction periods. These limitations also can be dependent on the type of visbreaker. Most often coil visbreakers are limited by asphaltene sediment in the visbreaker tar at conversions just below where coke formation begins. This is because the converted asphaltenes are more soluble at reaction temperatures than at room temperature. However, there can be many exceptions because the reactor contains the total liquid product, and the visbreaker tar does not contain the distillable liquids, nonsolvents for asphaltenes. If soaker visbreakers have some backmixing, coke can form before asphaltene sediment become insoluble in the tar. Even if a visbreaker is within the limits of coke formation and tar sediment, the converted asphaltenes in the visbreaker tar can precipitate when blended with other heavy fuel oils. In terms of chapter 5, the oils can be incompatible.

Because all three visbreaker conversion limitations are related to asphaltene solubility, they can be diagnosed and solved using the Oil Compatibility Model of chapter 5. Thus, one should get the samples of the heavy fuel oils that the heavy fuel oil customer plans to blend with the visbreaker tar and measure the range of solubility blending numbers. The insolubility number of the visbreaker tar should be monitored and kept by a safety factor below the lowest solubility blending number of the fuel oils that it might be blended. If the visbreaker tar contains asphaltene sediment, it is a self-incompatible oil (section 5.3.10) and only the toluene equivalence needs to be measured (equal to both the insolubility number and the solubility blending number). If this is below the insolubility number for compatibility with other heavy fuel oils, one can use the Oil Compatibility Model to select other oils to blend with the visbreaker tar to redissolve the asphaltenes. Otherwise, one will need to reduce the visbreaker conversion for that feed to get within the hot filtration limit. Finally, if the visbreaker is limited by coke formation, the coke induction period (residence time) can be predicted for each visbreaker feed using the methods in chapter 5, section 5.3.13.2 without actually coking the visbreaker. This requires determining the solubility blending number of the total visbreaker liquid product, not just the visbreaker tar, by sampling and testing each of the liquid products and using the volumetric mixing rule for the solubility blending rule. Naturally, the insolubility number is measured on the visbreaker tar.

9.4 VISBREAKING PROCESS INNOVATIONS

Although most of the visbreaking innovations that are discussed here have been presented previously in this book, it is worthwhile summarizing them in one section. Again, this demonstrates that a mature technology, such as visbreaking, can be significantly improved beyond what is currently practiced.

9.4.1 Optimizing Visbreaker Conversion for Each Feed

It is common for refineries to keep visbreaker conditions constant even though the feed is often changed. This requires a refinery to operate at the mildest visbreaker conditions for the feed having the most restrictive limits. However, it was learned in chapter 4 that resids can vary greatly in thermal reaction rate constant, hydrogen donor quantity and reactivity, and hydrogen content (Conradson carbon residue). As a result, this mode of visbreaker operation reduces the conversion of all feeds except the limiting feed. Instead, by starting with the mildest visbreaker conditions, one can measure the solubility blending number and insolubility number of the visbreaker liquid products and determine if there is room to increase the visbreaker severity for any given feed and still stay within limits as discussed in chapter 9, section 9.3. If so, the linear relationships of solubility blending number and insolubility blending number of the feed, and the total liquid product with thermal severity allows one to estimate how much to increase the visbreaker severity for each feed and conservatively stay within the limits. As a result, the visbreaker conditions may be optimized for each feed.

9.4.2 REMOVE VOLATILE NONSOLVENT

If visbreaker conversion is limited by coke formation, methods for increasing the coke induction period were discussed in chapter 4. As coke is triggered by the phase separation of converted asphaltenes, one needs to improve the solubility of converted asphaltenes in the reacting resid. Probably, the best way is to strip out the distillable liquids (nonsolvents for asphaltenes) more effectively by adding steam or some other inert gas to the visbreaker. As the vapor products typically have half the residence time of the liquid phase in a coil visbreaker and less than one-tenth the residence time in soaker visbreaker,[4] this effect is expected to be greater for a coil visbreaker than a soaker visbreaker. Yet, the Eureka process (chapter 8, section 8.3.4) shows that if the distillable liquids are stripped effectively enough even in a soaker, the resid conversions may be in the range of coking (more than twice visbreaking) and still the tar can flow as a liquid at high temperatures. Although this would make the tar incompatible with other heavy fuel oils, it demonstrates that there is significant room to increase the coke induction period of most visbreaker feeds by stripping out the distillable liquid product. However, particularly for the coil visbreaker, one must be sure that the stripping gas does not cause an undesirable two-phase flow in the visbreaker, such as slug flow.

9.4.3 ADD HYDROGEN DONOR SOLVENT

Another method for increasing the coke induction period discussed in chapter 4 is to coprocess, with the visbreaker feed, a heavy oil that is a better solvent for asphaltenes (higher solubility blending number) than the feed, but has no or very soluble asphaltenes (low insolubility number). We learned that this was insufficient, but the coprocessing oil also needs to have at least as high a concentration of hydrogen donors as the feed. These requirements greatly limit the choice of coprocessing oil, but the naphthenic vacuum resid in Table 4.13, chapter 4 shows that some resids can generate hydrogen donors during thermal conversion and thereby meet the hydrogen donor requirement. The hydrogen donor concentration of the coprocessing oil can also be increased above that in the feed by hydrogenating an aromatic oil and obtaining the benefit of reducing of the amount of asphaltenes formed from heptane solubles and the amount of asphaltene cores formed from asphaltenes. Although hydrogen donor visbreaking was discussed in the patent and technical literature[5] of Exxon and the Exxon Donor solvent process for liquefaction of coal was demonstrated on a plant scale in the 1970s, this process has been rarely, if ever, practiced. Nevertheless, with the understanding in chapter 5 as a starting point, adding a hydrogen donor solvent to a visbreaker has the greatest potential for increasing the conversion without forming coke. However, one needs to be alerted that reactive hydrogen donors reduce the cracking rate. Therefore, one may need to raise the visbreaker temperature.

The best hydrogen donors for coprocessing with the resid would be three- and four-ring aromatics that have been partially hydrogenated. The example in Table 9.1 is for heavy cat cycle oil that is without hydrogenation in a first case,

TABLE 9.1

Comparison of Coke Yields in Tubing Bomb Reactors at Different Conversions for Arabian Heavy Vacuum Resid (AHVR) Alone and When Coprocessed with Equal Volumes of Heavy Cat Cycle Oil (HCCO) at Three Levels of Hydrogenation

Oil or Coprocessing Oil	DDQ Donor H(g/100 g Oil)	Donor H Relative Reactivity	Percentage of Conversion AHVR	Coke, wt% Toluene Insolubility
100% AHVR	1.9	2.7	28	1.7
100% AHVR	1.9	2.7	35	3.9
100% AHVR	1.9	2.7	44	9.2
50% HCCO	1.3	6.0	53	16.2
50% Lt Hyt HCCO	1.7	10.6	62	4.3
50% H Hyt HCCO	2.2	15.4	66	1.9

lightly hydrogenated in a second case, and heavily hydrogenated in a third case. When these were blended separately with an equal volume of Arab Heavy vacuum resid and thermally reacted, the amount of toluene-insoluble coke formed was decreased with increasing hydrogenation of heavy cat cycle oil even though the conversion was increased in the same direction. As compared to the cases of thermally reacting Arabian Heavy vacuum resid alone, the untreated heavy cat cycle oil with a lower concentration of hydrogen donors than the resid did not seem to suppress the coke formation at all.

Coprocessing the lightly hydrotreated heavy cat cycle oil with almost as high concentration of hydrogen donors as the resid, but about four times as reactive, obtained about the same coke yield as the resid alone, but at 27 wt% absolute higher conversion (yield of volatile liquids). Coprocessing the heavily hydrotreated heavy cycle oil with higher concentration of hydrogen donors as the resid and 5.7 times as reactive allowed for 38 wt% absolute higher conversion than the resid alone at about the same yield of coke. Of course, no measurable yield of coke is allowed in commercial visbreaking and not many will be willing to cut the resid feed rate in half to coprocess 50% of hydrogen donor. However, these results demonstrate that significant increases in visbreaker conversions are possible by coprocessing even 10 to 20% of a hydrotreated aromatic oil, such as heavy cat cycle oil, particularly if this oil were going to be blended into heavy fuel oil with the visbreaker tar anyway.

9.4.4 PREVENT ADHESION OF COKE TO THE VISBREAKER SURFACE

In chapter 6, section 6.7, we have determined that the carbonaceous mesophase does not wet the surface of an oxidized metal, such as stainless steel. However, in a visbreaker hydrogen sulfide is formed that converts the oxide layer to an iron sulfide surface. The carbonaceous mesophase wets and coats the iron sulfide surface with a liquid film that hardens into a strong adhering solid. This forms the "glue" that holds the solid coke from the bulk to the stainless steel surface. Instead, if the steel tube is coated with a material that is not wetted by the carbonaceous mesophase and is not sulfided, the coke will not adhere to the surface, but be easily removed by the flowing oil. Although promising, this method of coke mitigation by coating the visbreaker surface has been underutilized and, as a result, is highly proprietary.

9.4.5 COUPLE WITH FUELS DEASPHALTING

This method is to visbreak a resid followed by deasphalting. The deasphalted oil is taken off as a light product, and the asphalt is taken as a heavy product. The intermediate resin stream from the deasphalter is recycled back to the visbreaker. As discussed in chapter 7, section 7.2.7, theoretically this process could get liquid yields similar to coking. The asphalt would have to be flaked or pelletized as it would be solid at room temperature. Although this resin recycle concept is theoretically sound and visbreaking has been commercially combined with deasphalting,[6] recycling of resins in visbreaking has not been demonstrated on the pilot plant or commercial scale. Resin recycle may be even more promising for hydroconversion as discussed in chapter 10.

REFERENCES

1. JH Gary, GE Handwerk, MJ Kaiser. *Petroleum Refining Technology and Economics.* 5th ed. Boca Raton: CRC Press, pp. 111–116, 2007.
2. S Raseev, *Thermal and Catalytic Processes in Petroleum Refining.* New York: Marcel Dekker, pp. 138–159, 2003.
3. DE Allan, CH Martinez, CC Eng, WJ Barton. Visbreaking gains renewer interest. *Chem Eng Progr* 85–89, January 1983.
4. M Akbar, H Geelen. Visbreaking uses soaker drum. *Hydrocarbon Proc* 60(5): 81–85, 1981.
5. AW Langer. Hydrogen donor diluent visbreaking of residua. *I&EC Proc Des and Dev* 1: 309–312, 1962.
6. MJ Humbach, RF Anderson, RT Penning, RR Shah. Deasphalting/visbreaking, A simplified black oil conversion scheme. Proceedings of AIChE, Spring National Meeting, Anaheim, Paper 20d, 1984.

10 Hydroconversion

10.1 PROCESS OBJECTIVES

Hydroconversion combines conversion by thermal cracking with hydrogenation of polynuclear aromatics (PNAs) in heavy oil to form additional hydrogen donors that greatly improve reaction selectivity. As discussed in chapter 4, section 4.7.4, additional reactive hydrogen donors greatly increase the coke induction period over that obtained in visbreaking by terminating free radicals and by reducing the frequency of aromatics combining to form larger PNAs. Thus, in catalytic hydroconversion, the positive effect of hydrogen donor dilute cracking is obtained without the negative effect of reducing the cracking rate because the concentration of reactive hydrogen donors is kept only as high as needed. As a result, when incompatibility is avoided, hydroconversion of vacuum resids to volatile liquids can be over 85 wt%, as opposed to 20 to 30 wt% for visbreaking and 50 to 60 wt% for commercial coking. However, one has to deal with the cost of hydrogen and catalyst, high-pressure vessels, poisoning of catalysts, the difficulty of asphaltenes to diffuse through small catalyst pores, and the intolerance to coke and sediment formation.

Because resids and heavy oils contain catalyst poisons, such as sulfur, basic nitrogen, vanadium, nickel, and polynuclear aromatics, as a primary upgrading process, hydroconversion should have methods to maintain catalyst activity while minimizing catalyst cost. First, the active catalyst should be a transition metal sulfide, otherwise, the high level of sulfur in resid feeds will poison other hydrogenation catalysts. The methods used to minimize the effect of catalyst poisons depends on the quality of the resid and the type of hydroconversion process—fixed bed, ebullating bed, or dispersed catalyst.

10.2 FIXED-BED RESID HYDROPROCESSING

Fixed-bed resid hydroprocessing units are usually more similar to hydrotreaters than hydroconversion units. They are most often run on good-to-moderate-quality resids that reduce sulfur and catalyst poisons—metals, basic nitrogen, and polynuclear aromatics (Conradason carbon residue)—before secondary catalytic processing, for example, fluid catalytic cracking. For instance, some atmospheric resids, such as Arabian Light, have catalyst poison levels just beyond the specifications for fluid catalytic cracking feed; they can be made into valuable fluid catalytic cracker feed after hydroprocessing. Likewise, hydroprocessing of deasphalted oil from low-quality vacuum resids can reduce catalyst poisons sufficiently to make fluid catalytic cracker (FCC) feed. Even if resid hydroprocessing units are operated for conversion, it is difficult to achieve more than 50% conversion before exceeding sediment specifications in heavy by-product or coke formation in the reactor (compatibility limit). It is also difficult to process vacuum resid for more than one year without having to shut down the unit and replacing the catalyst because of deactivation. Resid hydroprocessing units are commonly run at high hydrogen pressures (1500 to 3000 psig or 1.04 to 20.7 MPa) and low space velocities (0.2 to 0.5 $vh^{-1}v^{-1}$).[1]

Because metals removal is one of the fastest reactions and the metals accumulate in the pores of supported catalysts, it is common to have a guard bed in front of the fixed bed. When the guard bed is deactivated because of the deposition of metals, the feed is switched to a second guard bed, while the catalyst is replaced in the first guard bed. In order to hydrogenate the largest macromolecules in the resid—the asphaltenes—some or all of the catalysts need to have pores 100 to 200 Å in diameter. The remainder of the catalyst may actually be spent hydrotreating catalysts from vacuum gas oil hydrotreating. As freshly sulfided catalysts deactivate, to a certain degree quickly in resid service, and then slowly, not much is lost by using a cheap, spent catalyst from vacuum gas oil service that has undergone only the first level of deactivation. Resid hydrogenation catalysts usually are different combinations of cobalt, molybdenum, and nickel on alumina or silica–alumina.

Fixed-bed resid hydroprocessing units are prone to fouling. In chapter 5, section 5.3.12.1, the case of the inlet of a hydroprocessing unit plugging with coke due to feed incompatibility was discussed. This cause is rare, but more often plugging is caused by inorganics from inefficient desalting or from corrosion. For this reason, atmospheric resids are often filtered before they enter the catalyst bed. If hydrogen sulfide is not effectively adsorbed out of recycled hydrogen before it is mixed with the feed and heated in a furnace coil, the hydrogen sulfide can react with the steel furnace tube to form iron sulfide and plug the bed inlet. This is even more likely if pigging is used to decoke the furnace tubes because it can remove the chromium oxide protective layer on the furnace tubes. Of course, the corroding of the furnace tube to spring a leak and cause a fire is an even more serious problem. In chapter 6, section 6.4, the fouling of a heat exchanger after resid hydrotreatment was discussed, showing that resid hydroprocessing can upset the delicate compatibility balance of a resid to cause fouling by insoluble asphaltenes on cooling.

10.3 EBULLATING BED HYDROCONVERSION

One way to deal with catalyst deactivation is to use a reactor that allows for the continuous addition of fresh catalyst and removal of spent catalyst. This is achieved by a recycle pump that fluidizes the catalyst with an up-flowing reacting resid, while adding fresh catalyst to the top of the expanded bed and removing spent catalyst from the bottom. A schematic of such a hydroconversion unit is shown in Figure 10.1, and that of an LC-Fining ebullating bed reactor is shown in Figure 10.2. H-Oil ebullating bed reactors are similar, except the recycle pump is

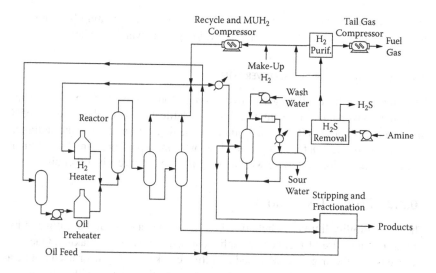

FIGURE 10.1 A schematic of an ebullating bed hydroconversion unit. (Reproduced with permission of Chevron Lummus Global LLC.)

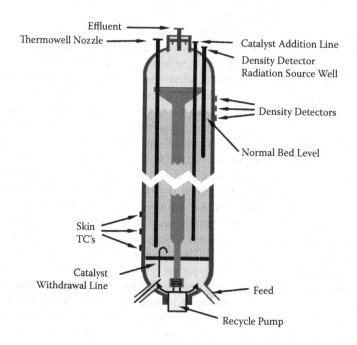

FIGURE 10.2 A schematic of an L-C Fining ebullating bed reactor. (Reproduced with permission of Chevron Lummus Global LLC.)

external to the reactor instead of internal. Nevertheless, ebullating bed units are the most popular choice for commercial resid hydroconversion today.

10.3.1 LIMITATIONS TO CONVERSION

The conversion by ebullating bed hydroconversion is greatly dependent on the feed, but conversions of vacuum resids to volatile liquids of the order of 65% are common.[1] These units can be limited by the deposition of coke and sediment downstream of the reactor in high- and low-pressure separators,[2] and by sediment in the bottoms, which interfere with it being blended into residual fuel oil.

10.3.2 PROCESS INNOVATIONS

The compatibility limits of ebullating bed hydroconversion can be mitigated by similar methods used for coke and asphaltene sediments in visbreaking. One difference is that the aromatic oil added to mitigate coke in the reactor need not be a hydrogen donor because the hydrogen pressure and the catalyst will hydrogenate an aromatic oil into a hydrogen donor. Indeed, heavy cat cycle oil and cat cracker bottoms are commonly added, when available, to the feed of ebullating bed units to mitigate coking and asphaltene sediments. For a different reason, Syncrude[2] recycled 30% light or heavy gas oil to quench the oil in high- and low-pressure separators. Their objective was to prevent thermal reactions from forming carbonaceous mesophase in the separators. This did extend the run length of an LC-Finer from one to two years by eliminating fouling in the low-pressure separator even though it only reduced the rate of fouling in the high-pressure separator. Thus, some carbonaceous mesophase may have been formed in the reactor.

Robert et al.[3] of IFP studied what affects the formation of asphaltene sediments in H-Oil bottoms. Catalysts with smaller pores are more prone to form sediments because they do not hydrogenate all of the asphaltenes. They followed changes up to sediment formation using the P-test (see chapter 5, section 5.3.7), but could not make sense of the data after the sediment formed. They probably were unfamiliar with experimental tests for self-incompatible oils (see chapter 5, section 5.3.10). Their SARA (saturates, aromatics, resins, and asphaltenes) separation results explained some but not all of their data. They found that more stable bottoms had the lowest ratio of vanadium to nickel. However, this is because asphaltenes usually have a higher ratio of vanadium to nickel than resins; thus, a higher ratio of vanadium to nickel implies a lower resin-to-asphaltene ratio (see chapter 6, section 6.4). Finally, they found that stability is more related to the amount of asphaltenes in the bottoms than the hydrogen-to-carbon (H/C) atomic ratio of the asphaltenes. As discussed in chapter 5, stability requires having a solubility blending number higher than the insolubility number. Although the insolubility number should increase as the asphaltene H/C atomic ratio decreases, instability only happens when the insolubility number equals or exceeds the solubility blending number. While one could correlate stability with only one variable—the P value, or the ratio of the solubility blending number to the insolubility number—the author recommends determining the solubility blending number and the insolubility number separately, and also looking for carbonaceous mesophase in the heavy liquid product. This enables one to understand better the mechanism of coke and asphaltene sediment formation to devise mitigation methods, as described in chapter 6.

Recently, Smith and Lott[4] of Headwaters have been recommending addition of their soluble catalyst to ebullating bed hydroconversion to obtain 8 to 12% greater yield, by more effectively hydrogenating asphaltenes, than can be achieved by supported catalysts. This essentially combines ebullating bed and dispersed-catalyst hydroconversion processes.

In sections 7.2.7 (chapter 7) and 9.4.5 (chapter 9), the advantage of coupling deasphalting with visbreaking was pointed out, in which the moderate Conradson carbon residue fraction (resins) was recycled back to the reactor to obtain much higher conversion (see Figure 7.14 in chapter 7) while removing the low and high Conradson carbon residue products. By replacing visbreaking with ebullating bed hydroconversion, one may achieve an even higher conversion. The removal of asphalt and recycle of resins should also improve the compatibility within the reactor. However, it is unlikely that the resulting asphalt would have an insolubility number low enough to be blended into heavy fuel oil, though it could be fed to a coker or a gasifier.

10.4 DISPERSED CATALYST HYDROCONVERSION PROCESSES

One way to avoid pore plugging of catalysts is to use finely dispersed catalysts. Instead of diffusing the asphaltenes to the catalysts, the catalysts are diffused to the asphaltenes. These dispersed catalysts can be formed from oil-soluble transition metal compounds, such as naphthanates, or lower-cost, water-soluble compounds, such as phosphomolybdic acid.[5] As illustrated in Figure 10.3, the thermal

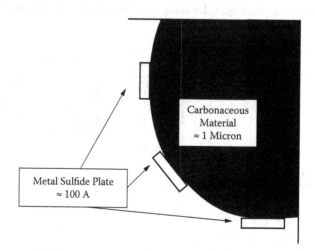

FIGURE 10.3 Dispersed catalysts are small metal sulfide plates on carbonaceous particles.

reaction of these precursors with resid in the presence of hydrogen and hydrogen sulfide forms small (~100 A) metal sulfide plates on (~1 μm) carbonaceous particles.[5] There can be as many as a billion of these catalytic particles in 1 mL of oil. Molybdenum is the most popular transition metal used as a compromise between cost and catalytic activity. However, the vanadium and nickel in the resid form part of the dispersed catalyst from hydrodemetalization reactions.

Although there have been many dispersed catalytic hydroconversion processes marketed, such as Microcat-RC,[5–7] Canmet,[8] VEBA-Combi-Cracking,[9] Eni Slurry Technology,[10,11] HDH-SHP,[12] and (HC)₃[13] or HCAT,[4] few have been commercialized. Dispersed catalysts can be continuously added in low enough amounts (i.e., 100 ppm) to consider them throwaway catalysts with the carbonaceous by-product. However, economics usually dictate some form of catalyst recycling to minimize catalyst cost. Nevertheless, the conversion of vacuum resids to gas and volatile liquids can be above 95 wt%, with more than 85 wt% volatile liquids (even over 100 vol% because of density decrease). An example schematic of the Microcat-RC process is shown in Figure 10.4, and example product yields[14] are shown in Table 10.1. Most dispersed catalyst hydroconversion processes use slurry bubble column reactors, as does the Microcat-RC process.

10.4.1 LIMITATIONS TO CONVERSION

The first limitation to conversion in dispersed-catalytic hydroconversion processes is the incompatibility within the reactor to form coke. If that is overcome, the only other limitation would be the conversion of PNA cores through hydrogenation, followed by cracking. This aromatic core limitation is really an economic barrier that often makes it unwise to increase resid conversion as high as 95 wt%. The problem is illustrated in Figure 10.5. If a ring of a PNA having more than four rings is hydrogenated, it can be thermally cracked only slowly, as it is competing with a much faster hydrogen-donation reaction. If the hydrogenated ring does

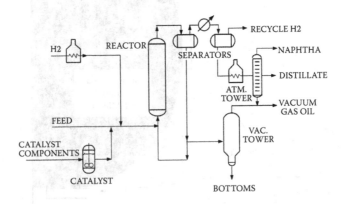

FIGURE 10.4 A schematic of the Microcat-RC process. (Used with permission from ExxonMobil Research & Engineering Co.)

TABLE 10.1

Yields of the Microcat-RC Process (524°C+ Cold Lake Vacuum Resid at 95% Conversion)

Product Objective:	FCC Feed Quality	High-Quality Diesel
C_1–C_4 gas, wt%	7.48	7.74
Liquid products, LV%		
C5–177°C naphtha	14.06	20.53
177–343°C distillate	53.91	49.74
343–565°C gas oil	33.59	32.61
Total liquids	101.56	102.88
565°C+ bottoms, wt%	2.14	2.14
Hydrogen cons., SCF/B	1663	2159

Source: CJ Mart, EG Ellis, DS McCaffrey, Jr. Exxon Mobil Resid Upgrading Technologies. PETROTECH 2005, 6th International Petroleum Conference and Exhibition, New Delhi, India, 2005. (Used with permission from ExxonMobil Research & Engineering Co.)

thermally crack, it only forms hydrocarbon gases and a polynuclear aromatic with methyl groups. When this is cracked down to a four-ring aromatic, it becomes part of the vacuum gas oil product. But, four-ring aromatics are the least valuable part of vacuum gas oil, as they need to be hydrogenated and cracked with acid catalysts to be converted into transportation fuels, with hydrocarbon gas by-products. Therefore, in the end, the conversion of large PNAs consumes an enormous amount of hydrogen, doubles the size of the hydroconversion reactor because of the slow kinetics, and produces a low-quality liquid product with a large quantity of low-value hydrocarbon gases. Thus, there is an economic optimum hydroconversion, which is in the range of 85 to 90% for Arabian Heavy vacuum resid because of this polyaromatic core limit.[15] As discussed in chapter 4, section 4.7.4, the super reactive hydrogen donor, THQ, could not convert the polyaromatic cores in either the asphaltenes or the heptane-soluble fraction of Arabian Heavy vacuum resid; however, it is questionable if such conversion is advisable even for hydroconversion. The economic optimum conversion depends on the feed and on the value and use of the unconverted carbonaceous liquid by-product (essentially, all PNA cores).

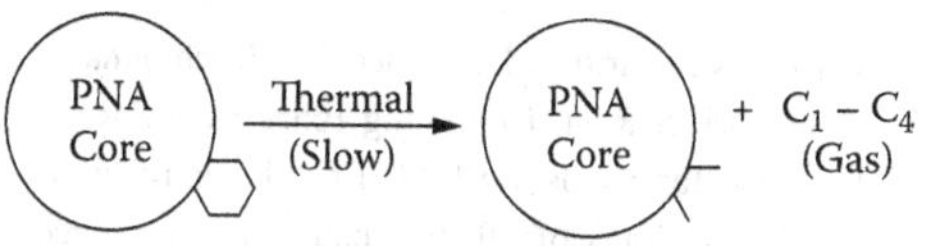

PNA Core → Thermal (Slow) → PNA Core + C_1 – C_4 (Gas)

FIGURE 10.5 Illustration of the problem of converting large polynuclear aromatics by hydroconversion.

10.4.2 PROCESS INNOVATIONS

To achieve high conversion, one needs to avoid the worst conversion range for compatibility, which varies with feed, but is often in the range of 70 to 80%.[5,16] This can be avoided by operating the reactor as two continuously stirred tank reactors, at lower conversion per pass, stripping out the volatile liquids by the flowing hydrogen, and recycling much of the separator bottoms. This allows the average conversion to be maintained below the incompatibility range in one reactor and above the incompatibility range in the second reactor. Alternatively, by making the bottoms recycle large compared with the feed, the average conversion can be kept above the incompatibility range in a single reactor. The bottoms can be checked for mesophase and sediment to make sure that compatibility is being achieved within the reactor. In addition, the insolubility number and solubility blending numbers of all the liquid products can be measured, and the oil compatibility model can be used to calculate the compatibility of the total liquid product as well as the bottoms. As high-conversion bottoms can have a high viscosity at near room temperature, it is often better to run the toluene equivalence at two concentrations than include the heptane dilution test. Also, high-conversion bottoms can achieve insolubility numbers more than 100 (toluene insoluble) without forming coke or sediment because the solubility blending number is still much higher than the insolubility number. As was discussed in chapter 5, section 5.3.12, in such cases, chlorobenzene should be substituted for toluene in the toluene equivalence test. Multiplying the chlorobenzene equivalence by 1.52 converts it to toluene equivalence values.

Eni (Dilbianco, Panariti et al.)[10,11] has been developing a dispersed catalyst, hydroconversion process that uses deasphalting to remove the deasphalted oil

---

TABLE 10.2
Yields of the Eni Slurry Process on Arabian Heavy Vacuum Resid

	Products (wt%)
Gas (HC + H$_2$S)	10.9
C$_5$–170°C naphtha	4.9
170–350°C distillate	30.6
350–500°C gas oil	24.4
500°C + DAO	24.4
Total liquids	89.1

Source: N Panariti, A Delbianco, M Marchionna, R Montanari, S Rosi. Upgrading Extra Heavy Residues and Tar Sands via EST: Eni Slurry Technology. Proceedings of 6th International Conference on Refining Processing, AIChE, 431–436, 2003.

---

product and recycles a major part of the asphalt with the dispersed catalyst. This Eni Slurry Technology satisfies the need to recycle the catalyst and maintain solubility of converted asphaltenes. By not requiring hydrogenated rings to be cracked, the reactor can be run at lower temperatures and still obtain reasonable kinetic rates. Meanwhile, secondary upgrading processes, such as fluid catalytic cracking and hydrocracking, can be used to crack the hydrogenated rings. The full vacuum distillation of the reactor product is not advisable because of dehydrogenation reactions (hydrogenation catalysts are dehydrogenation catalysts without hydrogen) and subsequent poisoning of the catalyst as well as duplication of the separation with deasphalting. Table 10.2 gives the yields of the Eni Slurry process on Arabian Heavy Vacuum Resid.

REFERENCES

1. JH Gary, GE Handwerk, MJ Kaiser. *Petroleum Refining Technology and Economics*. 5th ed. Boca Raton, FL: CRC Press, pp. 183–190, 2007.
2. VJ Nowlan, NS Srinivasan. Control of coke formation from hydrocracked athabasca bitumen. *Fuel Sci. Technol Int* 14: 41–54, 1996.
3. EC Robert, et al. Contribution of analytical tools for the understanding of sediment formation: Application to H-oil process. *Pet. Sci Technol* 21: 615–627, 2003.
4. L Smith, R Lott. HCAT heavy oil hydrocracking technology. 2nd AIChE-SPE Workshop on Exploiting the Value of Heavy Oil, Houston, 2007.
5. R Bearden. Microcat-RC, Technology for hydroconversion upgrading of petroleum residua. *ACS Div. Pet. Chem.* Meeting, ACS Award in Petroleum Chemistry, San Francisco, 1997.
6. R Bearden, CL Aldridge. Novel catalyst and process for upgrading residua and heavy crudes. *Energy Prog* 1: 44–48, 1981.
7. CL Aldridge, R Bearden. Hydrocracking of Hydrocarbons. U.S. Patent 4,192,735, 1980.
8. DJ Patmore, CP Khulbe, K Belinko. Residuum and heavy oil upgrading with the canmet hydrocracking process. ACS Preprint, *Div Pet Chem* 26: 431–439, 1981.
9. W Doehler, DIK Kretschmar, L Merz, L Niemann. VEBA-Combi-Cracking—A technology for upgrading of heavy oils and bitumen, ACS Preprint, *Div Pet Chem* 32: 484–489, 2001.
10. A Delbianco, N Panariti, S Correra, L Montanari, M Marchionna, R Montanari, S Rosi. Petroleum residues upgrading: Towards 100 percent conversion. Proceedings of 3rd International Conference on Petroleum Phase Behavior and Fouling, AIChE, 520–525, 2002.
11. N Panariti, A Delbianco, M Marchionna, R Montanari, S Rosi. Upgrading extra heavy residues and tar sands via EST: Eni slurry technology. Proceedings of 6th International Conference on Refining Processing, AIChE, 431-436, 2003.
12. RG Tailleur, B Solari. Upgrading orinoco heavy oil with hdh-shp technology. Proceedings of 6th International Conference on Refining Processing, AIChE, 421–430 2003.
13. R Lott, LK Lee. Upgrading of heavy crude oils and residues with the (HC)$_3$ hydrocracking process. Proceedings of 6th International Conference on Refining Processing, AIChE, 331–347, 2003.

14. CJ Mart, EG Ellis, DS McCaffrey, Jr. ExxonMobil resid upgrading technologies. PETROTECH 2005, 6th International Petroleum Conference and Exhibition, New Delhi, India, 2005.
15. KA Gould, IA Wiehe. Natural hydrogen donors in petroleum resids. *Energy and Fuels* 21: 1199–1204, 2007.
16. IA Wiehe. Mechanisms of coke and sediment resulting from resid conversion. ACS Div. Pet. Chem. Meeting, Symposium Honoring Roby Bearden, ACS Award in Petroleum Chemistry, San Francisco, 1997.

11 Future Processing of Petroleum Macromolecules

11.1 PROS AND CONS OF PETROLEUM PROCESS IMPROVEMENTS

The petroleum process industry is very conservative in accepting new technology. This is partially caused by the huge scale of processes, which result in major consequences for even minor problems. Another reason is that the products of refining are not differentiated by the consumer. A third cause is that today, with refining going on at full capacity, all refineries make a reasonable profit as long as they remain running at high rates, whatever the quality of their processes. In such an environment, why would anyone want to try a different process for the first time? Brand new processes always are at risk for failure and have unanticipated problems that need to be worked out. These are competing with tried-and-true processes that have been running for 20 years or more. As a result, the petroleum industry, which takes tremendous risks in looking and drilling for petroleum in increasingly challenging regions of the world, does not see much security in gambling on processing petroleum differently. This is what makes petroleum processing a mature industry—not that the technology has no room for improvement. It is a far cry from the electronics industry that follows Moore's Law in that the number of transistors that can be inexpensively placed on an integrated circuit is doubling approximately every two years. In electronics, those processes that do not continuously improve at a fast rate, do not survive.

Of course, even in the petroleum process industry there are drivers that force change. One major driver is the environmental factor, including restrictions on air and water quality from refineries and upgraders, and on transportation fuel products. In the near future, this will include reductions in carbon dioxide emissions. Another driver is the high cost of crude oil and of energy, the two largest direct costs for petroleum processing. They are encouraging refineries to process opportunity crudes, to add resid conversion processes, to use cogeneration, to optimize heat integration, and to mitigate fouling. A third driver is the need (because of full capacity) to minimize the process downtime caused by fouling, corrosion, and equipment failures. Finally, there is the imperative to increase the yields of good quality transportation fuels from a given feed. The way to accomplish yield improvement with minimum risk is to change the catalyst without changing the process.

However, if significant yield improvements are demonstrated at a few refineries, such as was the case of fluid catalytic cracking with short-contact-time riser reactors, then most refineries will want to make the process change.

As the petroleum processing industry continues to increase the processing of heavy and extra heavy oils relative to light oils with depleting supply, increasing the yield of transportation fuels from petroleum macromolecules will receive continued emphasis. As a result, these processes will be pushed to the limits imposed by real barriers and will capture much of the possible improvements pointed out in the previous chapters.

11.2 GASIFICATION TO BE EMPHASIZED

The barrier to the conversion of petroleum macromolecules that is stressed throughout this book is the presence of polynuclear aromatics that require all conversion processes on petroleum resids to have a heavy aromatic by-product. Meanwhile, the environment friendly trend of decreasing the sulfur content of fuels allowed for electric power generation will eventually not allow these heavy aromatic by-products to be used for this application without bringing about either major changes in power plants or large reduction in the sulfur content of heavy aromatic by-products.

One solution is the integrated gasification combined cycle (IGCC)[1-7] in which the fuel is first gasified with steam and oxygen to produce a syngas containing carbon monoxide, hydrogen, hydrogen sulfide, and excess oxygen. The heat is recovered by steam waste heat boilers to generate electricity with steam turbines. The hydrogen sulfide is adsorbed out of the syngas and converted to sulfur. This desulfurized syngas is then combusted in combustion turbines to generate more electricity.

Although the capital cost of IGCC is about 20% more than conventional pulverized coal power plants and it requires a steady supply of oxygen separation from air, it has many advantages for a refinery or upgrader in addition to eliminating sulfur dioxide emissions. It has much higher thermal efficiency, and thus releases less carbon dioxide for the same electric power produced than conventional power plants. Without the release of large volumes of nitrogen, IGCC power plants could also capture and store carbon dioxide more easily and cheaply when carbon dioxide reduction is required. In addition, the carbon monoxide in syngas can be converted by a water–gas shift to hydrogen, and satisfy the high requirements for refining petroleum macromolecules. Finally, syngas can be used as the feed to manufacture chemicals, as is already being done by Eastman Chemicals, through gasification of coal or by Fischer-Tropsch process to hydrocarbons for lube oils, or even for diesel fuel as is being done for the conversion of natural gas and coal to hydrocarbon liquids.

Thus, in the end, gasification will be an extremely flexible process that can be used to produce the best mix of electric power, hydrogen, energy, chemicals, and hydrocarbons that fits the time and location of production, also utilizing the heavy aromatic by-products. These gasification plants make a lot of sense if they are located near several refineries/chemical plants and are partially owned by power companies. This can assure a cheap, reliable supply of fuel to the power plant, and a market for hydrogen and syngas. Similarly, if located near an extra heavy oil production and upgrading site, the gasifier can supply hydrogen for the upgrader and either steam or syngas for the energy required in production, such as cyclic steam or steam assisted gravity drainage. A synthetic hydrocarbon diluent for pipeline transportation of extra heavy oil could also be produced from the syngas.

11.3 DEASPHALTING OF FUELS TO BECOME COMMON

In chapter 7, it has been stressed that deasphalting does a much better job of separating petroleum macromolecules based on quality than conventional vacuum distillation, and that secondary upgrading can accept resid feeds if the quality is sufficient. As a result, one can separate the high quality part of a heavy oil or resid to send on to secondary processing without degrading in a thermal process, such as coking. Meanwhile, conversion is needed to break the covalent bonds between the good and bad parts of the poorer quality petroleum macromolecules. If deasphalting is used to separate the parts after conversion, the conversion process may be milder than if the separation was made on the basis of boiling point. The Eni Slurry Technology for hydroconversion described in section 10.4.2 is an example.

However, an even better result is obtained if deasphalting separates the feed and products into three streams on the basis of Conradson carbon residue (CCR; hydrogen content). One stream (deasphalted oil [DAO] of low CCR) meets the minimum standard for secondary processing, another stream (converted asphaltenes of high CCR) is low enough in quality to be rejected as the heavy aromatic by-product, and the remaining stream (resins of moderate CCR) is recycled back for more conversion. As discussed in sections 7.2.7, 9.4.5, and 10.3.2, this type of conversion-deasphalting process will give the maximum conversion while favoring compatibility. It also enables setting specifications on both the DAO and reacted asphaltene streams even though the feed to the process varies greatly in quality. As discussed in chapter 7, there is not much room for improving the selectivity of the separation. This encourages one to focus on lowering the cost of deasphalting rather than looking for alternative separation processes.

11.4 YIELDS FROM COKING PROCESSES TO INCREASE

In chapter 8, it was shown that for a feed of Arabian Heavy vacuum resid, the potential improvement over Fluid Coking for the yield of liquids with the same hydrogen content is 24 wt% absolute with 15 wt% absolute demonstrated on the small laboratory scale that reduced the coke yield from 1.2 CCR to 0.8 CCR. These are incentives too large for the petroleum industry to ignore. This can be accomplished by deasphalting and coking the asphalt in a coker with short vapor residence time and a long resid residence time. Alternatively, a reactor with short vapor residence time and a long resid residence time can be designed to flash off the lighter part of the resid, with a process, such as Distact distillation, to replace the deasphalting step.

Admittedly, to get the benefit of prior separation, the vacuum resid feed needs to have a significant fraction of saturates and aromatics (low CCR). Thus, it is questionable if a heavily biodegraded feed, such as Athabasca vacuum resid, would yield much benefit by prior separation, reducing the yield advantage closer to 8 wt% absolute of a coker with a short vapor residence time and a long resid residence time but without prior separation. Nevertheless, with either of these incentives, a new coker design, such as the cross-flow coker of ETX Systems,[8] can be expected to be developed and demonstrated in the near future. This should encourage most refineries to use a new generation coker to meet the need for significantly increasing resid conversion to liquids while greatly lowering the yield of coke.

11.5 DISPERSED CATALYST TO OVERTAKE EBULLATING BED HYDROCONVERSION

As the world moves from light crudes to extra-heavy crudes, refineries will be pressured into producing even more transportation fuels. Part of this increase will be from capacity expansion, but a significant part will have to result from greatly increasing the yield from resid conversion. Table 11.1 compares the yields on Arabian Heavy vacuum resids[9] from existing processes, processes under development, and processes only demonstrated in the laboratory. Obviously this comparison in various stages of development is not fair; also, the volume yield would be a better measure and the hydrogen content of the liquid product should be reported. However, as these measures are not available, the reader should consider it a very rough comparison. Nevertheless, most will agree that dispersed catalyst

TABLE 11.1
Liquid Yields by Resid Conversion Processes on Arabian Heavy Vacuum Resid

Process	State of Development	Liquid Yield (wt%)
Visbreaking	Commercial	28
Visbreaking + cyclohexane nonpolars	Small laboratory	62
Delayed coking	Commercial	59
Once-through delayed coking	Commercial	61
Fluid Coking	Commercial	61
Once-through Fluid Coking	Demonstrated	66
Short vapor residence time coking	Small laboratory	69
Separation + short vapor residence time coking	Small laboratory	76
Ebullating bed hydroconversion	Commercial	65–70
Eni Slurry Technology	Pilot plant	89

hydroconversion processes (Eni Slurry Technology in Table 10.2) give the highest yields of the resid conversion processes. Therefore, it is logical that the petroleum industry will finally be drawn to this technology. Certainly, the present yields from ebullating bed hydroconversion are too low, as they are barely above the present coking levels and would not compete with a new generation coker.

In chapter 10, it was shown that although the yield of 100% by volume liquid products has been demonstrated for dispersed catalyst hydroconversion (Table 10.1 for Microcat-RC), it is probably beyond the economic optimum. Still, a process that produces an equal volume, or near equal volume, of volatile liquids from a vacuum resid could be quite attractive. In addition, coupling a dispersed catalyst hydroconversion process with gasification is synergistic. The liquid state of the heavy aromatic by-product of a dispersed catalyst hydroconversion process is the preferred feed (better than coke) to a gasifier, while the high yield of cheap hydrogen from the gasifier would benefit the high hydrogen consumption of the dispersed catalyst hydroconversion process and subsequent heavy hydrotreating of the lighter products.

11.6 FOULING TO BE NEARLY ELIMINATED

With the strong future incentives to reduce carbon dioxide emissions from refineries and upgraders, the energy efficiency of petroleum processing will be greatly increased. This will be the first carbon dioxide emission reduction done by refineries and upgraders because it is the only one that also reduces costs. In addition, with refineries at full capacity there are large rewards for refineries to minimize the time they are shut down. While energy conservation can be accomplished by optimizing heat integration[10] and by fouling mitigation, fouling and corrosion mitigation reduce shut-down time.

The use of the Oil Compatibility Model of chapter 5 and the fouling mitigation strategy in chapter 6 provide a good background for mitigating fouling, particularly organic fouling. As discussed in chapter 6, mitigating inorganic fouling requires reducing corrosion and improving the effectiveness of desalters. As discussed in section 7.1, the scientific understanding of stable emulsion formation in desalters has been rapidly advancing recently. However, a concerted effort is needed that combines those working on mitigating petroleum fouling with those working on mitigating petroleum corrosion to make the next significant advance, such as by devising coatings that prevent both corrosion and fouling adhesion. Once that is accomplished, both organic and inorganic fouling could soon nearly be eliminated in petroleum processing.

11.7 INFLUENCE OF OTHER CHANGING TECHNOLOGIES

The future processing of petroleum macromolecules will certainly be altered and influenced by future changes in other technologies. One might anticipate some of these, but many others will come as a surprise. Certainly, dramatic changes in transportation fuels will require changes in petroleum processing.

11.7.1 HYDROGEN AND FUEL CELLS

Many dream of the use of hydrogen as a fuel using fuel cells. This is certainly a nice dream because the fuel is converted only to water with no carbon dioxide emission, and it takes advantage of the much higher theoretical efficiency of fuel cells. Although progress in the development of fuel cells for vehicles has been slow, if not discouraging, the hydrogen fuel will need to be supplied from other sources. If hydrogen is stored in the vehicle under pressure, it will be supplied from reforming methane (natural gas) in which the carbon dioxide by-product is captured and stored. If a greater supply is required, the IGCC (section 11.2) run on coal or petroleum resids is the next most probable source. However, transportation vehicles carrying hydrogen under high pressure will be an unacceptable safety hazard. The fuel will have to be a hydrocarbon liquid that can be easily reformed onboard to produce hydrogen as needed, such as naphthenes. If so, it is possible that it will be produced from petroleum or from Fischer-Tropsch hydrocarbons using syngas. Petroleum macromolecules can be converted directly or used as a fuel for syngas, depending on what hydrocarbons are selected.

11.7.2 BIOFUELS

Biofuels were discussed in chapter 1. They have the advantage of being formed from a renewable source that uses as much carbon dioxide in its formation as it produces in its combustion. Certainly, as much biofuels should be used as possible. However, it will be difficult to completely replace fossil fuels with biofuels. This is especially true because the world is expected to run increasingly short of food in the future and all agricultural resources will be engaged in food production. So, it seems likely in the future that biofuel production will be sourced from waste by-products of agriculture.

11.7.3 GAS CONVERSION LIQUIDS

Syngas can also be formed from natural gas and be converted by Fischer-Tropsch to hydrocarbons. This product already competes with liquefied natural gas for transporting from remote locations to population centers. The hydrocarbons can be used for production of lubricating oils or transportation fuels, especially diesel fuel, containing no sulfur. If this becomes economical enough to compete with petroleum, petroleum macromolecules could also be used as the feed for producing syngas.

REFERENCES

1. Wikipedia Web site, http://en.wikipedia.org/wiki/Combined_cycle, 2007.
2. C Schrader, L Shah. Integration of Texaco gasification and Fischer-Tropsch for improved economics and a cleaner environment. Proceedings of the 3rd International Conference on Refinery Processing, AIChE, 357–364, 2000.

3. JD de Giraaf, R van der Berg, PL Zuideveld. Shell gasification processes. Proceedings of the 3rd International Conference on Refinery Processing, AIChE, 365–374, 2000.
4. J Sadhukhan, XX Zhu. Integration of gasification technologies in the context of overall refinery operation, Proceedings of the 3rd International Conference on Refinery Processing, AIChE, 375–384, 2000.
5. CF Penrose, A Rodarte, PS Wallace, WE Preston. Gasification of refinery bottoms enhances profitability. Proceedings of the 3rd International Conference on Refinery Processing, AIChE, 385–406, 2000.
6. JF McGehee. Residue upgrading by solvent deasphalting and pitch gasification—A review. Proceedings of the AIChE Spring National Meeting, Paper 87c, 2007.
7. J Plunkett, C White, S Salerno, DG Noblis, G Tomlinson. IGCC: Current status and future potential. Proceedings of the AIChE Spring National Meeting, Paper 9a, 2007.
8. R Pinchuk, W Brown, G Monaghan, W McCaffrey, O Asprino, P Gonzalez. Tailoring RTDs in fluidized beds for improved coker yields—The Envision Technologies\Corp. cross-flow coker. Proceedings of 8th International Conference on Refinery Processing, AIChE, Paper 66a, 2005.
9. IA Wiehe. The processing of resids and heavy oils. Proceedings of the AIChE Spring National Meeting, Paper 87b, 2007.
10. XXF Zhu. Energy optimization studies to enhance both new projects and return on existing assets. Proceedings of the AIChE Spring National Meeting, Paper 109b, 2007.

Index

Printed in the United States
by Baker & Taylor Publisher Services